Application and Behavior of Nanomaterials in Water Treatment

Application and Behavior of Nanomaterials in Water Treatment

Special Issue Editors
Protima Rauwel
Erwan Rauwel
Wolfgang Uhl

MDPI • Basel • Beijing • Wuhan • Barcelona • Belgrade

Special Issue Editors
Protima Rauwel
Estonian University of Life Sciences
Estonia

Erwan Rauwel
Estonian University of Life Science
Estonia

Wolfgang Uhl
Norwegian Institute for Water Research
(NIVA) and Norwegian University of
Science and Technology (NTNU)
Norway

Editorial Office
MDPI
St. Alban-Anlage 66
4052 Basel, Switzerland

This is a reprint of articles from the Special Issue published online in the open access journal *Nanomaterials* (ISSN 2079-4991) from 2018 to 2019 (available at: https://www.mdpi.com/journal/nanomaterials/special_issues/nano_water)

For citation purposes, cite each article independently as indicated on the article page online and as indicated below:

LastName, A.A.; LastName, B.B.; LastName, C.C. Article Title. *Journal Name* **Year**, *Article Number*, Page Range.

ISBN 978-3-03921-171-5 (Pbk)
ISBN 978-3-03921-172-2 (PDF)

Cover image courtesy of Protima Rauwel.

© 2019 by the authors. Articles in this book are Open Access and distributed under the Creative Commons Attribution (CC BY) license, which allows users to download, copy and build upon published articles, as long as the author and publisher are properly credited, which ensures maximum dissemination and a wider impact of our publications.

The book as a whole is distributed by MDPI under the terms and conditions of the Creative Commons license CC BY-NC-ND.

Contents

About the Special Issue Editors vii

Protima Rauwel, Wolfgang Uhl and Erwan Rauwel
Editorial for the Special Issue on 'Application and Behavior of Nanomaterials in Water Treatment'
Reprinted from: *Nanomaterials* **2019**, *9*, 880, doi:10.3390/nano9060880 1

Protima Rauwel and Erwan Rauwel
Towards the Extraction of Radioactive Cesium-137 from Water via Graphene/CNT and Nanostructured Prussian Blue Hybrid Nanocomposites: A Review
Reprinted from: *Nanomaterials* **2019**, *9*, 682, doi:10.3390/nano9050682 4

Zhongchuan Wang, Pengfei Fang, Parveen Kumar, Weiwei Wang, Bo Liu and Jiao Li
Controlled Growth of LDH Films with Enhanced Photocatalytic Activity in a Mixed Wastewater Treatment
Reprinted from: *Nanomaterials* **2019**, *9*, 807, doi:10.3390/nano9060807 25

Xiaoze Shi, Shuai Zhang, Xuecheng Chen and Ewa Mijowska
Evaluation of Nanoporous Carbon Synthesized from Direct Carbonization of a Metal–Organic Complex as a Highly Effective Dye Adsorbent and Supercapacitor
Reprinted from: *Nanomaterials* **2019**, *9*, 601, doi:10.3390/nano9040601 36

Rizwan Khan, Muhammad Ali Inam, Sarfaraz Khan, Du Ri Park and Ick Tae Yeom
Interaction between Persistent Organic Pollutants and ZnO NPs in Synthetic and Natural Waters
Reprinted from: *Nanomaterials* **2019**, *9*, 472, doi:10.3390/nano9030472 51

Yuelong Xu, Bin Ren, Ran Wang, Lihui Zhang, Tifeng Jiao and Zhenfa Liu
Facile Preparation of Rod-like MnO Nanomixtures via Hydrothermal Approach and Highly Efficient Removal of Methylene Blue for Wastewater Treatment
Reprinted from: *Nanomaterials* **2019**, *9*, 10, doi:10.3390/nano9010010 66

Ha Eun Shim, Jung Eun Yang, Sun-Wook Jeong, Chang Heon Lee, Lee Song, Sajid Mushtaq, Dae Seong Choi, Yong Jun Choi and Jongho Jeon
Silver Nanomaterial-Immobilized Desalination Systems for Efficient Removal of Radioactive Iodine Species in Water
Reprinted from: *Nanomaterials* **2018**, *8*, 660, doi:10.3390/nano8090660 82

Rong Guo, Ran Wang, Juanjuan Yin, Tifeng Jiao, Haiming Huang, Xinmei Zhao, Lexin Zhang, Qing Li, Jingxin Zhou and Qiuming Peng
Fabrication and Highly Efficient Dye Removal Characterization of Beta-Cyclodextrin-Based Composite Polymer Fibers by Electrospinning
Reprinted from: *Nanomaterials* **2019**, *9*, 127, doi:10.3390/nano9010127 93

Cuiru Wang, Juanjuan Yin, Ran Wang, Tifeng Jiao, Haiming Huang, Jingxin Zhou, Lexin Zhang and Qiuming Peng
Facile Preparation of Self-Assembled Polydopamine-Modified Electrospun Fibers for Highly Effective Removal of Organic Dyes
Reprinted from: *Nanomaterials* **2019**, *9*, 116, doi:10.3390/nano9010116 110

Jun Yang, Taiping Xie, Chenglun Liu and Longjun Xu
Dy(III) Doped BiOCl Powder with Superior Highly Visible-Light-Driven Photocatalytic Activity for Rhodamine B Photodegradation
Reprinted from: *Nanomaterials* **2018**, *8*, 697, doi:10.3390/nano8090697 127

Yi Liu, Yumin Huang, Aiping Xiao, Huajiao Qiu and Liangliang Liu
Preparation of Magnetic Fe_3O_4/MIL-88A Nanocomposite and Its Adsorption Properties for Bromophenol Blue Dye in Aqueous Solution
Reprinted from: *Nanomaterials* **2019**, *9*, 51, doi:10.3390/nano9010051 139

Taiping Xie, Hui Li, Chenglun Liu, Jun Yang, Tiancun Xiao and Longjun Xu
Magnetic Photocatalyst $BiVO_4$/Mn-Zn ferrite/Reduced Graphene Oxide: Synthesis Strategy and Its Highly Photocatalytic Activity
Reprinted from: *Nanomaterials* **2018**, *8*, 380, doi:10.3390/nano8060380 152

About the Special Issue Editors

Protima Rauwel received her PhD in 2005 from University of Caen, France, in condensed matter physics and material science. After her PhD, she continued working with nanomaterials, and their characterization and applications, through postdoctoral positions at the University of Aveiro, Portugal. She has also held a Researcher position at the Center for Materials Science and Nanotechnology, at the University of Oslo, Norway. Presently, she works as a Senior Researcher at the Estonian University of Life Sciences. She is an expert in transmission electron microscopy and develops hybrid nanomaterials for photovoltaic applications. She also applies nanomaterials to water remediation. She has 80 publications, 2 patents, 4 book chapters, 1 book, and an H-index of 22. She is also CEO of PRO-1 NANOSolutions, a startup company that applies nanotechnology to water remediation and nanomedicine.

Erwan Rauwel received his PhD degree from Univ. of Caen in Materials Science in 2003. He continued with postdoctoral studies at Minatec, Grenoble, France, and as Marie Curie Fellow at the Univ. of Aveiro, Portugal and then as Senior Researcher at University of Oslo. He is now Professor at the Institute of Technology of the Estonian University of Life Science in Estonia, where his team investigates the properties of nanoparticles for water purification and biomedical applications and hybrid nanocomposites for photocurrent generation and energy harvesting. He has more than 60 peer-reviewed publications with a h-index of 20, 4 book chapters, and 5 patents. He is also Chief Scientist of his start-up company specializing in nanomaterials (PRO-1 NANOSolutions).

Wolfgang Uhl is the research manager of Systems Engineering and Technology at the Norwegian Institute for Water Research (NIVA) and Adjunct Professor at the Norwegian University of Science and Technology (NTNU), Department of Civil and Environmental Engineering in Trondheim, Norway. He received his MS in Chemical Engineering from the University of Karlsruhe, Germany, a MS in Biotechnology from the University of Lund, Sweden, and his PhD in Mechanical Process Engineering from the University of Duisburg-Essen, Germany. He has about 30 years of experience in research and consulting in the water business. He has been working and he published regarding all aspects of water quality, water treatment and water distribution with respect to drinking water, wastewater, and process water, including water reclamation and reuse.

Editorial

Editorial for the Special Issue on 'Application and Behavior of Nanomaterials in Water Treatment'

Protima Rauwel [1,*], Wolfgang Uhl [2,3] and Erwan Rauwel [1]

1. Institute of Technology, Estonian University of Life Sciences, Kreutzwaldi 56/1, 51014 Tartu, Estonia; erwan.rauwel@emu.ee
2. Norwegian Institute for Water Research (NIVA), Gaustadalléen 21, N-0349 Oslo, Norway; wolfgang.uhl@niva.no
3. Department of Civil and Environmental Engineering, Norwegian University of Science and Technology (NTNU), S. P Andersens Vei 5, 7491 Trondheim, Norway
* Correspondence: protima.rauwel@emu.ee

Received: 29 May 2019; Accepted: 10 June 2019; Published: 14 June 2019

The simultaneous population explosion and the growing lack of clean water today requires disruptively innovative solutions in water remediation. The last decade has witnessed the emergence of various nanomaterials capable of bridging the gap between the demand for and supply of clean water. Accelerated research on finding suitable nanomaterials in water treatment is therefore fueled by the need of the hour. The main asset of nanomaterials is their highly specific surfaces due to their size reduction, which in turn promotes enhanced catalytic activity, subsequently bringing about a more efficient degradation of dyes and organic pollutants. Nanomaterials such as oxide nanoparticles, nanocarbons, doubled layered hydroxides, and other nanosorbents offer enormous advantages in heavy metal capture and extraction from aqueous media.

This Special Issue compiles eleven articles dedicated to nanomaterials for water treatment: ten research articles and one review article. Together they constitute an interesting and a multi-disciplinary approach to pollution elimination in aqueous media. The papers present different nanomaterials such as layered double hydroxides [1]; nanoporous carbon [2]; oxide nanoparticles, i.e., ZnO [3] and MnO [4]; Ag metal nanoparticles [5]; polymer fibers [6,7]; and inorganic BiOCl doped Dy^{+3} powders [8]. Hybrid materials combining metal organic frameworks (MOF) such as MIL-88A [9] and Prussian blue combined with graphene and carbon nanotubes (CNT) [10], along with magnetic nanoparticles, i.e., magnetite (Fe_3O_4) and ferrite (Mn-Zn) [11], are also featured. These nanomaterials have been applied to the degradation of dyes and pharmaceuticals, along with heavy metal ion and radioactive ion extraction. The purpose of this Special Issue is to communicate the most recent advances in the application and behavior of nanomaterials in water treatment. It targets a broad readership of physicists, chemists, materials scientists, catalysis researchers, water researchers, environmentalists, and nanotechnologists. In the paragraphs that follow, we, the guest editors of this Special Issue, provide a brief overview of the individual articles published and hope to incite the interest of potential readers.

We open the discussion on the published articles with the paper on silver metal nanoparticles by Shim et al. [5]. Their work focuses on desalination via the extraction of radioactive iodine from water. Their methodology combines silver nanoparticles immobilized on a cellulose-based membrane reinforced with *Deinococcus radiodurans*, which is a radiation-resistant bacterium. Ag nanoparticles capture iodine complexes, whereas the bacteria serve to bio-remediate the produced slurry. Metal oxide nanoparticles have also been presented in this compilation for the degradation of organic species, dyes, and pharmaceuticals. The study by Khan et al. assesses the effects of ZnO nanoparticles in various contaminated aqueous media [3]. Their study is of importance, as ZnO nanoparticles are employed in various applications, and therefore their concentrations in wastewaters are increasing. They more specifically study the stability of ZnO in the presence of persistent organic pollutants, i.e., polybrominated diphenyl ethers. The latter behaves as a surfactant tending to increase the colloidal stability of ZnO nanoparticles, which could prove detrimental if consumed. Other inorganic nanomaterials for the degradation of methylene blue were studied by Xu et al. [4]. They synthesized MnO nanomaterials with rod-like morphologies, which have the potential to be reused in successive cycles against methylene blue degradation with a high efficiency of 99.8%. Inorganic powders of BiOCl were doped with Dy^{+3} by Yang et al. [8]. Their paper describes the synthesis of the rare-earth doped inorganic nanopowders and their photocatalytic activity towards Rhodamine B degradation. The authors have worked on the structural, optical, and adsorption properties of the nanopowder. They have also provided a schematic of the energy band diagram and electron transfer mechanism in the Dy^{+3} doped and undoped BiOCl powders.

Organic materials such as polymer fibers also demonstrate efficacy in degradation of dyes and pharmaceuticals. The report by Guo et al. describes a cost-effective and facile fabrication of electrospun ε-polycaprolactone- and β-cyclodextrin-based composite polymer fiber [7]. These fibers exhibit high surface areas, improved mechanical strength, and excellent uptake of methylene blue azo dye. Wang et al. have also used polycaprolactone fibers modified by polydopamine [6]. The nanocomposite had a roughened microstructure, implying a higher specific surface with more active sites for the extraction of dyes. They exhibited efficiency against both methylene blue and methylene orange. Other organic materials include activated carbons, which have also presented efficiency against dyes such as methylene blue. Shi et al. have synthesized nanoporous carbon from metal organic complexes [2]. Their tunable pore sizes and uniform pore distribution allow diffusion of the methylene blue molecules through them. Nanoporous carbons are excellent electrode materials and exhibit supercapacitance properties in aqueous electrolytes. Due to their higher anion exchange capacity, layered double hydroxides (LDH) are considered to be promising nanomaterials for the extraction of organic and inorganic anions [1]. Wang et al. have synthesized Ni-Al-Fe LDH, which exhibited a higher photocatalytic activity than pure LDH. Their study demonstrates a catalytic effect of the captured heavy metals on the surface of LDH towards the degradation of organic contaminants in wastewater.

Magnetic extraction has the advantage of reclaiming the spent sorbent. Fe_3O_4-MIL-88A is one such magnetic MOF presented by Liu et al., capable of degrading phenolic dyes, i.e., bromophenol blue [9]. They tested their material's efficiency on nine dyes in all, out of which eight contained sulphonyl groups. Their study therefore brings insights into magnetic extraction of dyes. Magnetic photocatalysts, i.e., $BiVO_4/Mn_{1-x}Zn_xFe_2O_4/RGO$, were studied by Xie et al. in their work [11]. They thoroughly investigated the photocatalytic activity of the composite. In addition, graphene played a very important role in enhancing the photocatalytic activity of the material towards the degradation of Rhodamine B dye. The last paper by Rauwel et al. is a contribution from the guest editors, and reviews the various hybrid nanomaterials studied by various groups [10]. They focus on the extraction of $^{137}Cs^+$ from aqueous media in the light of the recent Fukushima Daiichi catastrophe. The paper mainly surveys the extraction of $^{137}Cs^+$ with nanocomposites of Prussian blue/graphene/CNT. The possibility of magnetic extraction when combining the hybrid material with Fe_3O_4 nanoparticles is also discussed.

Author Contributions: P.R. wrote this Editorial Letter. E.R. and W.U. provided their feedback, which was assimilated into the Letter.

Funding: P.R. and E.R. acknowledge the Centre of Excellence project EQUiTANT (F180175TIBT) for financial support. W.U. thanks the Norwegian Research Council and NIVA's internal publication fund for support

Acknowledgments: The guest editors thank all the authors for submitting their work to the Special Issue and for its successful completion. A special thank you to all the reviewers participating in the peer-review process of the submitted manuscripts and for enhancing their quality and impact. We are also grateful to Yueyue Zhang and the editorial assistants who made the entire Special Issue creation a smooth and efficient process.

Conflicts of Interest: The authors declare no conflict of interest.

References

1. Wang, Z.; Fang, P.; Kumar, P.; Wang, W.; Liu, B.; Li, J. Controlled Growth of LDH Films with Enhanced Photocatalytic Activity in a Mixed Wastewater Treatment. *Nanomaterials* **2019**, *9*, 807. [CrossRef] [PubMed]
2. Shi, X.; Zhang, S.; Chen, X.; Mijowska, E. Evaluation of Nanoporous Carbon Synthesized from Direct Carbonization of a Metal–Organic Complex as a Highly Effective Dye Adsorbent and Supercapacitor. *Nanomaterials* **2019**, *9*, 601. [CrossRef] [PubMed]
3. Khan, R.; Inam, M.A.; Khan, S.; Park, D.R.; Yeom, I.T. Interaction between Persistent Organic Pollutants and ZnO NPs in Synthetic and Natural Waters. *Nanomaterials* **2019**, *9*, 472. [CrossRef] [PubMed]
4. Xu, Y.; Ren, B.; Wang, R.; Zhang, L.; Jiao, T.; Liu, Z. Facile Preparation of Rod-like MnO Nanomixtures via Hydrothermal Approach and Highly Efficient Removal of Methylene Blue for Wastewater Treatment. *Nanomaterials* **2018**, *9*, 10. [CrossRef] [PubMed]
5. Shim, H.E.; Yang, J.E.; Jeong, S.-W.; Lee, C.H.; Song, L.; Mushtaq, S.; Choi, D.S.; Choi, Y.J.; Jeon, J. Silver Nanomaterial-Immobilized Desalination Systems for Efficient Removal of Radioactive Iodine Species in Water. *Nanomaterials* **2018**, *8*, 660. [CrossRef] [PubMed]
6. Wang, C.; Yin, J.; Wang, R.; Jiao, T.; Huang, H.; Zhou, J.; Zhang, L.; Peng, Q. Facile Preparation of Self-Assembled Polydopamine-Modified Electrospun Fibers for Highly Effective Removal of Organic Dyes. *Nanomaterials* **2019**, *9*, 116. [CrossRef] [PubMed]
7. Guo, R.; Wang, R.; Yin, J.; Jiao, T.; Huang, H.; Zhao, X.; Zhang, L.; Li, Q.; Zhou, J.; Peng, Q. Fabrication and Highly Efficient Dye Removal Characterization of Beta-Cyclodextrin-Based Composite Polymer Fibers by Electrospinning. *Nanomaterials* **2019**, *9*, 127. [CrossRef] [PubMed]
8. Yang, J.; Xie, T.; Liu, C.; Xu, L. Dy(III) Doped BiOCl Powder with Superior Highly Visible-Light-Driven Photocatalytic Activity for Rhodamine B Photodegradation. *Nanomaterials* **2018**, *8*, 697. [CrossRef] [PubMed]
9. Liu, Y.; Huang, Y.; Xiao, A.; Qiu, H.; Liu, L. Preparation of Magnetic Fe3O4/MIL-88A Nanocomposite and Its Adsorption Properties for Bromophenol Blue Dye in Aqueous Solution. *Nanomaterials* **2019**, *9*, 51. [CrossRef] [PubMed]
10. Rauwel, P.; Rauwel, E. Towards the Extraction of Radioactive Cesium-137 from Water via Graphene/CNT and Nanostructured Prussian Blue Hybrid Nanocomposites: A Review. *Nanomaterials* **2019**, *9*, 682. [CrossRef] [PubMed]
11. Xie, T.; Li, H.; Liu, C.; Yang, J.; Xiao, T.; Xu, L. Magnetic Photocatalyst BiVO4/Mn-Zn ferrite/Reduced Graphene Oxide: Synthesis Strategy and Its Highly Photocatalytic Activity. *Nanomaterials* **2018**, *8*, 380. [CrossRef] [PubMed]

© 2019 by the authors. Licensee MDPI, Basel, Switzerland. This article is an open access article distributed under the terms and conditions of the Creative Commons Attribution (CC BY) license (http://creativecommons.org/licenses/by/4.0/).

Review

Towards the Extraction of Radioactive Cesium-137 from Water via Graphene/CNT and Nanostructured Prussian Blue Hybrid Nanocomposites: A Review

Protima Rauwel * and Erwan Rauwel

Institute of Technology, Estonian University of Life Sciences, Kreutzwaldi 56/1, 51014 Tartu, Estonia; erwan.rauwel@emu.ee
* Correspondence: protima.rauwel@emu.ee

Received: 5 April 2019; Accepted: 24 April 2019; Published: 2 May 2019

Abstract: Cesium is a radioactive fission product generated in nuclear power plants and is disposed of as liquid waste. The recent catastrophe at the Fukushima Daiichi nuclear plant in Japan has increased the ^{137}Cs and ^{134}Cs concentrations in air, soil and water to lethal levels. ^{137}Cs has a half-life of 30.4 years, while the half-life of ^{134}Cs is around two years, therefore the formers' detrimental effects linger for a longer period. In addition, cesium is easily transported through water bodies making water contamination an urgent issue to address. Presently, efficient water remediation methods towards the extraction of ^{137}Cs are being studied. Prussian blue (PB) and its analogs have shown very high efficiencies in the capture of ^{137}Cs$^+$ ions. In addition, combining them with magnetic nanoparticles such as Fe_3O_4 allows their recovery via magnetic extraction once exhausted. Graphene and carbon nanotubes (CNT) are the new generation carbon allotropes that possess high specific surface areas. Moreover, the possibility to functionalize them with organic or inorganic materials opens new avenues in water treatment. The combination of PB-CNT/Graphene has shown enhanced ^{137}Cs$^+$ extraction and their possible applications as membranes can be envisaged. This review will survey these nanocomposites, their efficiency in ^{137}Cs$^+$ extraction, their possible toxicity, and prospects in large-scale water remediation and succinctly survey other new developments in ^{137}Cs$^+$ extraction.

Keywords: carbon nanotubes; graphene; Prussian blue; 137-Cesium; water remediation; magnetic extraction; ^{137}Cs$^+$ selectivity; radioactive contamination

---

1. Introduction

Cesium-137 is a radioactive element with a half-life of 30.4 years emitting both beta and gamma radiations. Under normal circumstances, its release is an outcome of radioactive testing of fission reactions, whereby making it the most abundant radioactive atmospheric pollutant capable of entailing health hazards. The ^{137}Cs isotope is produced when uranium and plutonium undergo fission after having absorbed neutrons in a nuclear reactor and is detectable and measurable by gamma counting. It decays into ^{137}Ba with a half-life of 2.6 min. Small amounts of ^{137}Cs$^+$ are released on a regular basis in spent fuel ponds through cracks in the fuel rod that reach the coolant and fuel reprocessing waters, which are all subsequently discharged into the sea as effluents [1]. Today, in view of the Fukushima-Daiichi catastrophe, the extraction of radioactive cesium from soil and water is a necessity and calls for the development of new technologies. The projected radiation effects of ^{137}Cs will remain at their maximum for the next 100 years, seconded by isotopes of strontium and plutonium. Other sources of ^{137}Cs contamination can be attributed to the Chernobyl accident in 1986; [2] it took 10 years for the levels to drop but their effects are still being perceived even today [3]. In general, ^{137}Cs is not very mobile and tends to accumulate on the soil surface and is subsequently absorbed by plants and also captured by fungi [4]. A soil cleanup of 40,000 km² reduced the radioactive contamination to 1/10th of the original value in Chernobyl.

^{137}Cs is the primary cause of water contamination since the recent Fukushima Daiichi catastrophe posing more far-reaching threats than via soil contamination [5], which tends to be more localized [6]. ^{137}Cs decontamination has now become a priority recommended by the International Atomic Energy Agency [7]. Mobility of ^{137}Cs$^+$ through moving water bodies displaces the contamination to other areas, thus widening the 'contamination reduction' zone. Therefore, aquatic life and fish-farming products are also contaminated, which are thereafter consumed by humans. Additionally, ^{137}Cs$^+$ behaves very similarly to K$^+$ and Na$^+$, thus facilitating its digestion and assimilation in living organisms [8]. Consumption of contaminated game and fish would introduce ^{137}Cs in the human body, which would then emit harmful radiations directly targeting the cell nucleus. Moreover, in aqueous media ^{137}Cs$^+$ ions tend to be rather robust and unaffected by changes in pH and redox conditions, whereby making them a menace [9].

Several techniques were developed for the extraction of ^{137}Cs$^+$ from water: reverse osmosis [10], coagulation-sedimentation [11], ion-exchange [12], nanofiltration [13], electro-dyalysis [14] and more recently, fibrous zeolite-polymer composites [15]. On the other hand, adsorption is a highly efficient and cost-effective process with very high ion selectivity provided that the right sorbents are used [16]. Various adsorbents: inorganic adsorbents, polymer-inorganic adsorbents [7] and bioadsorbents [17], have shown efficiency in ^{137}Cs$^+$ extraction. Bioadsorbents however, have several disadvantages such as low sorption efficiency, especially in the presence of other salts in the aqueous medium viz., Na$^+$ and K$^+$. They also suffer from degradation in extreme conditions (high temperature and low pH). On the other hand, inorganic sorbents have shown large-scale applicability with high cation sorption capacities (clays viz., zeolites, bentonite and coal). However, activation of sorption sites is necessary through a surface functionalization treatment, which conversely, increases their production cost. For nanozeolites, the Q_{max} or maximum adsorption capacity values could reach up to 69 mg/g, but on the other hand, it is difficult to recover the exhausted material. Moreover, the K_d value, which is the ratio of the equilibrium adsorption of the sorbent to the equilibrium concentration of the solute, is low, implying an ineffectiveness in high solute concentrations [18]. Finally yet importantly, they do not withstand harsh aqueous environments. Synthetic polymers combined with inorganic materials seem to be more robust with a better sorption capacity than inorganic sorbents; they also show stability in harsh environments and have been already applied on a large scale in Fukushima [7].

Among the various methods available for the extraction of ^{137}Cs$^+$ from aqueous solutions, hybrid materials manifest enormous potential in selectively targeting and extracting ^{137}Cs$^+$ ions [19]. Inorganic ligands such as macrocyclic o-benzo-p-xylyl-22-crown-6-ether (OBPX22C6) ligand bonded to the hydroxyl groups of the mesoporous silica, exhibited a yield of 60% attributed to the Cs-π interaction of the OBPX22C6 benzene. Zeolite-Poly(ethersulfone) composite fiber having 30 wt % loading showed excellent properties for the decontamination of radioactive ^{137}Cs$^+$ [7]. The decontamination with such composites has also been demonstrated for a contamination of 823Bq/L with pH = 12. Other silicates have also been employed for their known selectivity to ^{137}Cs$^+$; these include crystalline silicotitanate [20,21], sodium mica [22] and sodium zirconium silicate [23]. Since they are usually in the form of very fine powders, they are therefore unsuitable for column loading. Moreover, they are difficult to separate from aqueous solutions by filtration or centrifugation therefore, reclaiming them once expended becomes problematic.

This review will focus on the recent developments in the extraction of ^{137}Cs$^+$ from water as depicted in the schematic outline of Figure 1. It will more specifically survey PB-CNT-Graphene based nanocomposite efficiencies in ^{137}Cs$^+$ extraction. Large-scale applicability in real case scenarios of such nanocomposites will be probed and their nanotoxicity issues will be discussed. A short summary of other new materials is also provided at the end, which opens-up new possibilities in combining these new materials with PB-CNT-graphene based nanocomposites.

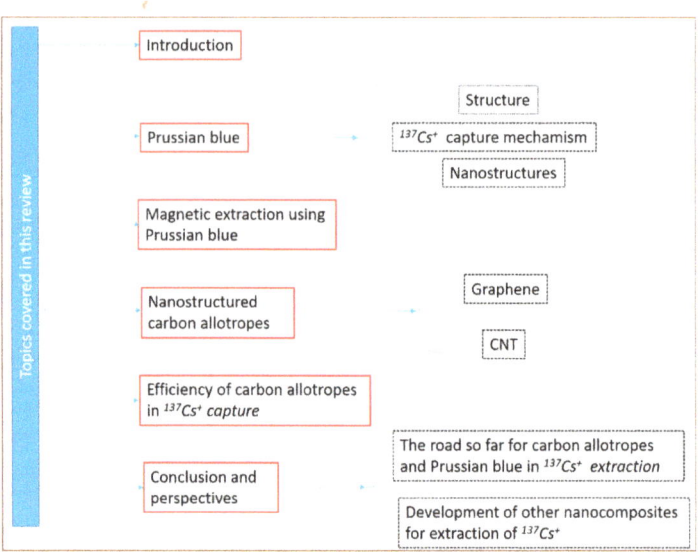

Figure 1. Topics covered in this review.

2. Prussian Blue

2.1. Structure

Prussian blue (PB) is a dark blue pigment synthesized by ferrous ferrocyanide salts with chemical formula $Fe_7(CN)_{18}$. It has a porous structure with the capacity to adsorb the $^{137}Cs^+$ ions into its pores and store them there. It is a metal organic framework (MOF) where the inorganic vertices, which donate electrons in the structure, are linked to each other via organic compounds. The complete chemical formula is $Fe^{III}_4[Fe^{II}(CN)_6]_3 \cdot xH_2O$. The compound has a face-centered cubic structure (FCC) structure (Figure 2) belonging to the Fm3m space group with a lattice parameter of 10.166 Å. Fe exists in two oxidation states within the structure: Fe^{3+} and Fe^{2+}. These ions form two different FCC lattices displaced by half a lattice parameter with respect to each other. However, the bi and tri-valent Fe are coordinated differently. Furthermore, they are linked to each other via cyanide groups (C ≡ N) i.e., C groups are linked to Fe^{2+} and N groups to Fe^{3+} with high and low spins respectively, in octahedral configurations. The Fe^{2+} and Fe^{3+} ratio of 3:4 implies that in order to obtain a charge neutrality within the structure a 25% vacancy of $[Fe^{+2}(CN)_6]^{4-}$ molecules is necessary [24]. Coordinated water molecules occupy the resulting octahedral cavities created by such vacancies; six water molecules are linked to Fe^{2+}. The other interstitial water molecules occupy the eight corners of the unit cell ($\frac{1}{4}, \frac{1}{4}, \frac{1}{4}$) and are essential for the insertion of the $^{137}Cs^+$ ions in the structure.

Fe^{+2} can be replaced by other transition metals with the same +2 oxidation states such as Ni, Mn, Cu and Co, coordinated exactly like Fe^{+2} in the structure and are called PB analogs. However, there are reports of Cd and Zn with slightly larger atomic radii also being incorporated into the structure owing to their +2 oxidation state [25]. The aim in including different species into the structure is to provoke a distortion of the PB lattice by producing vacancies of the high spin state molecule along with distortions in the vacant cages, in order to facilitate the capture and sequestration of the $^{137}Cs^+$ [26].

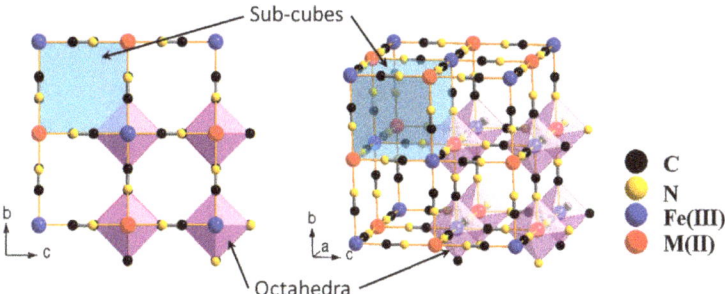

Figure 2. Framework of Prussian blue analogues. Adapted with permission from [27], Copyright RSC, 2012.

2.2. $^{137}Cs^+$ Ion Capture Mechanism in Prussian Blue

The compound is insoluble in water and the basic mechanism consists of ion exchange of $^{137}Cs^+$ and H^+ with the former occupying hydrophilic vacancies [28]. PB analogs have very different mechanisms of ion exchange or capture depending upon the anionic and alkali metal cation concentrations. Since PB and its analogs contain large amounts of interstitial and coordinated water, $^{137}Cs^+$ is captured by a defect created by a $[Fe^{+2}(CN)_6]$ vacancy, which creates a spherical cavity whose size is equivalent to the hydration radius of $^{137}Cs^+$. Nevertheless, recent calculations have demonstrated that a completely dehydrated $^{137}Cs^+$ ion can be incorporated into the structure with the release of a water molecule from the interstitial sites [29]. This is similar to certain clays, where on dehydrating the interlayers the $^{137}Cs^+$ selectivity increases [30]. On the other hand, water soluble analogs such as metal hexacyanoferrates (HCF) consisting of a alkali metal cation with a $[Fe^{+2}(CN)_6]$ anion, used for the extraction of $^{137}Cs^+$ have shown less efficiency. In such compounds Na^+ or K^+ are incorporated during the synthesis of the MOF in order to render them water-soluble [31]. In addition to $^{137}Cs^+$ capture mechanisms for non-soluble analogs; the water-soluble analogs mainly depend on the Na^+ or K^+ ion exchanges with Cs^+. Takahashi et al., have studied the $^{137}Cs^+$ uptake in KCuHCF PB analog in order to understand their lower adsorption capacity [31]. Three main mechanisms governed the $^{137}Cs^+$ ion exchange according to them, with the $^{137}Cs^+$-K^+ ion exchanges being predominant, as also stipulated by other research groups. In case of low anionic vacancies, the percolation of $^{137}Cs^+$ through the vacancies was prevalent. Finally, for low K^+ incorporation in the structure, proton exchange between $^{137}Cs^+$ and K^+ ions was evidenced. Ayrault et al., report a degradation in the crystal structure of the KCuHF soluble compound after $^{137}Cs^+$ adsorption which was not observed in the non-soluble counterpart [32].

2.3. Nanostructured Prussian Blue

$^{137}Cs^+$ adsorption in PB crystals is a very slow process. Fujita et al., have demonstrated that after two weeks of adsorption experiments, the depth of $^{137}Cs^+$ adsorption was at most between 1–2 nm, irrespective of the crystal size [33]. This implies that most of the adsorption occurs on the surface of the crystallites. This low diffusion depth is mainly attributed to the blocking of the vacancies by captured $^{137}Cs^+$ ions, which in turn hinders further $^{137}Cs^+$ diffusion. Since the diffusion depth appears to be a constant, increasing the specific surface would therefore be a solution to increasing the $^{137}Cs^+$ uptake. One way of augmenting the specific surface is by synthesizing nanoparticles of PB. The surface to volume ratio of crystallites increases as their size decreases; therefore nanoparticles have an extremely large surface to volume ratio. For example, a 3 nm nanoparticle will have 50% of its atoms on its surface. This would also imply that in the case of PB nanocrystals most of the vacancies and sites responsible for $^{137}Cs^+$ adsorption would be available on the surface, thus enhancing its specific surface. To this end, different research groups have produced various PB analogs of type Metal(M)-Co, where the nature of M defines the efficiency of the uptake. Liu et al., have demonstrated that Zn assisted Fe-Co PB analogs present high $^{137}Cs^+$ uptake efficiency [34]. They also observed that the size of the PB analog particle reduced with the reduction of Fe in the structure; for pure Zn-Co analogs, a crystallite size of ~73 nm was calculated, displaying the highest ^{137}Cs$^+$ adsorption as depicted in Figure 3.

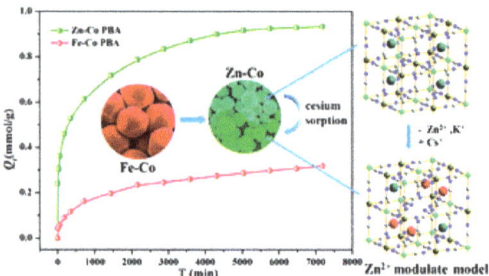

Figure 3. Adsorption efficiency of Fe-Co and Zn-Co prussian blue (PB) analogs as a function of time. The Zn-Co PB analog exhibits a higher Q_t (Q_t is the adsorption capacity per unit gram of the sorbent at a given time t) owing to the reduction in size of the nanoparticles. Adapted from [34] under the Creative Commons agreement from RSC, 2017.

Considering that the ^{137}Cs$^+$ adsorption depth is only about 1–2 nm, hollow PB nanoparticles may offer many more advantages. They not only have a very active surface area due to their large surface to volume ratio but their hollow interior is also capable of capturing and storing ^{137}Cs$^+$ [35]. A surfactant polyvinylpyrrolidone (PVP) was used to stabilize the nanoparticle and increase their dispersion in aqueous solutions. Figure 4 compares the efficiency of solid and hollow PB cubes of ~200 nm, in the capture of ^{137}Cs$^+$. The elemental mapping of Figure 4B depicts a higher concentration of captured ^{137}Cs$^+$ ions within the hollow structures than the filled ones in Figure 4A. Nevertheless, ^{137}Cs$^+$ diffusion depth greater than 2 nm would require higher activation energy at room temperature. Other methods are required to determine the exact diffusion depth in such structures, as these results are mainly qualitative. Besides, it is well known that the use of a surfactant shields the active sites and prevents the capture of ^{137}Cs$^+$. In order to avoid such shielding effects, PB could be coated onto support materials instead through hydroxyl bonds that anchor the PB to the support material. Carboxylic groups also tend to immobilize the PB particles in a sturdier manner. Wi et al., used a polyvinyl support surface functionalized with acrylic acid. This allowed converting the OH groups to COOH and providing a better adhesion of the PB [36]. The PB nanoparticles were immobilized on the PVP sponge and an increase in ^{137}Cs$^+$ uptake efficiency by five times was reported, compared to the hydroxyl bond functionalization.

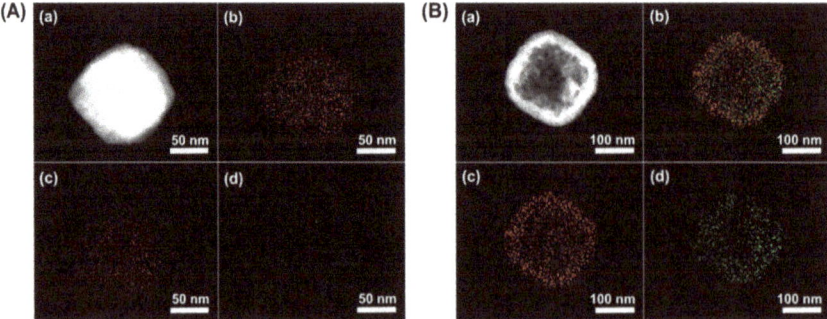

Figure 4. Elemental mapping images of solid (**A**) and hollow PB (**B**) nanoparticles of 190 nm in diameter. (**a**) Dark-field TEM image, (**b**) elemental mapping of both Fe and Cs (**c**) elemental mapping of Fe, and (**d**) elemental mapping of Cs. Adapted with permission from [35], Copyright RSC, 2012.

3. Magnetic Extraction Using Prussian Blue

There are reports of photo-induced magnetism where an electromagnetic radiation induces a residual magnetization even after the excitation is turned off, [37] due to low and high spin combinations of the transition metals in PB analogs. On their own PB and its analogs exhibit ferromagnetism at a Curie temperature of 11 K with a saturation magnetization of 3.4 emu/g as obtained by Tokoro et al., for Mn-Rb-Fe PB analogs [38]. The presence of Mn in the structure creates a Jahn-Teller distortion [39] by changing the M–CN–M bond angle and deviating it from 180°, whereupon inducing ferromagnetism. Among the various PB analogs, Mn based ones have shown the highest saturation magnetization [40]. One method of decreasing the Curie temperature of PB is by synthesizing nanoparticles of PB. Uemura et al., have demonstrated a decrease in Tc from 5.5 K in bulk PB to 4 K for PB nanoparticles protected by PVP [41]. Nevertheless, finding practical applications involving magnetic extraction would require having a Tc at around room temperature. Also, humidity increases the Curie temperature for Co-Cr PB analogs, [38] thus making it difficult for their direct application in aqueous media. This implies that most methods using PB for $^{137}Cs^+$ extraction, do not prescribe any efficient approach to recover the exhausted adsorbent.

Literature on nanostructured PB alone is very scarce as they are generally combined with magnetic nanomaterials like superparamagnetic iron oxide nanoparticles (SPIONs) i.e., Fe_3O_4 or γ-Fe_2O_3 nanoparticles. In the past, there have been reviews briefly describing $^{137}Cs^+$ adsorption employing magnetic PB-Fe_3O_4 nanoparticles [16]. However, this review goes further as it not only describes more recent developments in the latter but also discusses the development of the magnetic nanocomposite in detail from a nanoscale point of view. In nanostructures, physical properties such as magnetic moment as well as adsorption vary as a function of the nanoparticle size and further depend upon the surfactants used to stabilize them during synthesis. In the paragraphs that follow, the efficiency of magnetic PB nanoparticles and their combination with carbon allotropes are assessed. To the best of our knowledge, in the literature such nanostructures have not been evaluated in detail.

Core-shell structures with the magnetite constituting the core and the PB active layer constituting the shell have been employed. Jang et al., have reported that the poly(diallyldimethylammoniumchloride) (PDDA)@Iron oxide nanoparticles can act as nucleation sites for the precipitated PB, resulting in the coating of a negatively charged PB on the PDDA@Iron oxide nanoparticle surface [42]. Furthermore, they have also studied the magnetic properties of Fe_3O_4 and have observed a decrease in the saturation magnetization from 56 emu/g for pure Fe_3O_4 to 12 emu/g for the PB-Fe_3O_4 nanocomposite. The reduction was mostly due to the shielding of the superparamagnetism of Fe_3O_4 by the PB capping. Nevertheless, successful magnetic extraction was achieved with the nanocomposite [43]. PB analog compounds tend to show degradation in various applications after successive cycles of reuse. Chang et al., Figure 5A, have synthesized Fe_3O_4 (shell)-PB (core) nanocomposites with sizes between 20–40 nm; [44] two different concentrations of $FeCl_3$ were used during the synthesis. The higher Fe concentrations produced Fe_3O_4 cores with a higher magnetic saturation moment (Figure 5B), which conversely had an adverse effect on the $^{137}Cs^+$ adsorption due to the higher shielding of the active sites in the PB core as depicted in Figure 5C.

Some researchers have also used different ferrimagnetic nanoparticles like $CoFe_2O_4$ combined with PB for the extraction of cesium and have indicated very high efficiencies compared to pure $CoFe_2O_4$, which has some interesting $^{137}Cs^+$ adsorption capacity by itself [45]. $CoFe_2O_4$ possesses the spinel structure and tends to be physically more robust but is a hard magnetic compound compared to Fe_3O_4. The latter is a soft magnet with a very small residual magnetization.

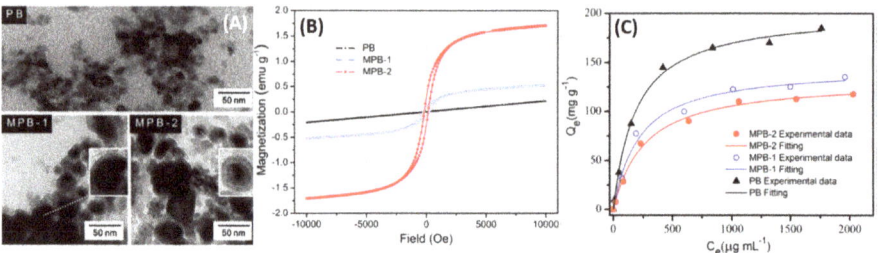

Figure 5. (**A**) TEM images of PB and magnetic PB (MPB) nanoparticles. (**B**) Field-dependent magnetization of PB and magnetic PB nanoparticles. (**C**) Adsorption capacity of samples under different Cs$^+$ concentration. $T = 25\ °C$; contact time = 6 h; [Cs$^+$]$_{initial}$ = 50–2500 µg/mL; $m_{adsorbent}/V_{solution}$ = 4 µg/mL. Adapted with permission from [44], Copyright RSC, 2016.

4. Nanostructured Carbon Allotropes in ^{137}Cs$^+$ Capture

Carbon based materials have a large surface area with good ion-exchange capabilities. These are further enhanced by the presence of functional groups such as hydroxyl or carboxyl groups on their surfaces, which can trap heavy metals from aqueous solutions. Carbon has various allotropes such as fullerenes, graphite, and diamond. The 2D carbon allotropes include carbon nanotubes and graphene [46]. The latter is the building block for graphite, which consists of stacked graphene sheets. On the other hand, a carbon nanotube (CNT) is a rolled-up graphene sheet and belongs to the fullerene or C60 family. The number of walls on a CNT is determined by the number of times the graphene sheet is rolled-up. The defects in graphene include folds and wrinkles operating as active sites towards their decoration. In a CNT, the edges, bends and breaks in the walls promote the attachment of functional groups that are subsequently linked to other organic or inorganic nanoparticles. Therefore, CNT and graphene are attractive nanoscale structures with large surface areas displaying high ion adsorption capacities and constituting versatile catalytic supports with tailored properties.

There are a few works based on activated carbon or carbonaceous materials in the extraction of low level ^{137}Cs$^+$ contamination [47]. Mesoporous carbon based magnetic nanoparticles for ^{137}Cs$^+$ sequestration are also being studied [48]. However, activated carbon is more efficient in extracting organic species than metal ions, [49] with an estimated adsorption efficiency between 0–10% for inorganic contaminants [50]. Activated carbon and PB analogs have shown moderate efficiency in ^{137}Cs$^+$ extraction with a Q_{max} = 63 mg/g [51]. PB acts as the active material while as activated carbon is the support material for the PB analog. On the other hand, nanoscale carbon-based materials are gaining importance as they can extract heavy metal ions as well as radioactive cesium ions on their own. PB and their analogs also show a higher efficiency and a greater adsorption capacity towards ^{137}Cs$^+$ when combined with carbon-based nanomaterials. Table 1 lists the various nanocomposites of CNT/graphene used in ^{137}Cs$^+$ extraction from aqueous solution. All of these nanocomposites are discussed in the paragraphs that follow.

4.1. Graphene Based Nanocomposites for ^{137}Cs$^+$ Extraction

There is not much work on pristine graphene in the adsorption of ^{137}Cs$^+$ from aqueous solutions. Only one paper reporting a very high efficiency of ^{137}Cs$^+$ extraction using pristine graphene has been published [52]. Graphene oxide and graphene functionalized with polyaniline (PANI) have also been used in radionuclide extraction, including ^{137}Cs$^+$ from aqueous solutions with the former showing a lower Q_{max} than the latter [53,54]. Combining graphene with nano PB further increases the efficiency as described below. Other combinations of graphene and PB with other bioabsorbants are also available in the literature viz. chitosan and biopolymers such as pectins [55,56]. For practical application, reclaiming the exhausted nanomaterials is an important aspect to consider. In addition, once the nano PB sorbent is exhausted it takes a very long time for a small quantity of it to precipitate, as PB has a tendency to disperse like a colloid in aqueous solutions. Combining them with graphene nanosheets cannot really solve the problem of recovering the exhausted materials either, as they too tend to form a colloidal suspension in aqueous solutions depending upon the pH. One way of overcoming the problem of reclaiming these PB nanoparticles is to pack them in a graphene oxide foam (Figure 6) that could be loaded directly into the filtration column [57]. This has already been applied to organic dyes such as methylene blue where wood was impregnated with graphene, intended for a continuous flow system. The sorbent showed potential of being regenerated and was reused multiple times [58].

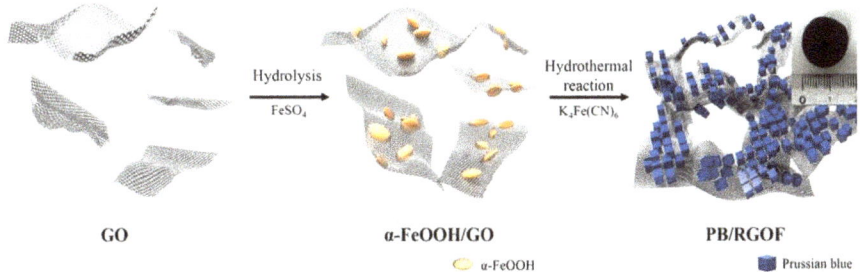

Figure 6. Schematics of PB and reduced graphene oxide nanocomposite synthesis in order to prepare foams. Adapted from [57] under Creative Commons agreement from Nature Research, 2015.

As mentioned earlier, PB nanoparticles have a tendency to agglomerate because of the magnetic moments generated by the various oxidation states of the transition metals [40]. In general, nanosize materials tend to agglomerate in order to reduce their surface energy [59]. The superparamagnetic properties of Fe_3O_4 allow easy recovery of the PB from the aqueous solution when combined together to form Fe_3O_4-PB nanocomposites. Nevertheless, the $^{137}Cs^+$ uptake efficiency of such nanocomposites tends to decrease with a higher magnetization of Fe_3O_4. This is mainly because when the magnetization increases, magnetic nanoparticles tend to show higher agglomeration, which conversely reduces their active surface area. To this end, PVP surfactant was employed, which increases the mono-dispersion of the nanomaterials and provides larger surface areas for adsorption. However, surfactants tend to reduce the adsorption efficiency as seen previously. In this regard, graphene based nanocomposites are therefore an alternative to the above-mentioned shortcomings. Firstly, Fe_3O_4 nanoparticles are immobilized on the graphene surface in a dispersed manner, whereby reducing their agglomeration. Secondly, these nanoparticles serve as spacers between the graphene sheets discouraging the graphene sheets to stick together due to Van der Waals interactions [60]. Thirdly, PB is then directly decorated onto the graphene without being linked to the Fe_3O_4 nanoparticles. Here again, PB hosts the active adsorption sites whereas Fe_3O_4 aids in magnetically extracting the exhausted nanocomposites. Graphene mainly serves as a support even though its synergistic interaction with PB cannot be neglected. Yang et al., obtained the nanocomposite by wet impregnation of graphene with $FeCl_3$ followed by $K_4[Fe(CN)_6]$ with constant magnetic stirring for 1 h at room temperature. The precipitates were subsequently recovered with a magnet by them [61].

4.2. Carbon Nanotubes for $^{137}Cs^+$ Extraction

Similarities in adsorption exist between graphene and CNT despite their opposite extreme aspect ratios. Like in the case of graphene, CNT also promises exciting $^{137}Cs^+$ extraction properties. Both have the capacity to be used either as colloids that can be magnetically extracted or as membranes which allow reclaiming them easily once exhausted [62]. The mechanism of adsorption is similar to graphene and relies on electrostatic, hydrogen and π-π interactions between the carbon allotrope and the contaminant [63]. Moreover, CNT have an additional advantage compared to graphene, which is the capacity to create mesopores when bundled together. Therefore, it presents several more sites for the adsorption of ions: within the walls of the CNT, between CNT or within the mesopores, on the walls of the CNT and in peripheral grooves [64].

In addition, nitric acid treated CNT have demonstrated efficacy in the extraction of heavy metals ions such as Cu^{2+}, Mn^{2+}, Co^{2+}, Zn^{2+}, Pb^{2+} owing to amine groups on the surface. This has motivated researchers to employ CNT for the extraction of other radioactive ions from aqueous solutions viz., U^{4+}, U^{6+}, Th^{4+} and radioactive Co^{2+} and Cu^{2+} ions [65]. Many groups have therefore carried out treatment or functionalization of CNT in order to make them more reactive to $^{137}Cs^+$ species. Several studies have shown that $^{137}Cs^+$ is attracted to OH groups [66]. CNT easily harbor different groups such as amine, hydroxyl, lactone, phenol or carboxyl groups on their surfaces post-functionalization [67]. They therefore are ideal candidates for the adsorption of $^{137}Cs^+$. Yang et al., have therefore prepared chitosan grafted CNT owing to the abundance of OH groups present in the former [68]. Their study demonstrated that the pH of the aqueous solution had a very important role to play on the $^{137}Cs^+$ adsorption by the OH groups of chitosan. They also manifested a higher adsorption than the bare CNT. Another study determined that increasing the amount of hydroxyl groups does not necessarily bring about an increase in $^{137}Cs^+$ adsorption. To this end, Yang et al., combined CNT with bentonite which are both known $^{137}Cs^+$ adsorbents linked to each other via chitosan [69]. They demonstrated that cation exchange is more effective than hydroxyl capture. More importantly, they established that hydroxyl capture depends upon the host matrix and increasing the number of OH groups will not increase the $^{137}Cs^+$ adsorption. This mainly suggests that the OH groups should be linked directly to the CNT matrix and are therefore limited by the number of active sites on the CNT walls. In this regard, single-walled carbon nanotube (SWCNT) could be more effective [70] than multi-walled carbon nanotube (MWCNT) [71], considering that the former possess a higher specific surface. In any case, the type of functional group also plays a role in $^{137}Cs^+$ selectivity.

Other groups have studied the effects of Prussian blue and their analogs on the adsorption efficiency of $^{137}Cs^+$ when combined with CNT. These tend to have a catalytic effect on the PB analogs by increasing the adsorption efficiency of the latter [72]. In one study by Li et al., Cu-Co-Ni PB analogs were combined with CNT [73]. Chitosan was wrapped around the CNT to increase their dispersion and also to link them to the PB analogs. Two studies have combined CNT with Cu cyanoferrates linked via different amine groups i.e., propargylamine (Figure 7) and dietheleneamine on SWCNT [74] and MWCNT [75], respectively. The Q_{max} increased from 150 mg/g (MWCNT) to 239 mg/g (SWCNT). These differences could be attributed to the use of different types of nanotubes and to variety of the amine ligands. In any case, the SWCNT study demonstrated that only 30% of $^{137}Cs^+$ ions were adsorbed by the PB nanoparticles. The rest were adsorbed onto other active sites of the functionalized SWCNT. Tsuruoka et al., have studied ferrocyanides embedded into diatomite nanoparticles encaged in both MWCNT and SWCNT which were impregnated in a polyurethane spongiform [76]. They have observed a synergistic effect owing to the porous structures created by both types of CNT. Zheng et al., have also demonstrated that it was possible to electrochemically clean the nanocomposite of $^{137}Cs^+$ ions and reuse them [77]. Water-soluble sodium CoHCF encapsulated in alginate beads reinforced with highly dispersed CNT was employed in the extraction of $^{137}Cs^+$. The beads were stable in a broad pH range of 4–10 and revealed potential for large scale applications [72]. A combination of graphene, carbon fiber and PB has also been used to extract $^{137}Cs^+$ directly from a lake in China [78]. Such a combination therefore allows an increased specific surface area with prospective scalable applications. A high efficacy of the graphene-CNT nanocomposites has been demonstrated for the extraction of aromatic compounds and Cu^{+2} heavy metal ions [79]. These hybrid carbon allotropes exhibit 25% higher adsorption capacities in both cases than their individual counterparts. They have also shown high desalination capacities along with adsorption of heavy metal ions [80]. This suggests that alkali metal ions would tend to have a higher affinity towards the hybrid carbon allotrope compared to PB, making $^{137}Cs^+$ extraction by PB more effective in a high salinity aqueous medium such as seawater. Combining them with Fe_3O_4 would then be a suitable method of reclaiming the exhausted sorbents.

Figure 8 provides the schematic of the $^{137}Cs^+$ extraction mechanism of graphene/CNT/magnetic PB nanocomposites.

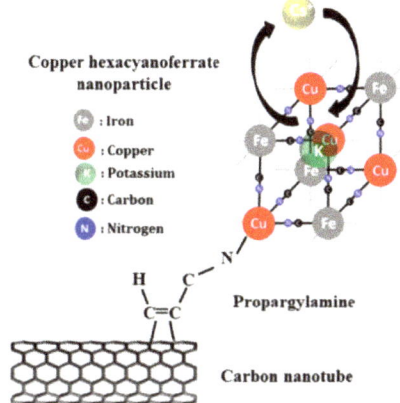

Figure 7. Mechanism of $^{137}Cs^+$ capture in propargylmine functionalized SWCNT decorated with copper hexanoferrates. Adapted with permission from [74], Copyright RSC, 2017.

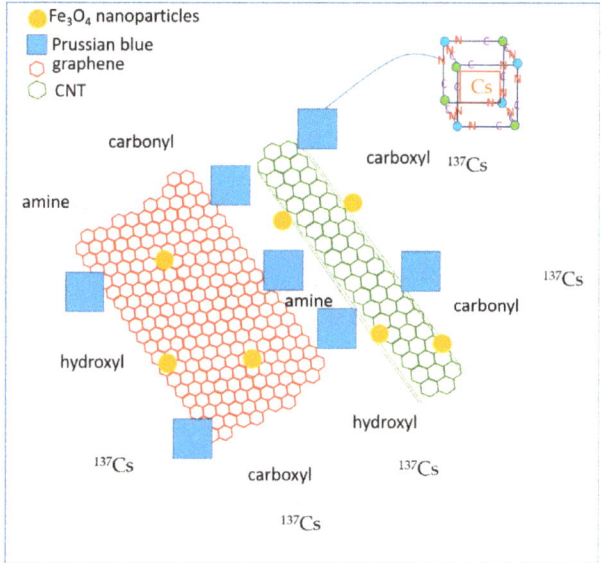

Figure 8. Mechanism of $^{137}Cs^+$ capture in graphene/CNT/PB nanocomposite. Fe_3O_4 is used for magnetic extraction and does not participate in the adsorption of $^{137}Cs^+$. PB is linked to CNT and graphene. Some of the possible functional groups that capture $^{137}Cs^+$ are depicted on the CNT and graphene surfaces. A sub-cube or unit-cell of PB capturing and sequestering the ^{137}Cs atom is also provided in the top right corner. The nanocomposite therefore harbors several active sites for the capture of $^{137}Cs^+$ ions.

Table 1. Various Graphene/carbon nanotubes (CNT)/PB/Fe$_3$O$_4$ based nanocomposites arranged in decreasing order of their Q_{max}. The table provides the ^{137}Cs$^+$ removal efficiency, K_d, initital concentration, temperature of extraction, isotherm and kinetic models used.

Nanocomposite	Removal Efficiency %	K_d L/g	pH	Initial Concentration mg/L	Temperature °C	Isotherm	Kinetic Model	Q_{max} mg/g
Pristine graphene [52]	41	0.115	1–12	10–10000	27	Langmuir	NA	465
Fe$_3$O$_4$-GO-pectin [56]	NA	50	1–11	9–822	30	Langmuir Freundlich Tempkin	PFO PSO	432
MWCNT-CuHCF [77]	95	568	7	1	28	Langmuir Fruendlich	NA	310
SWCNT-DMAD [70]	NA	NA	NA	NA	25	NA	NA	250
SWCNT-CuHCF [74]	NA	NA	NA	NA	25	NA	NA	230
Chitosan-PB-CNT [73]	NA	42.5	1–9	200	20	Langmuir Freundlich Redlich Peterson	PFO PSO Boyd Webbe Morris Elovich	219
GO-PANI [53]	100	94.5	2–10	1–100	25	Langmuir Freundlich	NA	185
Bentonite [69]	89	34.2	3–10	10	25	Langmuir Freundlich	NA	182.7
MWCNT-Cu ferrocyanide [75]	NA	NA	1–12	0.027–55	25	Langmuir	NA	151
MWCNT-Na-CoHCF-alginate [72]	53	23	2–10	200	25	Langmuir Freundlich	PFO PSO	133.33
MWCNT-Amino functionalized [71]	95	7–13	1–9	4.5×10^{-6}–2.28×10^{-4}	15–35	Langmuir Freundlich Tempkin	PFO PSO	117–136
PB-GO-Carbon fiber [78]	85.48	159.8	2–8	5–80	25	Langmuir Freundlich	PFO PSO	81.25
MWCNT [75]	NA	NA	1–12	0.027–55	25	Langmuir	NA	46
Chitosan-MWCNT [68]	25–70	26.7	3–10	1–42	25	Langmuir Freundlich	NA	45
Fe3O4-PB-GO- alginate beads [82]	80	48.7	7	25–150	0–30	Langmuir Freundlich	PSO	43.5
MWCNT-betonite [69]	80	23	3–10	10	25	Langmuir Freundlich	NA	159.4
Fe$_3$O$_4$-PB-hydrogels [81]	99.5	0.4	4–10	100–500	25	Langmuir	PSO	41.5
GO [54]	55	481	1–13	10	23	Langmuir Freundlich	PFO PSO	40
MWCNT [69]	35	27.8	3–10	10	25	Langmuir Freundlich	NA	45.6
Raw MWCNT [68]	12–55	55.6	3–10	1–42	25	Langmuir Freundlich	NA	29
RGO-PB [57]	99.5	6.455	NA	0.2	NA	Langmuir Freundlich	NA	18.67
MWCNT [67]	45	35	1–10	5–20	25	Langmuir Freundlich	NA	12.75
Pristine MWCNT [67]	NA	14	1–10	5–20	25	Langmuir Freundlich	NA	1.63

NA-not available.

5. Efficiency of Various Carbon Based Materials in $^{137}Cs^+$ Capture

Various batch tests performed on the different sorbents have shown increased efficiency of the graphene or CNT based sorbents (Table 1). Adsorption kinetics and equilibrium studies provide the efficacy of an adsorbent in a given adsorption system and are necessary to understand the underlying adsorption mechanisms. Most of the equilibrium adsorption studies that are provided in Table 1, employ the Langmuir and Freundlich adsorption models [67,77]. In the Langmuir model, the adsorption of $^{137}Cs^+$ from an aqueous solution onto a graphene-based nanocomposite is mainly used to obtain Q_{max} values. The model would hold for graphene-based nanocomposites as only a single monolayer of adsorption is considered, as defined by the active sites on the graphene/CNT surfaces. In the Langmuir model, all sites are considered equally favorable. This is contrary to the Freundlich model, which considers that the active sites possess varying adsorption energies and such is the case of graphene-PB based nanocomposites. In the Tempkin model it is assumed that the adsorption energy reduces as and when the sites become saturated [71,77]. The Brunauer-Emmet-Teller model assumes multilayer adsorption by considering the Langmuir model for each adsorbed layer [75]. Figure 9 provides examples of some of the isotherm models used in CNT/graphene-PB nanocomposites surveyed in this work. However, graphene and CNT based PB nanocomposites can be regarded as monolayer adsorption considering that the active sites lie on the surface of CNT and graphene, implying that all three models, Langmuir, Freundlich and Tempkin are appropriate. These equilibrium models are seconded by kinetic models such as, the pseudo first order (PFO) and second order kinetic models (PSO), which define the adsorption rate limiting factors [70,78,81,82]. The Elovich kinetic model has also been applied to chitosan-CNT-PB composites in consideration of their elemental heterogeneity, as the solid surfaces under study are energetically heterogeneous also [74].

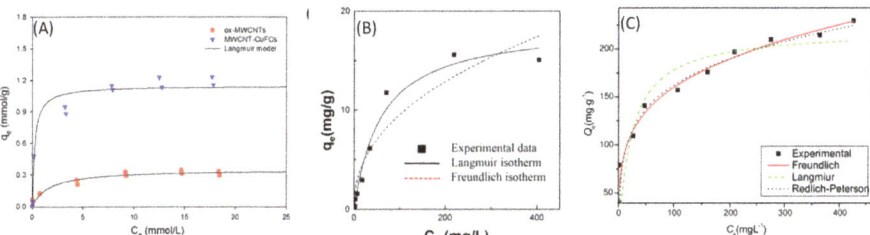

Figure 9. Examples of Langmuir, Freundlich and Redlich-Peterson adsorption isotherms models applied to (**A**) oxidized MWCNT and CuFC-MWCNT, adapted with permission from [75], Copyright Springer Nature, 2015, (**B**) Graphene foam-PB composite, adapted from [57] with permission under Creative Commons agreement from Nature Research, 2015 and (**C**) Chitosan-PB analog-CNT, adapted with permission from [73], Copyright Springer Nature, 2016.

Table 1 below therefore provides Q_{max} values obtained from batch tests of several different CNT, graphene, Fe_3O_4 and PB combinations. Q_{max} gives a direct indication of the available active sites or the maximum monolayer capacity for $^{137}Cs^+$ adsorption. Moreover, the Langmuir equation is valid over a wide concentration range. The table indicates that PB analogs along with CNT statistically provide the highest Q_{max} values reaching 310 mg/g. Pristine CNT or graphene in general do not show much efficiency compared to their hybrid counterparts. However, when they are decorated with amine groups their Q_{max} reaches 240 mg/g. This value is higher than CNT attached with hydroxyl groups of chitosan. However, chitosan when used as a linker between CNT and PB has a Q_{max} of 219 mg/g. Nevertheless, the value of the Q_{max} depends on various factors such as pH, temperature of the solution, presence of competing ions and the initial concentration of $^{137}Cs^+$. The particle size of the sorbent is also an important consideration but since we are only considering nanosize sorbents in this review, we can neglect this parameter in our discussions. Various studies on graphene or CNT based materials in this study show that the temperature of the solvent is instrumental in increasing the Q_{max} of the sorbent [54,71,78]. This is mainly because the process becomes endothermic with temperature increase and ^{137}Cs$^+$ in the solution is therefore more mobile due to which they attain active sites more efficiently on the sorbent. This further implies that ^{137}Cs$^+$ capture efficiency and can increase if the number of active sites obtained by various treatment or functionalization on the carbon-based materials are increased to their maximum capacity. The adsorption of ^{137}Cs$^+$ is also highly dependent on the initial pH of the solution, which has mostly to do with the functional groups of the carbon-based materials becoming protonated at low pH [67,72]. Since an OH group is necessary for the capture of ^{137}Cs$^+$ as discussed above a reduction in the former due to protonation would decrease the Q_{max} of the sorbent. At higher pH, a deprotonation of the functional group and an increase in OH$^-$ ions in the solution makes the ^{137}Cs$^+$ capture process more efficient.

The presence of other ions or salts such as Mg, Na, Li and K also affects the ^{137}Cs$^+$ adsorption due to the inherent competition between them. They all manifest similar affinities to the active sites on the carbon-PB based nanocomposites. In the case of PB or its analog, a higher affinity to K$^+$ and Na$^+$ has been observed with a given order of competition: K$^+$ > Ca^{2+} > Mg^{2+} > Na$^+$. When combined with MWCNT, the Q_{max} value decreased only slightly but showed a higher selectivity towards ^{137}Cs$^+$ [75]. Lin et al., have used CNT-PANI-NiHCF composites for electrically switching ion exchange studies and have demonstrated a higher affinity towards ^{137}Cs$^+$ than Na$^+$ ions [83]. Such nanocomposites are also employed for the extraction of Sr^{2+} at the same time. However, they appear to have a greater affinity towards ^{137}Cs$^+$ than ^{81}Sr^{2+}, which may have to do with the difference of oxidation states of the two cations [73]. The contact time between the sorbent and the sorbate are also necessary parameters towards understanding the adsorption of ^{137}Cs$^+$. Optimal adsorption times are required in order to evaluate the feasible scalability of the developed methods. This means that the contact time should be calculated in such a way that optimum adsorption should be obtained with minimum leaching of the adsorbed ^{137}Cs$^+$ as well as the transition metals and alkali metals constituting the PB framework. Both adsorption and equilibrium times should be short for large-scale applications. The rate of adsorption depends on the temperature, number of adsorption sites, and diffusion of the ^{137}Cs$^+$ within the porous matrix, which in turn depends on the density of the packed material. In the case of MWCNT, a contact time of 80 min was required to attain equilibrium [67]. A study with graphene oxide required a long contact time of 24 h [54] similarly to stirred pectin stabilized magnetic PB-GO nanocomposites [56]. For PB-graphene-carbon fiber composite the equilibrium was reached after 8 h [78].

6. Toxicity of Nanomaterials

Toxicity of nanomaterials is a key issue that is under debate [84]. If used in water remediation systems, their release and disposal need to be properly planned and evaluated. Nanomaterials are small enough to pass the skin barrier and enter the blood stream. In addition, they can be inhaled causing lung damage or ingested causing kidney damage. In the case of heavy metals, their ions can be fatal and can engender life-threatening maladies. Both CNT and graphene have been studied to evaluate their toxicity and environmental impact. Studies on zebra fish have shown that both their growth and cardiac rates are affected by these 2D nanocarbon allotropes [85]. The toxicity of CNT is an important enough issue to incite the development of related counter small-molecule-drugs [86]. CNT present risks during their entire lifecycle and mainly during occupational exposure. This implies that their manipulation needs to be regulated. CNT could also interact with biomolecules in the water and produce toxic effects [87]. The stability of PB needs to be thoroughly evaluated in various aqueous conditions. Studies have demonstrated the possibility of leaching of cyanide into ground water on decomposition of the ferrocyanides [88]. This also implies that ^{137}Cs$^+$ could leach out and contaminate ground water. One method to stabilize PB analogs viz., titanium ferrocyanide and curb leaching of cyanide and ^{137}Cs$^+$ includes transforming the spent material into lithium titanate, which consequently immobilizes the ^{137}Cs [89]. In some studies, the release of iron from PB in Fenton type reactions were studied by Doumic et al. [90]. They studied the catalytic effect in a pH = 3 environment and have observed that insoluble PB nanocomposites were more stable and showed negligible Fe release (10%

after 13 cycles). Kim et al., also have observed negligible leaching of Fe during the $^{137}Cs^+$ adsorption process in PB-cellulose hydrogel composites [91]. Yang et al., have also carried out a systematic study of the possible leaching of Fe from PB-Fe_3O_4-graphene nanocomposites in a wide pH range from 4 to 10 [61]. Even in the highest ionic strength seawater, the amount of Fe leaching ranged from 0.95% to 0.61% while in natural water, Fe leaching as low as 0.0026% was observed. In another study with PB embedded magnetic hydrogels, leaching of Fe was studied over a period of two weeks and only a 0.3% increase in Fe was observed in the solution [81]. Graphene is also considered a chemically stable material. Nevertheless, changes in pH affect the protonation of the carbonyl or hydroxyl groups. At high pH graphene dissolves like a salt while at low pH, it tends to form aggregates and at neutral pH it stays suspended in the solution [92]. This also implies that the stability of graphene oxide depends upon the functional groups on its surface and their particle sizes [93]. In such a scenario, the dose related toxicity could vary. Overall, the formation of colloids should be absolutely avoided. However, the effect of graphene on the human body is still unknown; on the other hand, graphene oxide has been known to accumulate in the lungs of mice when inhaled, but no pathological outcome was further reported. Nano-graphite and Fe_3O_4 were also tested for Fe leaching. A very low pH < 3 showed an exponential increase of Fe release. Fe_3O_4 are linked to graphene via electrostatic forces or via functional groups on their surfaces similar to CNT therefore, the risk of nanoparticles being loosely bonded does exist. Nevertheless, in general, many studies have shown that linking Fe_3O_4 to graphene and CNT exhibit exceptionally high stability with very little leaching of Fe [94]. In any case, the exhausted nanoadsorbent has to be removed from the water and presently; magnetic extraction appears to be the most adapted solution.

7. Conclusions and Perspectives

7.1. The Road so Far for $^{137}Cs^+$ and the Prospects of Graphene/CNT-PB Based Nanocomposites Towards Its Extraction

Research on $^{137}Cs^+$ contamination has taken giant leaps since the Chernobyl catastrophe. Cesium is known to damage soft tissue, bones and provoke bone cancer. Today, eight years after the Fukushima Daiichi disaster, cesium still seems to be prevalent in soil and water; a large-scale cleanup at both fronts is still being pursued. Radioactive $^{137}Cs^+$ has also been detected in tap waters of Tokyo. Salty ground water containing $^{137}Cs^+$ is one artery for land contamination to reach the sea. Even though there is some on-site storage capacity at Fukushima Daiichi of the contaminated water, a large amount of the least contaminated water has already been rejected into the sea. There has been some progress in the cleansing of waters, however the problem is still at large. Tokyo Electric Power Company (TEPCO) forecasts the decontamination procedure to continue for the next 30–40 years. Therefore, new materials and methods towards the capture and sequestration of $^{137}Cs^+$ are being developed at an accelerated pace in order to quickly reply to the concern.

Adsorption and ion exchange have been found to be efficient methods in the capture of $^{137}Cs^+$. The contamination in Fukushima has been cleaned by the Kurion process, which consisted of effective capture of the ions ($^{81}Sr^{2+}$ and $^{134,137}Cs^+$) via an extremely porous aluminosilicate based zeolite [95]. Areva, the French nuclear company then further treated the water with a nickel ferrocyanide (NiHFC) and sand-polymer mixtures [96], which reduced the contamination by at least more than 1000 times. Advanced liquid processing system (ALPS) technology uses different combinations of resins, zeolites, titanates, activated carbons and hexacyanoferrates with a selectivity towards 62 radionuclides but is only applicable after ^{137}Cs, ^{134}Cs and ^{90}Sr quantities are reduced by the Kurion and Areva methods [97]. Therefore, the Prussian blue analog or hexacyanoferrates have already shown efficiency at a large scale and a total of 482,000 m^3 of water has been decontaminated so far. The major drawback of the methods was the similarity between Na^+ and $^{134,137}Cs^+$ affinities thus, making the whole process less efficient. The storage of sludge obtained via such processes was initially planned in large underground tanks on the Fukushima-Daiichi premises with no radioactive leaching detected. The waste was then immobilized by cementation treatment [98]. The major disadvantage of cementation was the increase in volume of the waste to be stored in a nuclear repository along with the risk of the cement cracking. To this end, vitrification has been proposed as a method that not only immobilizes ^{137}Cs more securely without the risk of breaking or cracking but also compresses the quantity of stored radioactive waste.

Graphene based nanomaterials seem to have a catalytic effect on PB as the Na$^+$ and K$^+$ selectivity decreases compared to that of ^{137}Cs$^+$. This selectivity towards ^{137}Cs$^+$ can be further enhanced on electrical switching as in the case of PB with CNT. In addition, ^{81}Sr^{2+} and ^{134}Cs$^+$ ions are also extracted simultaneously with ^{137}Cs$^+$. Nevertheless, PB does show a higher affinity to ^{134,137}Cs$^+$ than ^{81}Sr^{2+}, which can be further enhanced by using high specific surface materials such as graphene, CNT or hollow nano PB and their analogs, which exhibit higher Q_{max} values than their bulk counterparts. These nanopowders can be dispersed into a water body to be decontaminated and recovered via magnetic extraction on attaching Fe$_3$O$_4$ nanoparticles to the graphene surfaces. The drawback of Fe$_3$O$_4$ is that the saturation magnetization is rather low for large-scale applications as they usually contain surfactants. This would imply a risk of leaving behind some nanomaterial uncaught by the applied magnetic field. Another scenario could consist of making membranes of functionalized graphene and CNT with PB, which could then be easily recovered once exhausted. The main drawback of using PB is an increase of alkali metal content in the water due to ion exchange. This would not have detrimental effects on seawater; however, increased salinity in ground water could render it unsafe to drink.

Graphene and CNT are resistant in harsh environments and further offer recycling capabilities. Moreover, carbon is the fourth most abundant element in the Earth's crust implying continuous availability for large-scale applications. However, no publications reporting large-scale use of graphene or CNT in water filtration are available. This suggests that new technologies and methods employing them are potential candidates. Nevertheless, large-scale applications require large cost-effective production and functionalization capacities. Moreover, a choice between fixed bed and batch adsorption has to be made when employing graphene. In certain cases, batch methods appear to be more efficient for contaminant adsorption than fixed bed methods. Such a choice would allow obtaining maximal adsorption with minimal amounts of absorbent, thereby reducing the quantities of radioactive wastes to be treated [99]. Further studies on the toxicity of these carbon-based materials are also required. The most important parameter is the contact time of the adsorbent with the adsorbate, which should be as low as possible for large-scale applications especially considering the thousands of cubic meter waiting to be remediated. Regeneration of graphene and CNT has been studied using acids and bases, making the recovery of the metallic contaminants feasible. However, from an industrial point of view, there is much work that needs to be conducted in this very promising field.

7.2. Development of Other Nanocomposites

Different materials need to be combined in order to increase the ^{137}Cs$^+$ selectivity and hence its extraction efficiency, as already proven by the Kurion-Areva-ALPS combinations. With regards to other potential materials, PB has also been combined with zeolites for the adsorption of ^{137}Cs$^+$ ions [100,101]. Zeolites were seen as promising materials [101] for ^{137}Cs$^+$ adsorption [102,103] and were subsequently combined with fibrous polymer adsorbents to enhance their ^{137}Cs$^+$ extraction capacity. These zeolite-polymer composite fibers were initially studied for heavy metal ion removal such as Pb^{2+}, Cu^{2+}, Ni^{2+}, and Cd^{2+} [104,105]. One advantage is that the surface to volume ratio could be tuned through the zeolite fibrous polymer ratio with a specific surface value of 145 m^2/g. The study suggested that with such zeolite polymer composite fibers, the Pb^{2+} extraction takes place by both ion exchange and inclusion mechanisms. Zeolite and poly(ethersulfone) were then combined for ^{137}Cs$^+$ extraction with 30 wt% of zeolite in porous fibrous polymer [15]. These composites were applied to the decontamination of radioactive ^{137}Cs$^+$ in the city of Fukushima for a period of 28 days. A total of 7700 Bq/kg of radioactive Cs was extracted during this period, while with the zeolite alone only 33 Bq/kg was extracted [7]. The main advantage of this composite is that water can flow unobstructed through the composite compared to the zeolite powders. Fiber based porous structures seem to be more adapted for large flow extraction than magnetic extraction that needs a specific set-up. The latter nevertheless has the advantage of limiting the quantity of nuclear waste produced during the radioactive $^{137}Cs^+$ extraction process. Functionalized graphene and CNT could very well be included in such porous structures without clogging the pores and increasing the $^{137}Cs^+$ selectivity during extraction.

Author Contributions: The publication was researched, conceptualized, written, reviewed, proof read and edited by P.R., E.R. acquired the funds for the project and contributed to the writing of two sections of the manuscript, and critically reviewing and proof reading it.

Funding: The Authors would like to acknowledge the Center of Excellence project EQUITANT number TK134 (F180175TIBT) for financial support.

Conflicts of Interest: The authors declare no conflict of interest.

References

1. Delmore, J.E.; Snyder, D.C.; Tranter, T.; Mann, N.R. Cesium isotope ratios as indicators of nuclear power plant operations. *J. Environ. Radioact.* **2011**, *102*, 1008–1011. [CrossRef]
2. Vakulovsky, S.M.; Nikitin, A.I.; Chumichev, V.B.; Katrich, I.Y.; Voitsekhovich, O.A.; Medinets, V.I.; Pisarev, V.V.; Bovkum, L.A.; Khersonsky, E.S. Cesium-137 and strontium-90 contamination of water bodies in the areas affected by releases from the chernobyl nuclear power plant accident: An overview. *J. Environ. Radioact.* **1994**, *23*, 103–122. [CrossRef]
3. Nesterenko, V.B.; Yablokov, A.V. Chapter I. Chernobyl Contamination: An Overview. *Ann. N. Y. Acad. Sci.* **2009**, *1181*, 4–30. [CrossRef]
4. Varskog, P.; Næumann, R.; Steinnes, E. Mobility and plant availability of radioactive Cs in natural soil in relation to stable Cs, other alkali elements and soil fertility. *J. Environ. Radioact.* **1994**, *22*, 43–53. [CrossRef]
5. Morino, Y.; Ohara, T.; Watanabe, M.; Hayashi, S.; Nishizawa, M. Episode Analysis of Deposition of Radiocesium from the Fukushima Daiichi Nuclear Power Plant Accident. *Environ. Sci. Technol.* **2013**, *47*, 2314–2322. [CrossRef]
6. Parajuli, D.; Tanaka, H.; Hakuta, Y.; Minami, K.; Fukuda, S.; Umeoka, K.; Kamimura, R.; Hayashi, Y.; Ouchi, M.; Kawamoto, T. Dealing with the Aftermath of Fukushima Daiichi Nuclear Accident: Decontamination of Radioactive Cesium Enriched Ash. *Environ. Sci. Technol.* **2013**, *47*, 3800–3806. [CrossRef]
7. Kobayashi, T.; Ohshiro, M.; Nakamoto, K.; Uchida, S. Decontamination of Extra-Diluted Radioactive Cesium in Fukushima Water Using Zeolite–Polymer Composite Fibers. *Ind. Eng. Chem. Res.* **2016**, *55*, 6996–7002. [CrossRef]
8. Staunton, S.; Dumat, C.; Zsolnay, A. Possible role of organic matter in radiocaesium adsorption in soils. *J. Environ. Radioact.* **2002**, *58*, 163–173. [CrossRef]
9. Lieser, K.H.; Steinkopff, T. Chemistry of radioactive cesium in the hydrosphere and in the geosphere. *Radiochim. Acta* **1989**, *46*, 39–47. [CrossRef]
10. Gibert, O.; Valderrama, C.; Peterková, M.; Cortina, J.L. Evaluation of Selective Sorbents for the Extraction of Valuable Metal Ions (Cs, Rb, Li, U) from Reverse Osmosis Rejected Brine. *Solvent Extr. Ion Exch.* **2010**, *28*, 543–562. [CrossRef]
11. Kosaka, K.; Asami, M.; Kobashigawa, N.; Ohkubo, K.; Terada, H.; Kishida, N.; Akiba, M. Removal of radioactive iodine and cesium in water purification processes after an explosion at a nuclear power plant due to the Great East Japan Earthquake. *Water Res.* **2012**, *46*, 4397–4404. [CrossRef]
12. Adabbo, M.; Caputo, D.; de Gennaro, B.; Pansini, M.; Colella, C. Ion exchange selectivity of phillipsite for Cs and Sr as a function of framework composition. *Microporous Mesoporous Mater.* **1999**, *28*, 315–324. [CrossRef]
13. Chitry, F.; Pellet-Rostaing, S.; Nicod, L.; Gass, J.-L.; Foos, J.; Guy, A.; Lemaire, M. Cesium/sodium separation by nanofiltration-complexation in aqueous medium. *Sep. Sci. Technol.* **2001**, *36*, 1053–1066. [CrossRef]
14. Mahendra, C.; Bera, S.; Babu, C.A.; Rajan, K.K. Separation of Cesium by Electro Dialysis Ion Exchange using AMP-PAN. *Sep. Sci. Technol.* **2013**, *48*, 2473–2478. [CrossRef]
15. Masaru Ooshiro, T.K. Shuji Uchida Fibrous zeolite-polymer composites for decontamination of radioactive waste water extracted from radio-Cs fly ash. *Int. J. Eng. Tech. Res.* **2017**, *7*, 6.

16. Liu, X.; Chen, G.-R.; Lee, D.-J.; Kawamoto, T.; Tanaka, H.; Chen, M.-L.; Luo, Y.-K. Adsorption removal of cesium from drinking waters: A mini review on use of biosorbents and other adsorbents. *Bioresour. Technol.* **2014**, *160*, 142–149. [CrossRef]
17. Olatunji, M.A.; Khandaker, M.U.; Mahmud, H.N.M.E.; Amin, Y.M. Influence of adsorption parameters on cesium uptake from aqueous solutions—A brief review. *RSC Adv.* **2015**, *5*, 71658–71683. [CrossRef]
18. El-Rahman, K.M.A.; El-Sourougy, M.R.; Abdel-Monem, N.M.; Ismail, I.M. Modeling the Sorption Kinetics of Cesium and Strontium Ions on Zeolite A. *J. Nucl. Radiochem. Sci.* **2006**, *7*, 21–27. [CrossRef]
19. Awual, M.R.; Yaita, T.; Miyazaki, Y.; Matsumura, D.; Shiwaku, H.; Taguchi, T. A Reliable Hybrid Adsorbent for Efficient Radioactive Cesium Accumulation from Contaminated Wastewater. *Sci. Rep.* **2016**, *6*, 19937. [CrossRef]
20. Solbrå, S.; Allison, N.; Waite, S.; Mikhalovsky, S.V.; Bortun, A.I.; Bortun, L.N.; Clearfield, A. Cesium and Strontium Ion Exchange on the Framework Titanium Silicate $M_2Ti_2O_3SiO_4 \cdot nH_2O$ (M = H, Na). *Environ. Sci. Technol.* **2001**, *35*, 626–629. [CrossRef]
21. Behrens, E.A.; Clearfield, A. Titanium silicates, $M_3HTi_4O_4(SiO_4)_3 \cdot 4H_2O$ (M=Na^+, K^+), with three-dimensional tunnel structures for the selective removal of strontium and cesium from wastewater solutions. *Microporous Mater.* **1997**, *11*, 65–75. [CrossRef]
22. Cho, Y.; Seol, B.N. A Study on Removal of Cesium and Strontium from Aqueous Solution Using Synthetic Na-Birnessite. *J. Korean Soc. Water Wastewater* **2013**, *27*, 155–164. [CrossRef]
23. El-Naggar, I.M.; Mowafy, E.A.; El-Aryan, Y.F.; Abd El-Wahed, M.G. Sorption mechanism for Cs^+, Co^{2+} and Eu^{3+} on amorphous zirconium silicate as cation exchanger. *Solid State Ion.* **2007**, *178*, 741–747. [CrossRef]
24. Kumar, A.; Yusuf, S.M.; Keller, L. Structural and magnetic properties of $Fe[Fe(CN)_6] \cdot 4H_2O$. *Phys. Rev. B* **2005**, *71*, 054414. [CrossRef]
25. Nie, P.; Shen, L.; Luo, H.; Ding, B.; Xu, G.; Wang, J.; Zhang, X. Prussian blue analogues: A new class of anode materials for lithium ion batteries. *J. Mater. Chem. A* **2014**, *2*, 5852–5857. [CrossRef]
26. Matsuda, T.; Kim, J.; Moritomo, Y. Control of the alkali cation alignment in Prussian blue framework. *Dalton Trans.* **2012**, *41*, 7620–7623. [CrossRef] [PubMed]
27. Lu, Y.; Wang, L.; Cheng, J.; Goodenough, J.B. Prussian blue: A new framework of electrode materials for sodium batteries. *Chem. Commun.* **2012**, *48*, 6544–6546. [CrossRef]
28. Ishizaki, M.; Akiba, S.; Ohtani, A.; Hoshi, Y.; Ono, K.; Matsuba, M.; Togashi, T.; Kananizuka, K.; Sakamoto, M.; Takahashi, A.; et al. Proton-exchange mechanism of specific Cs^+ adsorption via lattice defect sites of Prussian blue filled with coordination and crystallization water molecules. *Dalton Trans.* **2013**, *42*, 16049–16055. [CrossRef]
29. Ruankaew, N.; Yoshida, N.; Watanabe, Y.; Nakano, H.; Phongphanphanee, S. Size-dependent adsorption sites in a Prussian blue nanoparticle: A 3D-RISM study. *Chem. Phys. Lett.* **2017**, *684*, 117–125. [CrossRef]
30. Eberl, D.D. Alkali Cation Selectivity and Fixation by Clay Minerals. *Clays Clay Miner.* **1980**, *28*, 161–172. [CrossRef]
31. Takahashi, A.; Tanaka, H.; Minami, K.; Noda, K.; Ishizaki, M.; Kurihara, M.; Ogawa, H.; Kawamoto, T. Unveiling Cs-adsorption mechanism of Prussian blue analogs: Cs^+-percolation via vacancies to complete dehydrated state. *RSC Adv.* **2018**, *8*, 34808–34816. [CrossRef]
32. Ayrault, S.; Jimenez, B.; Garnier, E.; Fedoroff, M.; Jones, D.J.; Loos-Neskovic, C. Sorption Mechanisms of Cesium on $Cu^{II}_2Fe^{II}(CN)_6$ and $Cu^{II}_3[Fe^{III}(CN)_6]_2$ Hexacyanoferrates and Their Relation to the Crystalline Structure. *J. Solid State Chem.* **1998**, *141*, 475–485. [CrossRef]
33. Fujita, H.; Miyajima, R.; Sakoda, A.J.A. Limitation of adsorptive penetration of cesium into Prussian blue crystallite. *Adsorption* **2015**, *21*, 195–204. [CrossRef]
34. Liu, J.; Li, X.; Rykov, A.I.; Fan, Q.; Xu, W.; Cong, W.; Jin, C.; Tang, H.; Zhu, K.; Ganeshraja, A.S.; et al. Zinc-modulated Fe–Co Prussian blue analogues with well-controlled morphologies for the efficient sorption of cesium. *J. Mater. Chem. A* **2017**, *5*, 3284–3292. [CrossRef]
35. Torad, N.L.; Hu, M.; Imura, M.; Naito, M.; Yamauchi, Y. Large Cs adsorption capability of nanostructured Prussian Blue particles with high accessible surface areas. *J. Mater. Chem.* **2012**, *22*, 18261–18267. [CrossRef]
36. Wi, H.; Kang, S.-W.; Hwang, Y. Immobilization of Prussian blue nanoparticles in acrylic acid-surface functionalized poly(vinyl alcohol) sponges for cesium adsorption. *Environ. Eng. Res.* **2019**, *24*, 173–179. [CrossRef]

37. Pajerowski, D.M.; Gardner, J.E.; Frye, F.A.; Andrus, M.J.; Dumont, M.F.; Knowles, E.S.; Meisel, M.W.; Talham, D.R. Photoinduced Magnetism in a Series of Prussian Blue Analogue Heterostructures. *Chem. Mater.* **2011**, *23*, 3045–3053. [CrossRef]
38. Tokoro, H.; Ohkoshi, S.-I. Novel magnetic functionalities of Prussian blue analogs. *Dalton Trans.* **2011**, *40*, 6825–6833. [CrossRef] [PubMed]
39. Buzin, E.R.; Prellier, W.; Mercey, B.; Simon, C.; Raveau, B. Relations between structural distortions and transport properties in $Nd_{0.5}Ca_{0.5}MnO_3$ strained thin films. *J. Phys. Condens. Matter* **2002**, *14*, 3951–3958. [CrossRef]
40. Nakotte, H.; Shrestha, M.; Adak, S.; Boergert, M.; Zapf, V.S.; Harrison, N.; King, G.; Daemen, L.L. Magnetic properties of some transition-metal Prussian Blue Analogs with composition $M_3[M'(C,N)_6]_2 \cdot xH_2O$. *J. Sci. Adv. Mater. Devices* **2016**, *1*, 113–120. [CrossRef]
41. Uemura, T.; Kitagawa, S. Prussian Blue Nanoparticles Protected by Poly(vinylpyrrolidone). *J. Am. Chem. Soc.* **2003**, *125*, 7814–7815. [CrossRef] [PubMed]
42. Jang, S.-C.; Kang, S.-M.; Kim, G.Y.; Rethinasabapathy, M.; Haldorai, Y.; Lee, I.; Han, Y.-K.; Renshaw, J.C.; Roh, C.; Huh, Y.S. Versatile Poly(Diallyl Dimethyl Ammonium Chloride)-Layered Nanocomposites for Removal of Cesium in Water Purification. *Materials* **2018**, *11*, 998. [CrossRef]
43. Jang, J.; Lee, D.S. Magnetic Prussian Blue Nanocomposites for Effective Cesium Removal from Aqueous Solution. *Ind. Eng. Chem. Res.* **2016**, *55*, 3852–3860. [CrossRef]
44. Chang, L.; Chang, S.; Chen, W.; Han, W.; Li, Z.; Zhang, Z.; Dai, Y.; Chen, D. Facile one-pot synthesis of magnetic Prussian blue core/shell nanoparticles for radioactive cesium removal. *RSC Adv.* **2016**, *6*, 96223–96228. [CrossRef]
45. Hassan, M.R.; Aly, M.I. Adsorptive removal of cesium ions from aqueous solutions using synthesized Prussian blue/magnetic cobalt ferrite nanoparticles. *Part. Sci. Technol.* **2019**, 1–11. [CrossRef]
46. Nasir, S.; Hussein, M.Z.; Zainal, Z.; Yusof, N.A. Carbon-Based Nanomaterials/Allotropes: A Glimpse of Their Synthesis, Properties and Some Applications. *Materials* **2018**, *11*, 295. [CrossRef] [PubMed]
47. Kimura, K.; Hachinohe, M.; Klasson, K.T.; Hamamatsu, S.; Hagiwara, S.; Todoriki, S.; Kawamoto, S. Removal of Radioactive Cesium (^{134}Cs plus ^{137}Cs) from Low-Level Contaminated Water by Charcoal and Broiler Litter Biochar. *Food Sci. Technol. Res.* **2014**, *20*, 1183–1189. [CrossRef]
48. Husnain, S.M.; Um, W.; Chang, Y.-Y.; Chang, Y.-S. Recyclable superparamagnetic adsorbent based on mesoporous carbon for sequestration of radioactive Cesium. *Chem. Eng. J.* **2017**, *308*, 798–808. [CrossRef]
49. Sweetman, M.J.; May, S.; Mebberson, N.; Pendleton, P.; Vasilev, K.; Plush, S.E.; Hayball, J.D. Activated Carbon, Carbon Nanotubes and Graphene: Materials and Composites for Advanced Water Purification. *C* **2017**, *3*, 18. [CrossRef]
50. Brown, J.; Hammond, D.; Wilkins, T. *Handbook for Assessing the Impact of a Radiological Incident on Levels of Radioactivity in Drinking Water and Risks to Operatives at Water Treatment Works: Supporting Scientific Report*; Health Protection Agency, Center for Radiation, Chemical and Environmental Hazards, Radiation Protection Division Oxford: Oxford, UK, 2008; p. 97.
51. Wang, L.; Feng, M.; Liu, C.; Zhao, Y.; Li, S.; Wang, H.; Yan, L.; Tian, G.; Li, S. Supporting of Potassium Copper Hexacyanoferrate on Porous Activated Carbon Substrate for Cesium Separation. *Sep. Sci. Technol.* **2009**, *44*, 4023–4035. [CrossRef]
52. Kaewmee, P.; Manyam, J.; Opaprakasit, P.; Truc Le, G.T.; Chanlek, N.; Sreearunothai, P. Effective removal of cesium by pristine graphene oxide: Performance, characterizations and mechanisms. *RSC Adv.* **2017**, *7*, 38747–38756. [CrossRef]
53. Sun, Y.; Shao, D.; Chen, C.; Yang, S.; Wang, X. Highly Efficient Enrichment of Radionuclides on Graphene Oxide-Supported Polyaniline. *Environ. Sci. Technol.* **2013**, *47*, 9904–9910. [CrossRef] [PubMed]
54. Tan, L.; Wang, S.; Du, W.; Hu, T. Effect of water chemistries on adsorption of Cs(I) onto graphene oxide investigated by batch and modeling techniques. *Chem. Eng. J.* **2016**, *292*, 92–97. [CrossRef]
55. Rethinasabapathy, M.; Kang, S.-M.; Lee, I.; Lee, G.-W.; Lee, S.; Roh, C.; Huh, Y.S. Highly stable Prussian blue nanoparticles containing graphene oxide–chitosan matrix for selective radioactive cesium removal. *Mater. Lett.* **2019**, *241*, 194–197. [CrossRef]
56. Kadam, A.A.; Jang, J.; Lee, D.S. Facile synthesis of pectin-stabilized magnetic graphene oxide Prussian blue nanocomposites for selective cesium removal from aqueous solution. *Bioresour. Technol.* **2016**, *216*, 391–398. [CrossRef]

57. Jang, S.-C.; Haldorai, Y.; Lee, G.-W.; Hwang, S.-K.; Han, Y.-K.; Roh, C.; Huh, Y.S. Porous three-dimensional graphene foam/Prussian blue composite for efficient removal of radioactive ^{137}Cs. *Sci. Rep.* **2015**, *5*, 17510. [CrossRef]
58. Goodman, S.M.; Bura, R.; Dichiara, A.B. Facile Impregnation of Graphene into Porous Wood Filters for the Dynamic Removal and Recovery of Dyes from Aqueous Solutions. *ACS Appl. Nano Mater.* **2018**, *1*, 5682–5690. [CrossRef]
59. Laurent, S.; Forge, D.; Port, M.; Roch, A.; Robic, C.; Vander Elst, L.; Muller, R.N. Magnetic Iron Oxide Nanoparticles: Synthesis, Stabilization, Vectorization, Physicochemical Characterizations, and Biological Applications. *Chem. Rev.* **2008**, *108*, 2064–2110. [CrossRef]
60. Devasenathipathy, R.; Mani, V.; Chen, S.-M.; Arulraj, D.; Vasantha, V.S. Highly stable and sensitive amperometric sensor for the determination of trace level hydrazine at cross linked pectin stabilized gold nanoparticles decorated graphene nanosheets. *Electrochim. Acta* **2014**, *135*, 260–269. [CrossRef]
61. Yang, H.; Sun, L.; Zhai, J.; Li, H.; Zhao, Y.; Yu, H. In situ controllable synthesis of magnetic Prussian blue/graphene oxide nanocomposites for removal of radioactive cesium in water. *J. Mater. Chem. A* **2014**, *2*, 326–332. [CrossRef]
62. Ihsanullah. Carbon nanotube membranes for water purification: Developments, challenges, and prospects for the future. *Sep. Purif. Technol.* **2019**, *209*, 307–337. [CrossRef]
63. Yan, H.; Wu, H.; Li, K.; Wang, Y.; Tao, X.; Yang, H.; Li, A.; Cheng, R. Influence of the Surface Structure of Graphene Oxide on the Adsorption of Aromatic Organic Compounds from Water. *ACS Appl. Mater. Interfaces* **2015**, *7*, 6690–6697. [CrossRef]
64. Das, R. Carbon Nanotube in Water Treatment. In *Nanohybrid Catalyst Based on Carbon Nanotube: A Step-By-Step Guideline from Preparation to Demonstration*; Springer International Publishing: Cham, Swtizerland, 2017; pp. 23–54. [CrossRef]
65. Yu, J.-G.; Zhao, X.-H.; Yu, L.-Y.; Jiao, F.-P.; Jiang, J.-H.; Chen, X.-Q. Removal, recovery and enrichment of metals from aqueous solutions using carbon nanotubes. *J. Radioanal. Nucl. Chem.* **2014**, *299*, 1155–1163. [CrossRef]
66. Dwivedi, C.; Kumar, A.; Ajish, J.K.; Singh, K.K.; Kumar, M.; Wattal, P.K.; Bajaj, P.N. Resorcinol-formaldehyde coated XAD resin beads for removal of cesium ions from radioactive waste: Synthesis, sorption and kinetic studies. *RSC Adv.* **2012**, *2*, 5557–5564. [CrossRef]
67. Yavari, R.; Huang, Y.D.; Ahmadi, S.J. Adsorption of cesium (I) from aqueous solution using oxidized multiwall carbon nanotubes. *J. Radioanal. Nucl. Chem.* **2011**, *287*, 393–401. [CrossRef]
68. Yang, S.; Shao, D.; Wang, X.; Hou, G.; Nagatsu, M.; Tan, X.; Ren, X.; Yu, J. Design of Chitosan-Grafted Carbon Nanotubes: Evaluation of How the –OH Functional Group Affects Cs$^+$ Adsorption. *Mar. Drugs* **2015**, *13*, 3116–3131. [CrossRef] [PubMed]
69. Yang, S.; Han, C.; Wang, X.; Nagatsu, M. Characteristics of cesium ion sorption from aqueous solution on bentonite- and carbon nanotube-based composites. *J. Hazard. Mater.* **2014**, *274*, 46–52. [CrossRef]
70. Kaper, H.; Nicolle, J.; Cambedouzou, J.; Grandjean, A. Multi-method analysis of functionalized single-walled carbon nanotubes for cesium liquid–solid extraction. *Mater. Chem. Phys.* **2014**, *147*, 147–154. [CrossRef]
71. Jang, J.; Miran, W.; Lee, D.S. Amino-functionalized multi-walled carbon nanotubes for removal of cesium from aqueous solution. *J. Radioanal. Nucl. Chem.* **2018**, *316*, 691–701. [CrossRef]
72. Vipin, A.K.; Ling, S.; Fugetsu, B. Sodium cobalt hexacyanoferrate encapsulated in alginate vesicle with CNT for both cesium and strontium removal. *Carbohydr. Polym* **2014**, *111*, 477–484. [CrossRef]
73. Li, T.; He, F.; Dai, Y.J. Prussian blue analog caged in chitosan surface-decorated carbon nanotubes for removal cesium and strontium. *J. Radioanal. Nucl. Chem.* **2016**, *310*, 1139–1145. [CrossRef]
74. Draouil, H.; Alvarez, L.; Causse, J.; Flaud, V.; Zaibi, M.A.; Bantignies, J.L.; Oueslati, M.; Cambedouzou, J. Copper hexacyanoferrate functionalized single-walled carbon nanotubes for selective cesium extraction. *New J. Chem.* **2017**, *41*, 7705–7713. [CrossRef]
75. Lee, H.-K.; Choi, J.W.; Oh, W.; Choi, S.-J. Sorption of cesium ions from aqueous solutions by multi-walled carbon nanotubes functionalized with copper ferrocyanide. *J. Radioanal. Nucl. Chem.* **2016**, *309*, 477–484. [CrossRef]
76. Tsuruoka, S.; Fugetsu, B.; Khoerunnisa, F.; Minami, D.; Takeuchi, K.; Fujishige, M.; Hayashi, T.; Kim, Y.A.; Park, K.C.; Asai, M.; et al. Intensive synergetic Cs adsorbent incorporated with polymer spongiform for scalable purification without post filtration. *Mater. Express* **2013**, *3*, 21–29. [CrossRef]

77. Zheng, Y.; Qiao, J.; Yuan, J.; Shen, J.; Wang, A.-J.; Niu, L. Electrochemical Removal of Radioactive Cesium from Nuclear Waste Using the Dendritic Copper Hexacyanoferrate/Carbon Nanotube Hybrids. *Electrochim. Acta* **2017**, *257*, 172–180. [CrossRef]
78. Chen, F.-P.; Jin, G.-P.; Peng, S.-Y.; Liu, X.-D.; Tian, J.-J. Recovery of cesium from residual salt lake brine in Qarham playa of Qaidam Basin with prussian blue functionalized graphene/carbon fibers composite. *Colloids Surf. A Physicochem. Eng. Asp.* **2016**, *509*, 359–366. [CrossRef]
79. Dichiara, A.B.; Sherwood, T.J.; Benton-Smith, J.; Wilson, J.C.; Weinstein, S.J.; Rogers, R.E. Free-standing carbon nanotube/graphene hybrid papers as next generation adsorbents. *Nanoscale* **2014**, *6*, 6322–6327. [CrossRef]
80. Sui, Z.; Meng, Q.; Zhang, X.; Ma, R.; Cao, B. Green synthesis of carbon nanotube–graphene hybrid aerogels and their use as versatile agents for water purification. *J. Mater. Chem.* **2012**, *22*, 8767–8771. [CrossRef]
81. Yang, H.-M.; Hwang, J.R.; Lee, D.Y.; Kim, K.B.; Park, C.W.; Kim, H.R.; Lee, K.-W. Eco-friendly one-pot synthesis of Prussian blue-embedded magnetic hydrogel beads for the removal of cesium from water. *Sci. Rep.* **2018**, *8*, 11476. [CrossRef]
82. Yang, H.; Li, H.; Zhai, J.; Sun, L.; Zhao, Y.; Yu, H. Magnetic prussian blue/graphene oxide nanocomposites caged in calcium alginate microbeads for elimination of cesium ions from water and soil. *Chem. Eng. J.* **2014**, *246*, 10–19. [CrossRef]
83. Lin, Y.; Cui, X. Novel hybrid materials with high stability for electrically switched ion exchange: Carbon nanotube–polyaniline–nickel hexacyanoferrate nanocomposites. *Chem. Commun.* **2005**, *17*, 2226–2228. [CrossRef] [PubMed]
84. Gautam, R.K.; Chattopadhyaya, M.C. Chapter 13 - Nanomaterials in the Environment: Sources, Fate, Transport, and Ecotoxicology. In *Nanomaterials for Wastewater Remediation*; Gautam, R.K., Chattopadhyaya, M.C., Eds.; Butterworth-Heinemann: Boston, MA, USA, 2016; pp. 311–326. [CrossRef]
85. Liu, X.T.; Mu, X.Y.; Wu, X.L.; Meng, L.X.; Guan, W.B.; Ma, Y.Q.; Sun, H.; Wang, C.J.; Li, X.F. Toxicity of Multi-Walled Carbon Nanotubes, Graphene Oxide, and Reduced Graphene Oxide to Zebrafish Embryos. *Biomed. Environ. Sci.* **2014**, *27*, 676–683. [CrossRef]
86. Qi, W.; Tian, L.; An, W.; Wu, Q.; Liu, J.; Jiang, C.; Yang, J.; Tang, B.; Zhang, Y.; Xie, K.; et al. Curing the Toxicity of Multi-Walled Carbon Nanotubes through Native Small-molecule Drugs. *Sci. Rep.* **2017**, *7*, 2815. [CrossRef]
87. Das, R.; Abd Hamid, S.B.; Ali, M.E.; Ismail, A.F.; Annuar, M.S.M.; Ramakrishna, S. Multifunctional carbon nanotubes in water treatment: The present, past and future. *Desalination* **2014**, *354*, 160–179. [CrossRef]
88. Fujikawa, Y.; Ozaki, H.; Tsuno, H.; Wei, P.; Fujinaga, A.; Takanami, R.; Taniguchi, S.; Kimura, S.; Giri, R.R.; Lewtas, P. Volume Reduction of Municipal Solid Wastes Contaminated with Radioactive Cesium by Ferrocyanide Coprecipitation Technique. In *Nuclear Back-End and Transmutation Technology for Waste Disposal: Beyond the Fukushima Accident*; Nakajima, K., Ed.; Springer: Tokyo, Japan, 2015; pp. 329–341. [CrossRef]
89. Bartoś, B.; Filipowicz, B.; Łyczko, M.; Bilewicz, A.J. Adsorption of ^{137}Cs on titanium ferrocyanide and transformation of the sorbent to lithium titanate: A new method for long term immobilization of ^{137}Cs. *J. Radioanal. Nucl. Chem.* **2014**, *302*, 513–516. [CrossRef]
90. Doumic, L.I.; Salierno, G.; Ramos, C.; Haure, P.M.; Cassanello, M.C.; Ayude, M.A. "Soluble" vs. "insoluble" Prussian blue based catalysts: Influence on Fenton-type treatment. *RSC Adv.* **2016**, *6*, 46625–46633. [CrossRef]
91. Kim, Y.; Kim, Y.K.; Kim, S.; Harbottle, D.; Lee, J.W. Nanostructured potassium copper hexacyanoferrate-cellulose hydrogel for selective and rapid cesium adsorption. *Chem. Eng. J.* **2017**, *313*, 1042–1050. [CrossRef]
92. Shih, C.-J.; Lin, S.; Sharma, R.; Strano, M.S.; Blankschtein, D. Understanding the pH-Dependent Behavior of Graphene Oxide Aqueous Solutions: A Comparative Experimental and Molecular Dynamics Simulation Study. *Langmuir* **2012**, *28*, 235–241. [CrossRef]
93. Kashyap, S.; Mishra, S.; Behera, S.K. Aqueous Colloidal Stability of Graphene Oxide and Chemically Converted Graphene. *J. Nanopart.* **2014**, *2014*, 640281. [CrossRef]
94. Namvari, M.; Namazi, H. Clicking graphene oxide and Fe_3O_4 nanoparticles together: An efficient adsorbent to remove dyes from aqueous solutions. *Int. J. Environ. Sci. Technol.* **2014**, *11*, 1527–1536. [CrossRef]
95. Tsukada, T.; Uozumi, K.; Hijikata, T.; Koyama, T.; Ishikawa, K.; Ono, S.; Suzuki, S.; Denton, M.S.; Keenan, R.; Bonhomme, G. Early construction and operation of highly contaminated water treatment system in Fukushima Daiichi Nuclear Power Station (I)—Ion exchange properties of KURION herschelite in simulating contaminated water. *J. Nucl. Sci. Technol.* **2014**, *51*, 886–893. [CrossRef]

96. Prevost, T.; Blase, M.; Paillard, H.; Mizuno, H. Areva's Actiflo trademark -Rad water treatment system for the Fukushima nuclear power plant. *ATW Int. Z. Fuer Kernenerg.* **2012**, *58*, 308–313.
97. Lehto, J.; Koivula, R.; Leinonen, H.; Tusa, E.; Harjula, R. Removal of Radionuclides from Fukushima Daiichi Waste Effluents. *Sep. Purif. Rev.* **2019**, *48*, 122–142. [CrossRef]
98. Saini, A.; Koyama, T. Cleanup technologies following Fukushima. *MRS Bull.* **2016**, *41*, 952–954. [CrossRef]
99. Dichiara, A.B.; Weinstein, S.J.; Rogers, R.E. On the Choice of Batch or Fixed Bed Adsorption Processes for Wastewater Treatment. *Ind. Eng. Chem. Res.* **2015**, *54*, 8579–8586. [CrossRef]
100. Chen, G.-R.; Chang, Y.-R.; Liu, X.; Kawamoto, T.; Tanaka, H.; Kitajima, A.; Parajuli, D.; Takasaki, M.; Yoshino, K.; Chen, M.-L.; et al. Prussian blue (PB) granules for cesium (Cs) removal from drinking water. *Sep. Purif. Technol.* **2015**, *143*, 146–151. [CrossRef]
101. Ikarashi, Y.; Mimura, H.; Nakai, T.; Niibori, Y.; Ishizaki, E.; Matsukura, M. Selective Cesium Uptake Behavior of Insoluble Ferrocyanide Loaded Zeolites and Development of Stable Solidification Method. *J. Ion Exchange* **2014**, *25*, 212–219. [CrossRef]
102. Gu, B.X.; Wang, L.M.; Ewing, R.C. The effect of amorphization on the Cs ion exchange and retention capacity of zeolite-NaY. *J. Nucl. Mater.* **2000**, *278*, 64–72. [CrossRef]
103. Abusafa, A.; Yücel, H. Removal of ^{137}Cs from aqueous solutions using different cationic forms of a natural zeolite: Clinoptilolite. *Sep. Purif. Technol.* **2002**, *28*, 103–116. [CrossRef]
104. Nakamoto, K.; Ohshiro, M.; Kobayashi, T. Mordenite zeolite—Polyethersulfone composite fibers developed for decontamination of heavy metal ions. *J. Environ. Chem. Eng.* **2017**, *5*, 513–525. [CrossRef]
105. Nakajima, L.; Yusof, N.N.M.; Kobayashi, T. Calixarene-Composited Host–Guest Membranes Applied for Heavy Metal Ion Adsorbents. *Arab. J. Sci. Eng.* **2015**, *40*, 2881–2888. [CrossRef]

© 2019 by the authors. Licensee MDPI, Basel, Switzerland. This article is an open access article distributed under the terms and conditions of the Creative Commons Attribution (CC BY) license (http://creativecommons.org/licenses/by/4.0/).

Article

Controlled Growth of LDH Films with Enhanced Photocatalytic Activity in a Mixed Wastewater Treatment

Zhongchuan Wang [1], Pengfei Fang [1], Parveen Kumar [2], Weiwei Wang [1,*], Bo Liu [1,2,*] and Jiao Li [1]

[1] School of Material Science and Engineering, Shandong University of Technology, Zibo 255000, China; wangzhongchuan1994@163.com (Z.W.); 17864373857@163.com (P.F.); haiyan9943@163.com (J.L.)
[2] Laboratory of Functional Molecular and Materials, School of Physics and Optoelectronic Engineering, Shandong University of Technology, Zibo 255000, China; kumar@sdut.edu.cn
* Correspondence: wangweiwei@sdut.edu.cn (W.W.); liub@sdut.edu.cn (B.L.);
Tel.: +86-15689078202 (W.W.); +86-533-2783909 (B.L.)

Received: 15 April 2019; Accepted: 17 May 2019; Published: 28 May 2019

Abstract: Due to multiple charge transport pathways, adjustable layer spacing, compositional flexibility, low manufacturing cost, and absorption of visible light, layered double hydroxides (LDHs) are a promising material for wastewater treatment. In this study, LDH films and Fe-doped LDH films with different metal ions (Ni, Al, Fe) on the surface of conductive cloth were successfully prepared and applied for the photocatalytic degradation of wastewater containing methyl orange and Ag ions under visible-light irradiation. The chemical state of Fe ions and the composition of LDHs on methyl orange photodegradation were investigated. The experimental results showed that LDH films exhibited high photocatalytic activity. The photocatalytic activity of LDH films on methyl orange improved in the mixed wastewater, and the Fe-doped NiAl–LDH films exhibited best visible-light photocatalytic performance. The analysis showed that Ag ions in the mixed wastewater were reduced by the LDH films and subsequently deposited on the surface of the LDH films. The Ag nanoparticles acted as electron traps and promoted the photocatalytic activity of the LDH films on methyl orange. Thus, we have demonstrated that prepared LDH films can be used in the treatment of mixed wastewater and have broad application prospects in environmental remediation and purification processes.

Keywords: LDHs; film; mixed wastewater; photocatalytic activity

---

1. Introduction

Heavy metal ions and organic compounds are often discharged together during many industrial processes such as metal finishing, petroleum refining, and leather tanning and finishing, which indicates that the coexistence of heavy metal ions and organic compounds in wastewater is a common phenomenon [1–3]. For example, heavy metal ions such as Ag^+, Cr^{6+}, Cu^{2+}, generated from rinsing of plated articles, often coexist with organic dyes such as methyl orange. Many studies have focused on single pollutant treatments, however, it is often difficult to treat mixed contaminants using photocatalysts designed for single pollutants [4]. Therefore, the design and preparation of photocatalysts which can be used to treat the mixed contaminants of heavy metal ions and organic compounds are very important for practical pollutant remediation.

Among the developed photocatalytic materials, layered double hydroxides (LDHs) can be considered ideal photocatalysts for the treatment of various types of pollutants due to multiple charge transport pathways, adjustable layer spacing, compositional flexibility, low manufacturing cost and absorption of visible light. Many studies have been reported on the photocatalytic degradation of heavy metal ions, organic compounds, and CO$_2$ using LDHs as the photocatalyst [5–8]. The composition of LDHs with metal ions of variable valences (such as Co, Ni, Fe, Cr) could offer an effective pathway for electron-hole transportation and help to capture heavy metal ions in solution. For example, the Co(II) in CoAl–LDHs helps to capture Pd(II) species through an in situ redox reaction, resulting in the formation of Pd nanoclusters monodispersed on the surface of CoAl–LDHs [9]. It has also been shown that Au/Cr-substituted hydrotalcite could be an efficient heterogeneous catalyst for aerobic alcohol oxidation, owing to the formation of the Cr^{3+}–Cr^{6+} redox cycle [10]. In addition, NiFe–LDHs have multiple electronic excitation pathways via metal-to-metal charge transfer through Ni^{2+}–O–Fe^{3+}, d–d transitions of Ni^{2+}, and ligand-to-metal charge transfer through O–Ni^{2+}/Fe^{3+} [11].

The effective properties of photocatalytic materials could further be improved by the modification of metal ions or the formation of innovative compositions. For example, Ag@Ag$_3$PO$_4$/g-C$_3$N$_4$/NiFe–LDH nanocomposites improved the photocatalytic ability for the reduction of Cr(VI) to Cr(III) [5]. Similarly, Ag nanoparticle-coated Zn/Ti–LDH composites showed higher photocatalytic activity for rhodamine B (RhB) and NO, owing to the formation of Schottky barriers between LDH and Ag nanoparticles and the surface plasmon resonance effect of Ag nanoparticles [12]. Furthermore, NiFe–LDHs modified through ion doping with [M(C$_2$O$_4$)$_3$]$^{3-}$ (M = Cr and Rh) exhibited higher magnetic properties [13]. In another study, Tb^{3+}-doped CaAl–LDHs showed an enhancement in fluorescence intensity [14], H$_2$SrTa$_2$O$_7$ with difference photocatalytic activities toward CO and H$_2$ evolution was obtained using various metals as cocatalysts (Ag, Pd, Au, and Cu$_2$O) [15], and the photocatalytic properties of ZnS were improved by modification with Ru nanoparticles [16]. Similar studies have shown that the formation of dual surface heterostructure by deposing Au and CuO on Cu$_2$O cubes improved the photocatalytic activity of Cu$_2$O due to the synergistic effect of CuO/Cu$_2$O and Au/Cu$_2$O [17]. However, in one study, the diffusion of K$^+$ ions from poly(heptazine imide) (PHIK) to a metal–organic framework (MOF) made PHIK more negatively charged and MOF more positively charged, which provided the strongest interaction between PHIK and MOF and resulted in superior photocatalytic activity through rhodamine B degradation [18]. It has also been shown that the formation of heterojunctions could increase photocatalytic properties, such as Cu$_2$S/ZnO, Cu$_2$O/TiO$_2$, and CuInS$_2$/TiO$_2$ [19–21], and that the presence of Na$_2$S hole scavengers increased the photoreduction of CO$_2$ to form (HCOO$^-$) on ZnS [22]. Finally, due to the different Fermi levels between heavy metals and LDHs, heavy metals deposited on the surface of LDHs could act as electron traps to prevent electron-hole recombination [5,9,10], which would improve the photocatalytic performance of LDHs.

According to the above discussion, LDHs could be used for the photocatalytic degradation of mixed wastewater. However, LDH powders exhibit the following drawbacks: low photocatalytic activity, formation of aggregates, and difficulties in subsequent separation processes. Using a two-dimensional structure with excellent carrier mobility as the substrate is an effective approach for enhancing the photocatalytic efficiency by preventing agglomeration and decreasing charge recombination [23,24]; besides, the formation of films on the substrate is beneficial for recovering the catalysts from the solution. Among those substrates, conductive cloth has the characteristics of good conductivity, flexibility, and low manufacturing cost. From this viewpoint, we used conductive cloth as the substrate and LDHs as the active sites for photocatalytic domains. Based on the photo-induced reduction and photocatalytic degradation of LDHs, LDH films were successfully used to treat the mixed wastewater containing heavy metal ions and organic compounds. The formation of heavy metals and LDHs on the surface of the conductive cloth led to an enhancement of the photocatalytic activity for organic pollutants degradation and, thereby, realized the photocatalytic treatment of mixed wastewater.

2. Materials and Methods

2.1. Materials

Al(NO$_3$)$_3$·9H$_2$O, Ni(NO$_3$)$_2$·6H$_2$O, Fe(NO$_3$)$_3$·9H$_2$O, AgNO$_3$, ferric ammonium citrate ((NH$_4$)$_3$·[Fe(Cit)$_2$]), urea, methyl orange, and ammonium fluoride (NH$_4$F) were purchased from

Sinopharm Chemical Reagent Co. Ltd. (Shanghai, China) and used as received without further purification. The conductive cloth was purchased from Zhongyang shielding material production company, Guangzhou, China.

2.2. Synthesis of LDH Films

NiAl–LDH films were prepared using the method previously reported with some minor modifications [25]. The conductive cloth (2 cm × 6 cm) was cleaned with a mixed solution of deionized water, ethanol, and acetone (volume ratio, 1:1:1) in an ultrasonic bath for 30 min. $Ni(NO_3)_2 \cdot 6H_2O$ (0.15 mol·L^{-1}), $Al(NO_3)_3 \cdot 9H_2O$ (0.05 mol·L^{-1}), NH_4F (0.2 mol·L^{-1}), and urea (0.5 mol·L^{-1}) were dissolved in deionized water under magnetic stirring to form a clear solution at room temperature. The conductive cloth was vertically placed in the solution without stirring and heated at 110 °C for 8 h to form a thin film on its bottom side, which was washed by deionized water. For NiFe–LDH films, $Fe(NO_3)_3 \cdot 9H_2O$ (0.05 mol·L^{-1}) was used instead of $Al(NO_3)_3 \cdot 9H_2O$. LDH powders were also prepared under the same experimental conditions as that of LDH films but without conductive cloth.

Fe-doped LDH films: LDH films were vertically placed in a $(NH_4)_3 \cdot [Fe(Cit)_2]$ solution (0.01 mol·L^{-1}) and heated at 75 °C for 12 h to form Fe-doped LDH films, which were cleaned with deionized water. For Fe-doped LDH powders, LDH powders were used instead of LDH films.

2.3. Characterization

X-ray powder diffraction (XRD) patterns were recorded using a D8 ADVANCE X-ray diffractometer (Karlsruhe, Germany) with Cu K_α radiation (λ = 0.15406 nm). The scanning electron microscopy (SEM) images were recorded on a FEI-Sirion 200 F field emission scanning electron microscope (Hongkong, China). The transmission electron microscopy (TEM) images, high-resolution transmission electron microscopy (HRTEM) images, and the energy dispersive spectroscopy (EDS) spectra were taken with a FEI-Tecnai G2 field emission transmission electron microscope (Hongkong, China). The samples were obtained by peeling off LDH films from the substrate. Photocatalytic reactions were carried out using a 300 W Xe lamp as the light source and the light intensity in the visible region was about 85 mW·cm^{-2}. The UV–Vis diffuse reflectance spectra were recorded by a UV–Vis spectrophotometer (UV-3900H, Hitachi, Tokyo, Japan). Fourier transform infrared (FTIR) spectra were recorded by a Thermo Nicolet 5700 (Waltham, MA, USA). The chemical state of LDHs was investigated by X-ray photoelectron spectroscopy (XPS) on a ESCALAB 250Xi photoelectron spectrometer (Waltham, MA, USA) with Al K (1486 Ev) as the excitation light source. The N_2 adsorption/desorption tests were measured by Brunauer–Emmett–Teller (BET) measurements using a NOVA2200e surface area analyzer (Boynton Beach, FL, USA).

2.4. Photocatalytic Property Measurement

LDH films (2 cm × 6 cm) were immersed in a mixed solutions of methyl orange (20 mg·L^{-1}) and $AgNO_3$ (0, 5, 10, 20, and 30 mg·L^{-1}) and kept in the dark for 30 min to ensure adsorption–desorption equilibrium. The pH value of the solution was about 6. Samples were removed from the solution after various irradiation times and analyzed using a UV–Vis spectrophotometer at 464 nm. All photocatalytic experiments were repeated three times. The weight of the photocatalyst was calculated from the formula $m = m_1 - m_0$, where m_0 and m_1 represent the weight of the substrate before and after LDH growth, respectively. The LDHs grown on conductive cloth were about 6.6, 6.9, 6.5, and 6.7 mg for NiAl–LDH films, NiFe–LDH films, Fe-doped NiAl–LDH films, and Fe-doped NiFe–LDH films, respectively. To test the photocatalytic performance of LDH powders, LDH powders (15 mg) were used instead of LDH films.

3. Results and Discussion

XRD analysis was performed to confirm the crystal structure and phase (Figure S1). Both the structure of NiAl–LDH films and NiFe–LDH films matched well to the typical LDH lamellar structure

(JCPDS File No. 48-0594 and 51-0463). According to the (003), (006), and (009) reflections, the basal spacing values were 0.756 nm (NiAl–LDHs) and 0.764 nm (NiFe–LDHs), which coincide well with the values for CO_3^{2-}-intercalated LDH materials [26]. The Fe-doped LDH films displayed a phase of lamellar structures with a slightly lower degree of crystallinity and basal spacing (0.755 and 0.762 nm for Fe-doped NiAl–LDH films and Fe-doped NiFe–LDH films, respectively) which was lower than that of Fe complex-intercalated LDHs (1.214 nm) [27]. This suggests that the interlayer ion of LDHs was CO_3^{2-} and that no Fe complex intercalated in the interlayer. According to the (003) crystal plane spacing, grain size and stacking numbers of LDHs in the direction of the c axis were calculated using the Scherrer formula. The grain size for NiFe–LDH films and NiAl–LDH films were 43.2 and 33.9 nm, respectively, whereas the stacking numbers for NiFe–LDH films and NiAl–LDH films were 57 and 45, respectively.

The chemical state of Fe ions in the LDH films was investigated by XPS spectra. The peaks at 726.28 and 712.78 eV are attributed to the 2p1/2 and 2p3/2 spin states of Fe(III) for LDH lamellar structure [28–32] as seen in Figure 1a. After Fe ion doping, the positions of Fe 2p3/2 were basically the same (within the experimental uncertainty [33]), while the peaks for Fe 2p1/2 in both LDH films showed a negative shift of ~0.4 and 0.6 eV, as shown in Figure 1b,c. The mole ratio of Fe/Ni for Fe-doped NiFe–LDH films and Fe-doped NiAl–LDH films were 0.58:1 and 0.38:1, respectively. The increase in the mole ratio of Fe/Ni after doping indicates successful doping of Fe ions. No peaks for Fe complexes (725.0 and 711.5 eV) were observed, confirming that no Fe complex intercalated into the LDH films [28].

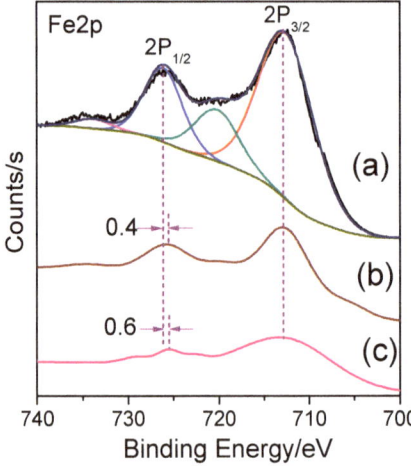

Figure 1. The high-resolution Fe 2p XPS spectra of (**a**) NiFe–LDH films, (**b**) Fe-doped NiFe–LDH films, and (**c**) Fe-doped NiAl–LDH films.

As seen in Figure 2, FTIR spectra of all LDH films provide evidence for the presence of intercalated CO_3^{2-} and interlayered water. After Fe(III) doping, the peak position of CO_3^{2-} moved toward high frequency (around 1388 cm^{-1}) due to the change in charge density by doping Fe(III) into the LDH films [34]. No vibration peak (1290 cm^{-1}) for an Fe complex appeared.

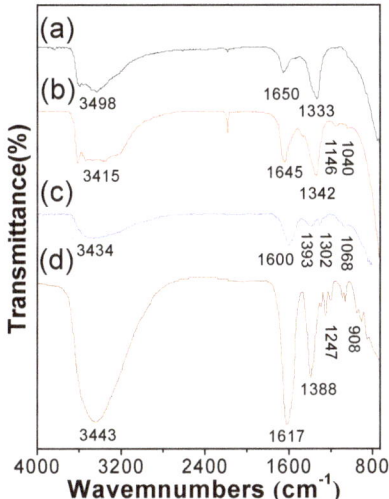

Figure 2. FTIR spectra of (**a**) NiFe–LDH films, (**b**) NiAl–LDH films, (**c**) Fe-doped NiAl–LDH films, and (**d**) Fe-doped NiFe–LDH films.

Figure 3 shows morphology images of the LDH films obtained using SEM analysis and suggests typical sheet-like structures uniformly distributed on the surface of the substrate, thus effectively solving the problem of powder agglomeration. Figure 3a shows NiAl–LDH sheets intersecting and aligned vertically on the conductive cloth under the orienting function of NH$_4$F [10]. NiFe–LDH films also exhibited a similar sheet-like morphology, as seen in Figure 3b, but the sheets were larger and more loosely distributed. The composition of the LDHs affected the sheet thickness, which was consistent with the results from XRD analysis. After Fe ion doping, NiAl–LDH and NiFe–LDH sheets displayed a lamellar structure of similar size. TEM investigation of LDH films revealed the almost transparent, thin, layered morphology of the sheets, as shown in Figure 4. All HRTEM images exhibited clear lattice fringes with a d-spacing value of 0.2593 nm corresponding to (012) reflection from the LDHs.

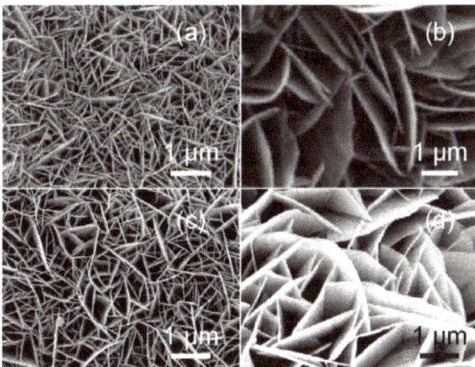

Figure 3. SEM images of (**a**) NiAl–LDH films, (**b**) NiFe–LDH films, (**c**) Fe-doped NiAl–LDH films, and (**d**) Fe-doped NiFe–LDH films.

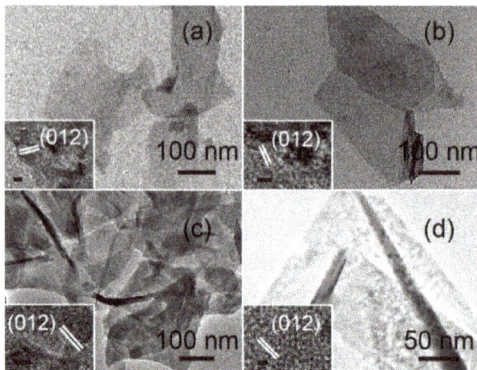

Figure 4. TEM images of (**a**) NiAl–LDH films, (**b**) NiFe–LDH films, (**c**) Fe-doped NiAl–LDH films, and (**d**) Fe-doped NiFe–LDH films. Insets are corresponding HRTEMs, scale bar: 2 nm.

The photocatalytic activity in methyl orange degradation under visible-light irradiation was evaluated, as shown in Figure 5a. All LDH films showed a weak adsorption effect after treatment in a dark place for 30 min. Figure 5b illustrates the blank experiment—in the absence of catalysts but under irradiation, showed that a small quantity of methyl orange was degraded Figure 5b. No obvious photosensitive degradation of methyl orange was observed. The degradation of methyl orange mainly depends on photocatalytic activity, which tends to increase with increasing irradiation time. In general, the photocatalytic activity depends on the amount of catalyst. Within a certain range, the use of more photocatalyst provides higher photocatalytic activity [35]. In our experiments, the degradation of methyl orange per milligram of LDHs was used to describe the photocatalytic activity. Under 120 min illumination, the degradation of methyl orange for NiAl–LDH and NiFe–LDH films was 9.1%·mg^{-1} and 8.4%·mg^{-1}, respectively. When Fe-doped LDH films were used as a photocatalyst, the degradation of methyl orange decreased.

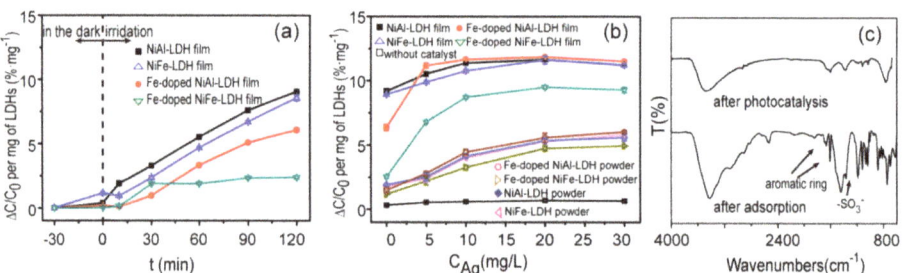

Figure 5. (**a**) Photocatalytic activity of methyl orange for LDH films without Ag ions. (**b**) Effect of Ag ion concentration on degradation of methyl orange under 120 min illumination. C_0: the equilibrium concentration of methyl orange before irradiation, ΔC: the change in the concentration of methyl orange after irradiation. (**c**) FTIR spectra of NiAl–LDH films after photocatalytic experiments and adsorption experiments.

The Fe(III) ionic radius (64.5 pm) is larger than Al(III) ionic radius (53.5 pm); hence, the replacement of Al(III) by Fe(III) could increase the distance between metal ions in the layer and reduce the charge density. This is unfavorable to electron-hole transfer throughout the layered framework, and thereby leads to a low level of photocatalytic activity as shown in Figure 6a [5,7]. Therefore, NiAl–LDH films showed better photocatalytic activity compared to NiFe–LDH films. The Figure 6b pathway 3 shows that after Fe doping, Fe(III) ions act as traps for photogenerated electron-hole pairs and accelerate electron-hole pair recombination [36–38], thus decreasing the photocatalytic degradation of methyl orange by Fe-doped LDH films.

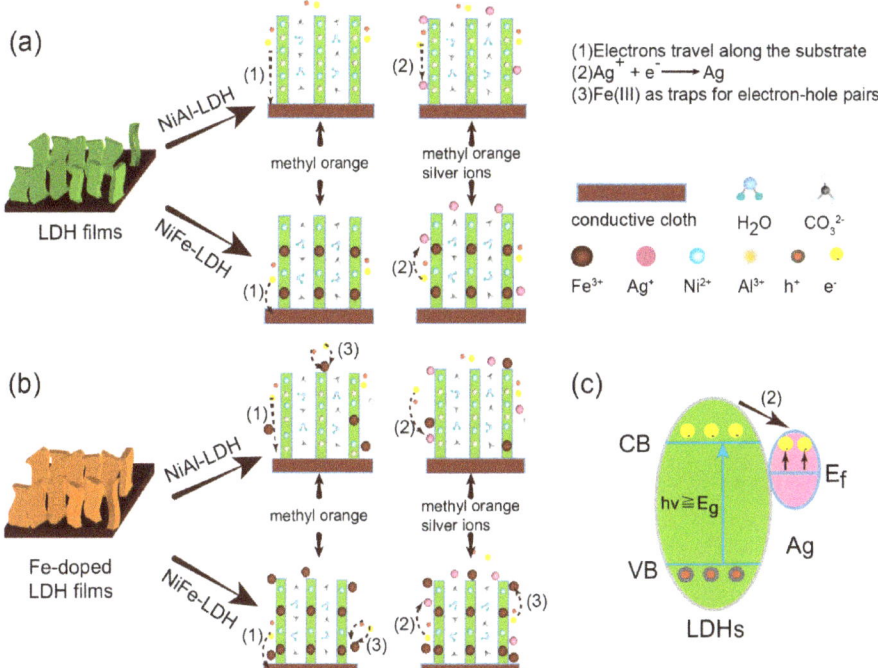

Figure 6. Schematic diagram of possible electron transport in (**a**) LDH films and (**b**) Fe-doped LDH films in degradation of methyl orange or a mixed solution of methyl orange and Ag ions. (**c**) Possible electron transport between LDHs and Ag nanoparticles.

The photocatalytic activity of LDH films was studied in mixed wastewater containing Ag ions and methyl orange. The degradation of methyl orange was gradually accelerated when the concentration of Ag ions increased from 0 to 20 mg·L^{-1}, as seen in Figure 5b. Fe-doped NiAl–LDH films exhibited the best photocatalytic activity with 20 mg·L^{-1} Ag ion concentration and 11.92%·mg^{-1} degradation rate of methyl orange, which is higher than reported values (degradation of methyl orange for xanthan gum/TiO$_2$ [35], MnO$_2$-M [39], and WO$_3$/g-C$_3$N$_4$ [40] were 2.3%, 9.5%, and 4.84%·mg^{-1}, respectively). When the concentration of Ag ions in solution was further increased from 20 mg·L^{-1}, the degradation of methyl orange decreased slightly due to the light shielding effect of Ag ions adsorbed on the surface of LDHs.

In the presence of Ag ions, the degradation of methyl orange by LDH films was higher as seen in Figure 5b. Figure 6, pathway 2 shows Ag nanoparticles were reduced from the solution and deposited on the surface of the LDH films, which received electrons and promoted the separation of photogenic electron-hole pairs [41], thus improving the photocatalytic activity of LDH films. That is, the electron traps of Ag nanoparticles prevented the recombination of electron-hole pairs on Fe(III) ions. The capture of electrons and holes by Ag nanoparticles and Fe(III) ions promoted the separation of electron-hole pairs. As a result, when the concentration of Ag ions increased from 0 to 20 mg·L^{-1}, there was a large increase in the degradation of methyl orange by Fe-doped LDH films, as shown in Figure 5b.

For comparison, corresponding LDH powders were also prepared. All of the LDH powders showed similar sheet-like morphology and lamellar structure to that of LDH films, as shown in Figures

S2 and S3. However, as seen in Figure 5b, the photocatalytic degradation of methyl orange by LDH powders was lower than that of LDH films, which confirms that the conductive cloth could facilitate the transportation of photogenerated charges shown in Figure 6a,b, pathway 1 [23]. However, this comparison is qualitative and we will further study the kinetics of excitons and free carriers in LDH films, for example, the charge transfer dynamics, to obtain the quantitative properties of LDH films [42–44]. In addition, LDH films could be easily separated from the solution, providing a simple method for the recovery of the catalyst.

We also used FTIR spectra to investigate LDH films after photocatalytic degradation and adsorption experiments, as shown in Figure 5c. After photocatalytic degradation of methyl orange, the peak intensity of CO_3^{2-} decreased and no obvious peaks for methyl orange were observed. While after adsorption experiments, some peaks for methyl orange were observed. The differences between photocatalytic degradation and adsorption experiments confirm that the degradation of methyl orange in the presence of LDH films was not through adsorption.

The LDH films after the photocatalytic reaction were examined by TEM and EDS. No noticeable change was observed in the morphology of LDH films before and after the photocatalytic reaction. LDH films exhibited good stability for the degradation of methyl orange. As shown in Figure 7, the LDH sheets were transparent to the electron beam, suggesting that they were very thin. The insets in Figure 7a–c show that after the photocatalytic reaction, the HRTEM images taken from one sheet can be indexed to the (0 0 10) plane, which matches well with the reported values of the LDH structure. Figure 7c shows that in the presence of Ag ions, some Ag nanoparticles were observed on the surface of LDH sheets. The lattice spacing of 0.2083 nm can be indexed to the (200) plane of Ag nanoparticles. EDS spectra also confirmed the formation of Ag after photocatalytic reaction, as seen in Figure S5.

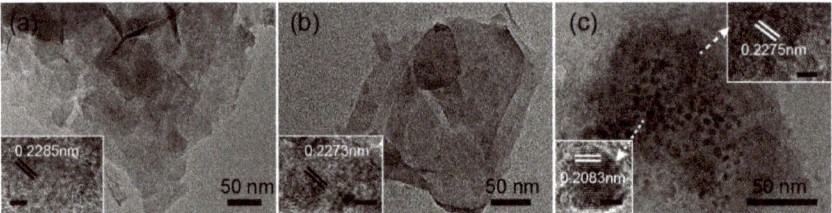

Figure 7. TEM images after photocatalytic reaction (**a**) NiAl–LDH films, (**b**) Fe-doped NiAl–LDH films, and (**c**) Fe-doped NiAl–LDH films in the presence of 5 mg·L^{-1} Ag ions. Insets are corresponding HRTEM images. Scales in insets are 2 nm.

4. Conclusions

LDH films and Fe-doped LDH films with different compositions (Ni, Al, and Fe) on the surface of conductive cloth were successfully prepared and applied for photocatalytic degradation of methyl orange in mixed wastewater. All LDH films showed layered structures and were distributed uniformly on the surface of the substrate. They exhibited high photocatalytic performance and could be easily separated from the solution, providing a simple method for the recovery of the catalyst. Benefiting from the electron trap of Ag nanoparticles, the photocatalytic activity of LDH films on methyl orange was improved, and the Fe-doped NiAl–LDH films presented the best visible-light photocatalytic performance. Our study indicates that LDH films can be used in the treatment of mixed wastewater and have broad application prospects for environmental remediation and purification processes. In order to better realize the application of LDH films in wastewater treatment, further research is needed, including 1) to explore different conductive substrates with excellent carrier mobility which could influence the photocatalytic performance of LDHs; and 2) to explore the reducing capacity of LDH films to other heavy metal ions with different Fermi levels in order to prevent hole-electron recombination, thereby enhancing photocatalytic activity. Based on such research, the factors affecting the photocatalytic performance of LDH films can be better understood. By adjusting the structure and composition of LDHs, different types of mixed wastewater containing heavy metal ions and organic compounds can be selectively treated. We envision that our findings will help in further development of new, improved, and more effective LDH films with enhanced photocatalytic activity for the treatment of mixed wastewater.

Supplementary Materials: The following are available online at http://www.mdpi.com/2079-4991/9/6/807/s1, Table S1: XPS Peak positions for Fe^{3+} obtained from LDH films and Fe-doped LDH films. Figure S1: XRD patterns. (a) NiAl–LDH films, (b) NiFe–LDH films, (c) Fe-doped NiAl–LDH films, and (d) Fe-doped NiFe–LDH films. Figure S2: SEM images. (a) NiAl–LDH powders, (b) NiFe–LDH powders, (c) Fe-doped NiAl–LDH powders, and (d) Fe-doped NiFe–LDH powders. Figure S3: XRD patterns. (a) NiAl–LDH powders, (b) Fe-doped NiAl–LDH powders, (c) NiFe–LDH powders, and (d) Fe-doped NiFe–LDH powders. Figure S4: EDS elements mapping for Fe-doped NiAl–LDH films after the photocatalytic degradation in the presence of methyl orange (20 mg·L^{-1}) and Ag ions (5 mg·L^{-1}). (a) Area without Ag particles, (b) area with Ag particles. Figure S5: EDS spectra for Fe-doped NiAl–LDH films after photocatalytic reaction in the presence of 5 mg·L^{-1} Ag ions. (a) Area with Ag nanoparticles, (b) area without Ag nanoparticles in Figure 7c. Figure S6: N_2 adsorption/desorption isotherms of (a) NiAl–LDH powders and Fe-doped NiAl–LDH powders, (b) NiFe–LDH powders and Fe-doped NiFe–LDH powders.

Author Contributions: Conceptualization, Z.W., W.W. and J.L.; Data curation, Z.W., P.F. and W.W.; Funding acquisition, W.W.; Methodology, Z.W. and P.F.; Project administration, W.W. and B.L.; Supervision, W.W.; Writing—Original Draft, Z.W., P.F. and W.W.; Writing—Review & Editing, P.K., W.W., B.L. and J.L.

Funding: This research was funded by Natural Science Foundation of Shandong Province (grant numbers ZR2015BM022).

Acknowledgments: Thanks Huan Yu for the useful discussion of the experiments.

Conflicts of Interest: The authors declare no conflict of interest.

References

1. Chen, S.; Huang, Y.; Han, X.; Wu, Z.; Li, C.; Wang, J.; Deng, Q.; Zeng, Z.; Deng, S. Simultaneous and efficient removal of Cr(VI) and methyl orange on LDHs decorated porous carbons. *Chem. Eng. J.* **2018**, *352*, 306–315. [CrossRef]
2. Yang, Y.; Wang, G.; Deng, Q.; Ng, D.H.L.; Zhao, H. Microwave-assisted fabrication of nanoparticulate TiO_2 microspheres for synergistic photocatalytic removal of Cr (VI) and methyl orange. *ACS Appl. Mater. Interfaces* **2014**, *6*, 3008–3015. [CrossRef]
3. Li, X.; Liu, Y.; Zhang, C.; Wen, T.; Zhuang, L.; Wang, X.; Song, G.; Chen, D.; Ai, Y.; Hayat, T.; et al. Porous Fe_2O_3 microcubes derived from metal organic frameworks for efficient elimination of organic pollutants and heavy metal ions. *Chem. Eng. J.* **2018**, *336*, 241–252. [CrossRef]
4. Zhao, G.; Sun, Y.; Zhao, Y.; Wen, T.; Wang, X.; Chen, Z.; Sheng, G.; Chen, C.; Wang, X. Enhanced photocatalytic simultaneous removals of Cr(VI) and bisphenol A over Co(II)-modified TiO_2. *Langmuir* **2019**, *35*, 276–283. [CrossRef] [PubMed]
5. Nayak, S.; Parida, K.M. Dynamics of charge-transfer behavior in a plasmon-induced quasi-type-II p–n/n–n dual heterojunction in Ag@Ag_3PO_4/g-C_3N_4/NiFe LDH nanocomposites for photocatalytic Cr(VI) reduction and phenol oxidation. *ACS Omega* **2018**, *3*, 7324–7343. [CrossRef]
6. Zhang, Q.; Li, S.; Jing, R.; Wu, M.; Zhao, S.; Liu, A.; Liu, Y.; Meng, Z. Superlattice assembly of two dimensional CoFe-LDHs nanosheets and titania nanosheets nanohybrids for high visible light photocatalytic activity. *Mater. Lett.* **2019**, *236*, 374–377. [CrossRef]
7. Salehi, G.; Abazari, R.; Mahjoub, A.R. Visible-light-induced graphitic-C_3N_4@nickel-aluminum layered double hydroxide nanocomposites with enhanced photocatalytic activity for removal of dyes in water. *Inorg. Chem.* **2018**, *57*, 8681–8691. [CrossRef]
8. Tonda, S.; Kumar, S.; Bhardwaj, M.; Yadav, P.; Ogale, S. g-C_3N_4/NiAl-LDH 2D/2D hybrid heterojunction for high-performance photocatalytic reduction of CO_2 into renewable fuels. *ACS Appl. Mater. Interfaces* **2018**, *10*, 2667–2678. [CrossRef]
9. Li, P.; Huang, P.; Wei, F.; Sun, Y.; Cao, C.; Song, W. Monodispersed Pd clusters generated in situ by their own reductive support for high activity and stability in cross-coupling reactions. *J. Mater. Chem. A* **2014**, *2*, 12739–12745. [CrossRef]

10. Liu, P.; Degirmenci, V.; Hensen, E. Unraveling the synergy between gold nanoparticles and chromium-hydrotalcites in aerobic oxidation of alcohols. *J. Catal.* **2014**, *313*, 80–91. [CrossRef]
11. Mohapatra, L.; Parida, K. A review on the recent progress, challenges and perspective of layered double hydroxides as promising photocatalysts. *J. Mater. Chem. A* **2016**, *4*, 10744–10766. [CrossRef]
12. Zhu, Y.; Zhu, R.; Zhu, G.; Wang, M.; Chen, Y.; Zhu, J.; Xi, Y.; He, H. Plasmonic Ag coated Zn/Ti-LDH with excellent photocatalytic activity. *Appl. Surf. Sci.* **2018**, *433*, 458–467. [CrossRef]
13. Coronado, E.; Martí-Gastaldo, C.; Navarro-Moratalla, E.; Ribera, A. Intercalation of $[M(ox)_3]^{3-}$ (M=Cr, Rh) complexes into $Ni^{II}Fe^{III}$-LDH. *Appl. Clay Sci.* **2010**, *48*, 228–234. [CrossRef]
14. Chen, Y.; Zhang, J.; Wang, X.; Wang, L. Fluorescence enhancement of Tb^{3+}-doped CaAl-LDH by cytosine. *J. Lumin.* **2018**, *204*, 42–50. [CrossRef]
15. Chen, W.; Wang, Y.; Shangguan, W. Metal (oxide) modified ($M_{\frac{1}{4}}$ Pd, Ag, Au and Cu) $H_2SrTa_2O_7$ for photocatalytic CO_2 reduction with H_2O: the effect of cocatalysts on promoting activity toward CO and H_2 evolution. *Int. J. Hydrogen Energ.* **2019**, *44*, 4123–4132. [CrossRef]
16. Baran, T.; Wojtyla, S.; Dibenedetto, A.; Aresta, M.; Macyk, W. Zinc sulfide functionalized with ruthenium nanoparticles for photocatalytic reduction of CO_2. *Appl. Catal. B-Enviro.* **2015**, *178*, 170–176. [CrossRef]
17. Jiang, D.; Zhang, Y.; Li, X. Synergistic effects of CuO and Au nanodomains on Cu_2O cubes for improving photocatalytic activity and stability. *Chinese J. Catal.* **2019**, *40*, 105–113. [CrossRef]
18. Rodríguez, N.A.; Savateev, A.; Grela, M.A.; Dontsova, D. Facile Synthesis of Potassium Poly(heptazine imide) (PHIK)/Ti-Based Metal−Organic Framework (MIL-125-NH_2) Composites for Photocatalytic Applications. *ACS Appl. Mater. Interfaces* **2017**, *9*, 22941–22949. [CrossRef]
19. Han, D.; Li, B.; Yang, S.; Wang, X.; Gao, W.; Si, Z.; Zuo, Q.; Li, Y.; Li, Y.; Duan, Q.; et al. Engineering Charge Transfer Characteristics in Hierarchical Cu_2S QDs @ ZnO Nanoneedles with p–n Heterojunctions: Towards Highly Efficient and Recyclable Photocatalysts. *Nanomaterials* **2019**, *9*, 16. [CrossRef] [PubMed]
20. Aguirre, M.E.; Zhou, R.; Eugene, A.J.; Guzman, M.I.; Grela, M.A. Cu_2O/TiO_2 heterostructures for CO_2 reduction through a direct Z-scheme: Protecting Cu_2O from photocorrosion. *Appl. Catal. B-Enviro.* **2017**, *217*, 485–493. [CrossRef]
21. Xu, F.; Zhang, J.; Zhu, B.; Yu, J.; Xu, J. $CuInS_2$ sensitized TiO_2 hybrid nanofibers for improved photocatalytic CO_2 reduction. *Appl. Catal. B-Enviro.* **2018**, *230*, 194–202. [CrossRef]
22. Zhou, R.; Guzman, M.I. CO_2 Reduction under Periodic Illumination of ZnS. *J. Phys. Chem. C* **2014**, *118*, 11649–11656. [CrossRef]
23. Lan, M.; Fan, G.; Yang, L.; Li, F. Significantly enhanced visible-light-induced photocatalytic performance of hybrid Zn-Cr layered double hydroxide/graphene nanocomposite and the mechanism study. *Ind. Eng. Chem. Res.* **2014**, *53*, 12943–12952. [CrossRef]
24. Chowdhury, P.R.; Bhattacharyya, K.G. Ni/Co/Ti layered double hydroxide for highly efficient photocatalytic degradation of Rhodamine B and Acid Red G: A comparative study. *Photochem. Photobiol. Sci.* **2017**, *16*, 835–839. [CrossRef] [PubMed]
25. Wu, X.; Ci, C.; Du, Y.; Liu, X.; Li, X.; Xie, X. Facile synthesis of NiAl-LDHs with tunable establishment of acid-base activity sites. *Mater. Chem. Phys.* **2018**, *211*, 72–78. [CrossRef]
26. Jiang, D.B.; Jing, C.; Yuan, Y.; Feng, L.; Liu, X.; Dong, F.; Dong, B.; Zhang, Y.X. 2D-2D growth of NiFe LDH nanoflakes on montmorillonite for cationic and anionic dye adsorption performance. *J. Colloid. Interf. Sci.* **2019**, *540*, 398–409. [CrossRef] [PubMed]
27. Parida, K.; Sahoo, M.; Singha, S. A novel approach towards solvent-free epoxidation of cyclohexene by Ti(IV)-Schiff base complex-intercalated LDH using H_2O_2 as oxidant. *J. Catal.* **2010**, *276*, 161–169. [CrossRef]
28. Huang, Z.; Wu, P.; Zhang, X.; Wang, X.; Zhu, N.; Wu, J.; Li, P. Intercalation of Fe(III) complexes into layered double hydroxides: Synthesis and structural preservation. *Appl. Clay Sci.* **2012**, *65–66*, 87–94. [CrossRef]
29. Rajeshkhanna, G.; Singh, T.I.; Kim, N.H.; Lee, J.H. Remarkable bifunctional oxygen and hydrogen evolution electrocatalytic activities with trace-Level Fe doping in Ni- and Co-layered double hydroxides for overall water-splitting. *ACS Appl. Mater. Interfaces* **2018**, *10*, 42453–42468. [CrossRef]
30. Zhou, T.; Cao, Z.; Zhang, P.; Ma, H.; Gao, Z.; Wang, H.; Lu, Y.; He, J.; Zhao, Y. Transition metal ions regulated oxygen evolution reaction performance of Ni-based hydroxides hierarchical nanoarrays. *Sci. Rep.* **2017**, *7*, 46154. [CrossRef]

31. Zhu, Y.; Zhao, X.; Li, J.; Zhang, H.; Chen, S.; Han, W.; Yang, D. Surface modification of hematite photoanode by NiFe layered double hydroxide for boosting photoelectrocatalytic water oxidation. *J. Alloys Compd.* **2018**, *764*, 341–346. [CrossRef]
32. Youn, D.H.; Park, Y.B.; Kim, J.Y.; Magesh, G.; Jang, Y.J.; Lee, J.S. One-pot synthesis of NiFe layered double hydroxide/reduced graphene oxide composite as an efficient electrocatalyst for electrochemical and photoelectrochemical water oxidation. *J. Power Sources* **2015**, *294*, 437–443. [CrossRef]
33. Bajnóczi, E.G.; Balázs, N.; Mogyorósi, K.; Srankó, D.F.; Pap, Z.; Ambrus, Z.; Canton, S.; Norén, K.; Kuzmann, E.; Vértes, A.; et al. The influence of the local structure of Fe(III) on the photocatalytic activity of doped TiO_2 photocatalysts-An EXAFS, XPS and Mósbauer spectroscopic study. *Appl. Catal. B-Enviro.* **2011**, *103*, 232–239. [CrossRef]
34. Ambrogi, V.; Perioli, L.; Nocchetti, M.; Latterini, L.; Pagano, C.; Massetti, E.; Rossi, C. Immobilization of kojic acid in ZnAl-hydrotalcite like compounds. *J. Phys. Chem. Solids.* **2012**, *73*, 94–98. [CrossRef]
35. Inamuddin. Xanthan gum/titanium dioxide nanocomposite for photocatalytic degradation of methyl orange dye. *Int. J. Biol. Macromol.* **2019**, *121*, 1046–1053. [CrossRef]
36. Reszczynska, J.; Grzyb, T.; Sobczak, J.W.; Lisowski, W.; Gazda, M.; Ohtani, B.; Zaleska, A. Visible light activity of rare earth metal doped (Er^{3+}, Yb^{3+} or Er^{3+}/Yb^{3+}) titania photocatalysts. *Appl. Catal. B-Enviro.* **2015**, *163*, 40–49. [CrossRef]
37. Boppella, R.; Choi, C.; Moon, J.; Kim, D.H. Spatial charge separation on strongly coupled 2D-hybrid of $rGO/La_2Ti_2O_7$/NiFe-LDH heterostructures for highly efficient noble metal free photocatalytic hydrogen generation. *Appl. Catal. B-Enviro.* **2018**, *239*, 178–186. [CrossRef]
38. Ambrus, Z.; Balázs, N.; Alapi, T.; Wittmann, G.; Sipos, P.; Dombi, A.; Mogyorósi, K. Synthesis, structure and photocatalytic properties of Fe(III)-doped TiO_2 prepared from $TiCl_3$. *Appl. Catal. B-Enviro.* **2008**, *81*, 27–37. [CrossRef]
39. Zhang, Y.; Zheng, T.X.; Hu, Y.B.; Guo, X.L.; Peng, H.H.; Zhang, Y.X.; Feng, L.; Zheng, H.L. Delta manganese dioxide nanosheets decorated magnesium wire for the degradation of methyl orange. *J. Colloid Interf. Sci.* **2017**, *490*, 226–232. [CrossRef]
40. Chen, G.; Bian, S.; Guo, C.Y.; Wu, X. Insight into the Z-scheme heterostructure $WO_3/g-C_3N_4$ for enhanced photocatalytic degradation of methyl orange. *Mater. Lett.* **2019**, *236*, 596–599. [CrossRef]
41. Lva, J.; Zhu, Q.; Zeng, Z.; Zhang, M.; Yang, J.; Zhao, M.; Wang, W.; Cheng, Y.; He, G.; Sun, Z. Enhanced photocurrent and photocatalytic properties of porous ZnO thin film by Ag nanoparticles. *J. Phys. Chem. Solids* **2017**, *111*, 104–109. [CrossRef]
42. Hoque, M.A.; Guzman, M.I. Photocatalytic activity: experimental features to report in heterogeneous photocatalysis. *Materials* **2018**, *11*, 1990. [CrossRef] [PubMed]
43. Schneider, J.; Bahnemann, D.W. Undesired role of sacrificial reagents in photocatalysis. *J. Phys. Chem. Lett.* **2013**, *4*, 3479–3483. [CrossRef]
44. Kamat, P.V.; Jin, S. Semiconductor photocatalysis: "tell us the complete story!". *ACS Energy Lett.* **2018**, *3*, 622–623. [CrossRef]

© 2019 by the authors. Licensee MDPI, Basel, Switzerland. This article is an open access article distributed under the terms and conditions of the Creative Commons Attribution (CC BY) license (http://creativecommons.org/licenses/by/4.0/).

Article

Evaluation of Nanoporous Carbon Synthesized from Direct Carbonization of a Metal–Organic Complex as a Highly Effective Dye Adsorbent and Supercapacitor

Xiaoze Shi [1], Shuai Zhang [1], Xuecheng Chen [1,2,*] and Ewa Mijowska [1,*]

1. Nanomaterials Physicochemistry Department, Faculty of Chemical Technology and Engineering, West Pomeranian University of Technology, Szczecin, Piastów Ave. 42, 71-065 Szczecin, Poland; xiaoze.shi@zut.edu.pl (X.S.); shuai.zhang@zut.edu.pl (S.Z.)
2. State Key Laboratory of Polymer Physics and Chemistry, Changchun Institute of Applied Chemistry, Chinese Academy of Science, Changchun 130021, China
* Correspondence: xchen@zut.edu.pl (X.C.); emijowska@zut.edu.pl (E.M.); Tel.: +48-91-449-6030 (X.C.); +48-91-449-4742 (E.M.)

Received: 13 March 2019; Accepted: 10 April 2019; Published: 11 April 2019

Abstract: The synthesis of interconnected nanoporous carbon (NPC) material from direct annealing of ultra-small Al-based metal–organic complex (Al-MOC) has been demonstrated. NPC presents a large accessible area of 1054 m^2/g, through the Methylene Blue (MB) adsorption method, which is comparable to the high specific surface area (SSA) of 1593 m^2/g, through an N$_2$ adsorption/desorption analysis. The adsorption properties and mechanisms were tested by various dye concentrations, pH, and temperature conditions. The high MB accessible area and the good electrical conductivity of the interconnected NPC, led to a large specific capacitance of 205 F/g, with a potential window from 0 to 1.2 V, in a symmetric supercapacitor, and a large energy density of 10.25 Wh/kg, in an aqueous electrolyte, suggesting a large potential in supercapacitors.

Keywords: nanoporous carbon; adsorption properties; dye; adsorption models; supercapacitor

1. Introduction

Nanoporous carbon (NPC) materials with large specific surface area (SSA), have been applied in different fields, especially, as electrode materials and dye adsorbents [1–3]. Dyes are applied in many different industrial fields, such as paper, textile, rubber, food, leather, cosmetics, plastic, and others. However, used dyes also become harmful, due to the fact that they can not only reduce light penetration and photosynthesis in water, but can also contain toxic and carcinogenic chemicals that might be a threat to human health [4]. Unfortunately, most of the wasted dyes are quite stable in the real environment [5]. Therefore, removing dyes from industrial waste water, through efficient technologies, is quite urgent. Adsorption is considered as an economical and efficient method. The adsorption method has the advantages of a simple design, easy operation, and possible regeneration. Methylene Blue (MB) has become a model cationic dye for dye adsorption investigations [6,7]. Various adsorbents, such as biomass materials [8–12], carbon nanotubes [13–16], graphene-based materials [17–20], and magnetic materials [21,22], have been investigated in the field of dye removal. Additionally, researchers always use the MB adsorption method to test the surface area of carbon samples, which can indicate the accessible surface of electrodes [23,24].

Carbon materials are also popular as electrode materials, due to their good electrical conductivity, high power density, and good cycling stability [25–29]. Many efforts have been made in these aspects to fabricate NPC materials with excellent properties, including a large SSA and high porosity. As it has been reported, NPC can be synthesized by various methods, such as chemical vapor decomposition

(CVD), chemical activation, and template methods [30–32]. Among these, the template method has drawn a lot of interest. Easily acquiring zeolites and mesoporous silica have been effectively used as templates, for the formation of NPC with different sizes and structures. Metal–organic frameworks (MOF) have become a new choice of precursors for NPC materials [33–35]. There are two types of MOF-derived NPC materials, depending on the production process. One induces a secondary carbon precursor and the other is a direct carbonization of the organic components of MOF. MOF-derived NPC materials possess a high SSA and a tunable pore size. The tunable pores can provide an appropriate size for the electrolyte ions and dye molecules diffusion, thus, increasing the accessible area and providing more physical active sites for the adsorption process. However, the specific capacitance of the MOF-derived NPC reported so far, is still not satisfactory. Due to the large size of the MOF template [36,37], it is difficult for electrolyte and ions to transport into the inner or the center of the micrometer-sized NPC, leaving many inaccessible active sites. Thus, the resultant specific capacitance is limited, especially under a high charging rate. In this respect, decreasing the MOF size is conducive to improving the electrochemical performance. Additionally, a proper pore size distribution (PSD) can also help to improve the capacitance [38].

Herein, we have prepared an interconnected NPC material from a direct carbonization of the ultra-small Al-MOC. NPC presents a high Brunauer-Emmett-Teller (BET) area, showing an excellent adsorption capacity for MB. NPC can adsorb MB with a maximum value of 415 mg/g, via a physical adsorption process, which suggests a large molecular accessible area of 1054 m^2/g, very close to the SSA from the N$_2$ adsorption/desorption analysis (1593 m^2/g). The large accessible surface area, together with the suitable pore distribution (micropores and mesopores) and high conductivity of the NPC, makes it a good candidate for an electrode material. Hence, a two-electrode supercapacitor has been designed with the NPC material, exhibiting a large specific capacitance of 226 F/g and an energy density of 10.25 Wh/kg, in an aqueous electrolyte.

2. Materials and Methods

2.1. Synthesis of NPC

NPC was fabricated from Al-MOC, according to our previous report [2]. In brief, 2 g of the Al-MOC was heated at 950 °C, for 3 h, in Ar atmosphere and then treated with 10 mL 17% HCl, to remove the metal components. The final product was dried at 105 °C and named as NPC. All the chemicals were purchased from MERCK (Darmstadt, Germany).

2.2. Characterization

Transmission electron microscopy (TEM), High-resolution transmission electron microscopy (HRTEM) (Tecnai F30, FEI, Eindhoven, The Netherlands), Scanning Electron Microscopy (SEM, Hitachi SU8020, Tokyo, Japan), X-ray diffraction using Copper K-α with an X-ray wavelength of 1.5406 Å (XRD, X'Pert PRO Philips diffractometer, Almelo, Holland), and Raman scattering (Renishawmicro, Renishaw, London, UK; λ = 785 nm) were performed, to characterize the structural properties. Thermogravimetric analysis (TGA, DTA-Q600 SDT TA, New Castel, DE, USA) was conducted to determine the composition. The N$_2$ adsorption/desorption analysis (Micromeritics ASAP 2010M, Boynton Beach, FL, USA) were used to calculate the SSA and the PSD. UV-vis spectrophotometer (Thermo Scientific, Waltham, MA, USA) was applied to calculate the concentration of MB.

2.3. Adsorption Equilibrium Isotherm

NPC (W: 0.02 g) was dissolved in a constant volume (V: 0.02 L) of MB aqueous solutions, with different initial concentrations (C_0: 100 to 900 mg/L), at 25 °C, for 24 h. The MB concentrations in the supernatant were measured as C_e (mg/L) and the adsorbed dye amounts by NPC (q_e, mg/g) were calculated as follows:

$$q_e = \frac{(C_0 - C_e)V}{W} \qquad (1)$$

2.4. Adsorption Kinetics

The adsorption kinetic measurements were carried out by continuously stirring the NPC and MB solutions (C_0: 200, 400, or 600 mg/L). The concentrations of MB (C_t: mg/L) in the supernatant were checked at preset time intervals (t: min), to calculate the adsorbed MB amount at time (q_t, mg/g):

$$q_t = \frac{(C_0 - C_t)V}{W} \qquad (2)$$

2.5. Adsorption Thermodynamics

To observe the influence of different temperatures on the adsorption capacity, NPC and MB solutions (500 or 700 mg/L) were placed under different temperatures (25, 35, 45, or 55 °C) and stirred for 24 h. The concentrations of MB and the amounts of MB adsorbed onto the NPC, at equilibrium, were measured, similar to that of adsorption equilibrium experiments.

2.6. Effect of pH Values

Five different pH values (3.62, 5.14, 7.21, 8.76, and 10.20) were investigated by adjusting the pH of MB (400 mg/L) solution, with NaOH or HCl (0.1 M).

2.7. Electrochemical Evaluation

NPC (80 wt.%) was mixed with carbon nanotubes (Sigma, Kawasaki, Kanagawa Prefecture, Japan; 10 wt.%) and polyvinylidenedifluoride (Solvay, Brussels, Belgium; 10 wt.%), and the electrode (d: 1 cm) was prepared by pressing the mixture at 10 MPa. A two-electrode system was used with active materials mass of 2 mg. The cyclic voltammetry (CV), galvanostatic charging/discharging (GCD), and electrochemical impedance spectroscopy (EIS) tests were performed on the EC-LAB VMP3 workstation (BioLogic Science Instruments, Seyssinet-Pariset, France), with 1 M Li_2SO_4 aqueous solution.

3. Results and Discussion

The morphology of NPC material is presented by SEM and TEM. Figure 1 shows the TEM images of the Al-MOC, before (Figure 1a) and after carbonization (Figure 1b). Al-MOC, which had an XRD pattern (Figure S1) comparable to the MIL-53 (Al) from the report [39], exhibited separated nanocubes with ultra-small size around 100 nm, while NPC turned into an interconnected three-dimensional bulk structure, after sintering under high temperatures (Figure 1b). SEM image of NPC also indicated a large size, after annealing (Figure S2). XRD pattern of the obtained NPC is shown in Figure 1c. The NPC powder presents the graphitic carbon (002) and (101) diffractions. The presence of (002) diffraction at $2\theta = 23°$ shifted to the left, compared to the perfect graphite diffraction at $2\theta = 26°$, suggesting that the NPC sample had a low crystallinity [40]. TGA data (Figure 1d) showed a negligible weight of the remaining NPC, after heating to 900 °C, suggesting a successful removal of alumina.

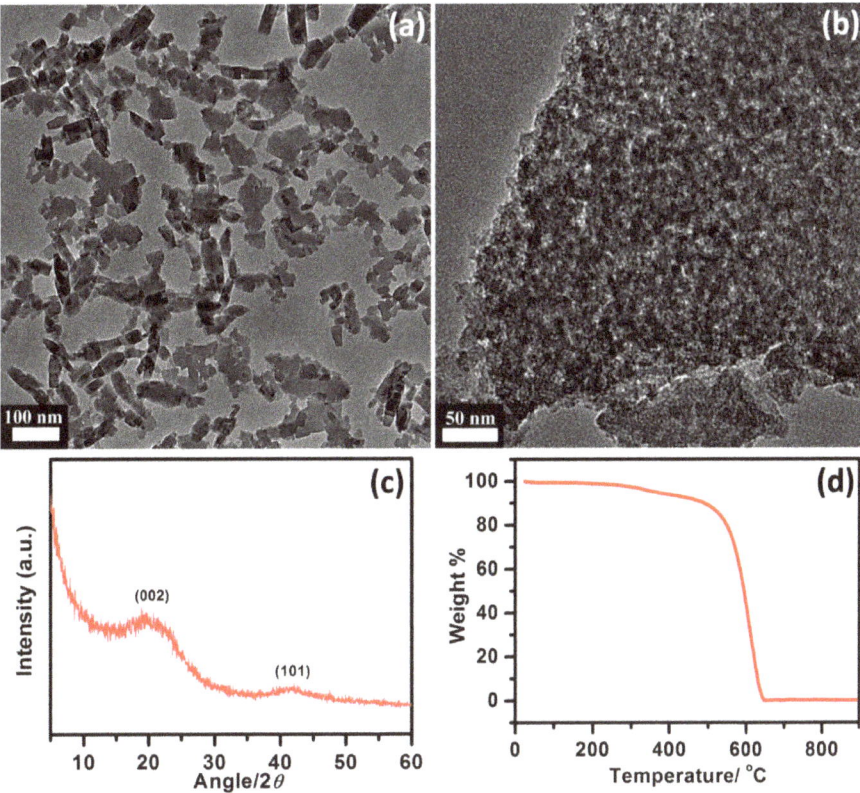

Figure 1. TEM images of Al-based metal–organic complex (Al-MOC) (**a**) and nanoporous carbon (NPC) (**b**); XRD pattern (**c**) and TGA curve (**d**) of NPC.

Although NPC showed an interconnected structure after high-temperature annealing, the HRTEM (Figure S3) images still exhibited an amorphous structure. Notably, irregular mesopores can be found from the agglomerate (Figure S3a–c). In order to achieve the detailed information, the N_2 adsorption–desorption isotherm within the relative pressure of 0–1 was measured. The isotherm in Figure 2a exhibits a typical type IV profile, suggesting the existence of micro-, meso- and macropores. It was calculated that the SSA of NPC was 1593 m^2/g and the total pore volume was 2.49 cm^3/g (Figure 2b), due to the new pores that appeared among the agglomerated complex in bulk NPC. Raman spectrum (Figure 2c) of the NPC sample showed two peaks at 1324 (D band) and 1580 1/cm (G band). The D and G bands were ascribed to the disordered carbons and ideal graphitic carbon, respectively [41]. The intensity ratio I_D/I_G for the NPC was 1.36, suggesting the disordered nature for NPC.

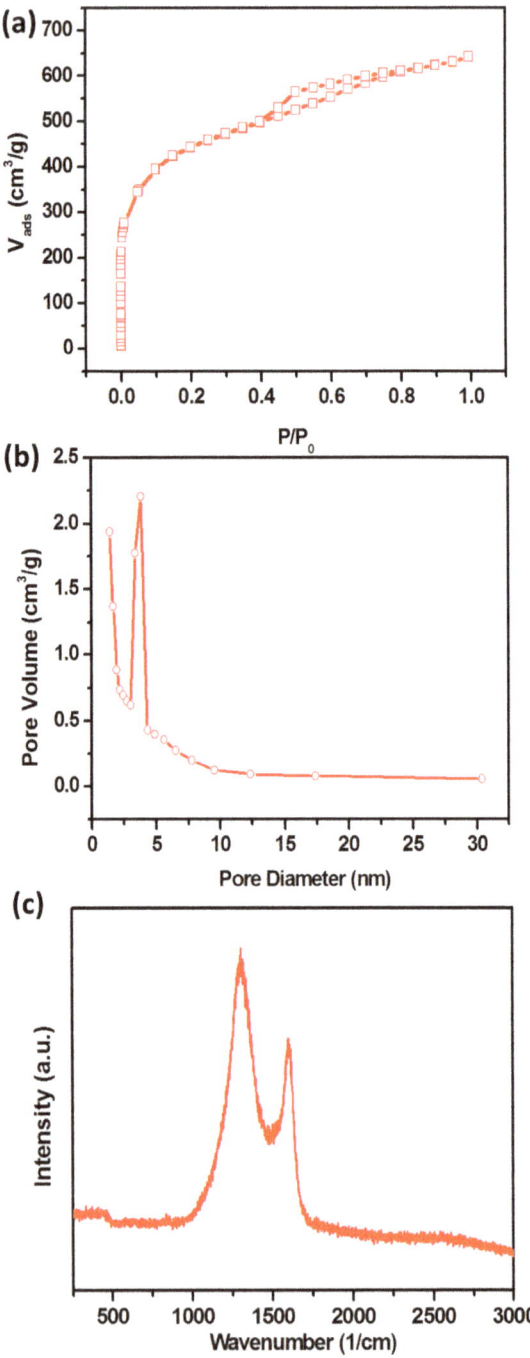

Figure 2. Nitrogen adsorption/desorption isotherms (**a**), pore size distributions (**b**), and Raman spectrum (**c**) of NPC.

Encouraged by the large SSA and the well-distributed porous structure, the adsorption properties of NPC were tested. First, the adsorption process was studied with different dye concentrations. A fast color fading could be observed with the low-concentration groups, and then the adsorption equilibrium, proceed slowly with the high-concentration groups. The fast color fading indicated a fast adsorption rate, which was contributed by the large SSA, as well as the large pore volume. Figure 3 shows that there was a significant increase of the adsorption capacities, with the first four concentrations, which grew slowly, with the latter five concentrations. This could be proved by the well-fitted Langmuir model [42], indicating a monolayer adsorption behavior, which meant that no further dye adsorption could happen at the occupied site of the NPC. The fitted Langmuir isotherm could be defined in the following form:

$$\frac{C_e}{q_e} = \frac{C_e}{q_{max}} + \frac{1}{K_L q_{max}} \quad (3)$$

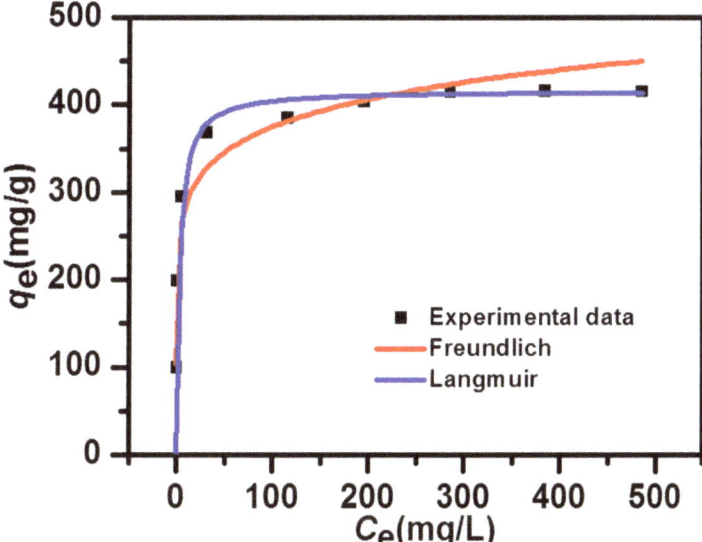

Figure 3. Non-linear fits of the Langmuir and Freundlich isotherm models, to the Methylene Blue (MB) adsorption.

From Figure 4a, q_{max} (mg/g) is the maximum amount of MB adsorbed onto the NPC and the constant (K_L: L/mg) had a relationship to the rate of adsorption. Additionally, the R_L parameter which indicated the adsorption process to be favorable ($0 < R_L < 1$) or unfavorable with the isotherm ($R_L > 1$), was calculated as follows [43]:

$$R_L = \frac{1}{1 + K_L C_0} \quad (4)$$

The Freundlich isotherm model which demonstrated a heterogeneous adsorption process [44], could be written in the following form:

$$\ln q_e = \ln K_F + \frac{1}{n} \ln C_e \quad (5)$$

where the constants (K_F and n) are related to the intercept and the slope of Figure 4b. The parameters for adsorption isotherms calculated from the above Equations (3)–(5) are shown in Table 1. The correlation coefficient (R^2) values indicated that the dye adsorption process was well-matched to both models, but was better matched to the Langmuir isotherm model. It revealed that the dye adsorption started on both the homogeneous and heterogeneous active sites of NPC. The heterogeneous sites were from the possible functional groups contained on the NPC surface. There were negligible differences between the maximum adsorption capacity (q_{max}) fitted by the Langmuir model (417 mg/g), and the experimental value (415 mg/g), which also proved to be a better match to the Langmuir isotherm. According to Table 1, the R_L value was near zero (0.0035), illustrating that the adsorption process was favorable and irreversible. Value n (8.656) in the range of 1–10, also illustrated that this adsorption was favorable [45,46].

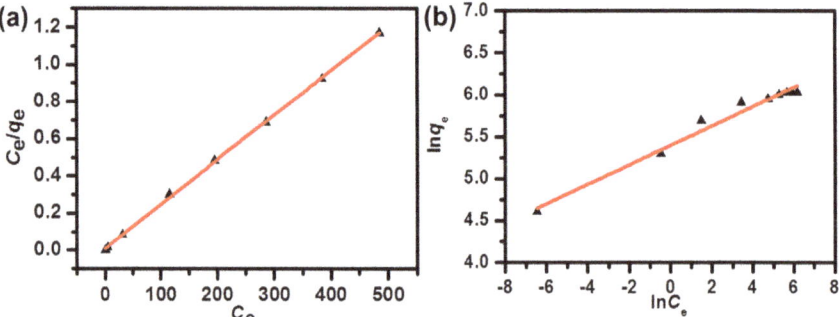

Figure 4. Linear fits of the Langmuir (a) and Freundlich (b) isotherms for the MB adsorption on the NPC samples.

Table 1. Parameters of the Langmuir and Freundlich isotherms for the adsorption of MB on the NPC samples.

	q_{max} (mg/g)	K_L (L/mg)	R_L	K_F	n	R_2
Langmuir	417	0.317	0.0035	-	-	0.999667
Freundlich	-	-	-	220.256	8.656	0.974407

To study the effect of time, adsorption kinetics was measured by choosing different time intervals. Adsorption kinetics could also help to quantify the adsorption rate and illustrate the mechanism of adsorption. The adsorption kinetics of the MB-NPC was obtained by three different MB concentrations (200, 400, and 600 mg/L). The MB adsorption was rapid in the first 5 min (Figure 5), and then the adsorption rate gradually slowed down, with the proceeding time. At last, the adsorption achieved equilibrium, within 2 h. Four kinetics models were applied to figure out the kinetic mechanism of the dye adsorption. First, the adsorption kinetics was examined by the pseudo-first-order model. The equation is described as follows [47]:

$$\ln(q_e - q_t) = \ln q_e - k_1 t \qquad (6)$$

Based on the equation, the plots of $\ln(q_e - q_t)$ versus t (Figure 6a) can be used to calculate k_1 and q_e. Second, the pseudo-second-order equation can be expressed as follows [48]:

$$\frac{t}{q_t} = \frac{1}{k_2 q_e^2} + \frac{1}{q_e} t \qquad (7)$$

Based on this equation, q_e and k_2 calculated from the plots of t/q_t versus t (Figure 6b) are shown in Table 2. The corresponding R^2 values in the pseudo-first-order kinetic model (<0,920) are relatively smaller than in the pseudo-second-order kinetic model (0.999). Moreover, the q_e values of the pseudo-second-order model from the fitted linear plots, are better agreed with the experimental data than those of the pseudo-first-order model, indicating that, in this work, it was more appropriate using the pseudo-second-order kinetic model, to describe the adsorption kinetics.

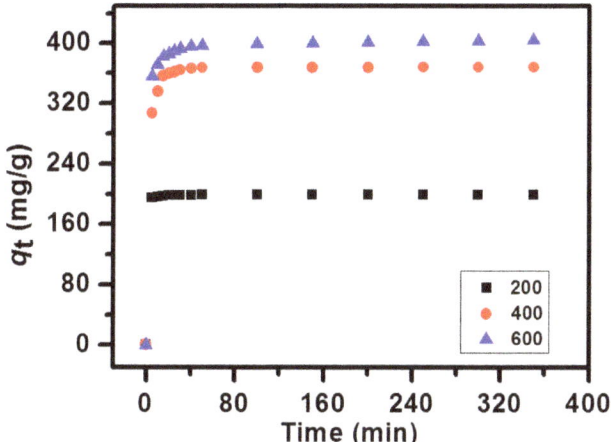

Figure 5. Effect of contact time on the adsorption capacity of MB on NPC samples.

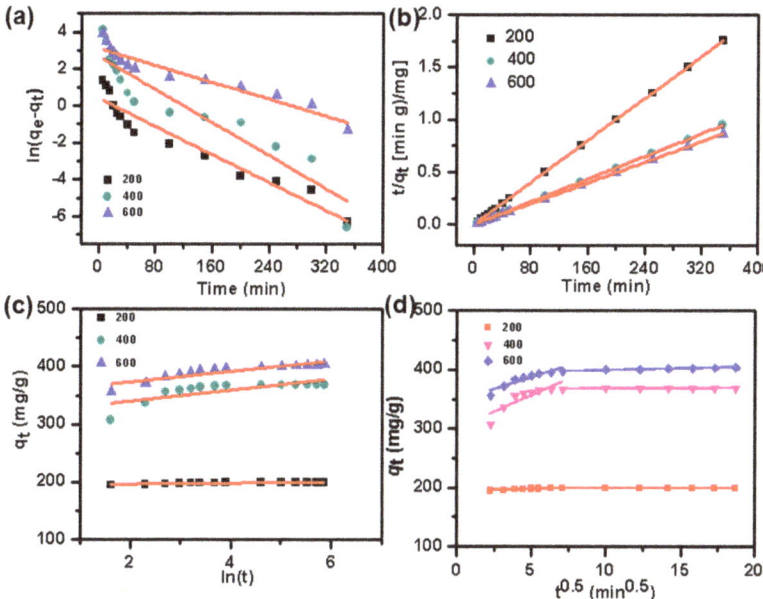

Figure 6. Linear fits of pseudo-first-order kinetics (a), pseudo-second-order kinetics (b), Elovich kinetic (c), and intraparticle diffusion (d) models for the adsorption of MB on the NPC samples.

The Elovich model, which can be used to identify chemical adsorption, is described as follows [49]:

$$q_t = \frac{1}{\beta} \ln(\alpha\beta) + \frac{1}{\beta} \ln(t) \tag{8}$$

where α and β values correspond to the initial adsorption and desorption rate, respectively. As shown in Table 2, the large α shows the viability of the MB-NPC adsorption, while the low value of β confirms that the MB-NPC adsorption is essentially irreversible [50,51]. This model, with a low value of R^2, has a poor linearity (Figure 6c), considering that the mechanism of adsorption did not occur via chemical adsorption.

Dye adsorption might involve the following three sequential steps—(1) external adsorption on the sorbent surface, (2) dye diffusion to the sorbent pore or intraparticle diffusion, and (3) chemical adsorption [7,52,53]. These steps can be investigated by the intraparticle diffusion model described as follows [54]:

$$q_t = k_i t^{0.5} + C \tag{9}$$

Based on this equation, the plot of q_t versus $t^{0.5}$ could be used to calculate the intraparticle diffusion rate constant (k_i: mg/g/min$^{0.5}$) and C (mg/g) (Figure 6d). The corresponding parameters are listed in Table 2. As shown in Figure 6d, there are two linear plots, by fitting q_t versus $t^{0.5}$, suggesting that intraparticle diffusion was involved but was not the predominant mechanism in the adsorption process. Thus, there might be other factors that could influence the adsorption kinetics [53,55]. The concentrations showed two defined stages (Figure 6d). For all concentrations, the first stage had a higher slope, indicating the external and boundary diffusion of the dye adsorption on the NPC surface. The second linear stage showed a drop in the slope, which could be assigned to the intraparticle diffusion. The C values could be used to evaluate the diffusion resistance, suggesting the thickness of the boundary layer. The C values showed an increasing trend towards the dye concentrations (Table 2), which meant that the boundary layer diffusion had a larger effect on the high initial dye concentration.

Table 2. Parameters of the kinetic models for the adsorption of MB on the NPC samples.

	C_0 (mg/L)		
	200	400	600
$q_{e,exp}$ (mg/g)	199.36	368.63	404.42
Pseudo-1st-order			
q_e (mg/g)	1.46	14.44	22.46
k_1 (1/min)	0.01887	0.02236	0.01148
R^2	0.91945	0.87247	0.90755
Pseudo-2nd-order			
q_e (mg/g)	199.6	369	404.9
k_2 (g/mg/min)	0.000005	0.004877	0.002537
R^2	0.999999	0.999996	0.999993
Elovich			
α (mg/g/min)	-	3.1×10^{15}	5.3×10^{17}
β (g/mg)	1.261	0.104	0.109
R^2	0.69778	0.53520	0.80215
Intraparticle diffusion			
K_d	0.01903	0.10303	0.58545
C	199.0359	366.7999	393.3279
R^2	0.859991	0.949224	0.983385

Temperature is another important factor for the dye adsorption. The temperature effect was investigated at 25, 35, 45, and 55 °C, with two MB initial concentrations (500 or 700 mg/L). The thermodynamic equations were as follows:

$$K_D = \frac{q_e}{C_e} \tag{10}$$

$$\Delta G = -RT \ln K_D \tag{11}$$

$$\ln K_D = \frac{\Delta S}{R} - \frac{\Delta H}{RT} \tag{12}$$

where K_D was the adsorption constant, T (K) is the temperature, R (8.314 J/mol/K) is the gas constant, and ΔG^θ is the change of the Gibbs free energy (kJ/mol), the change of enthalpy ((ΔH^θ, kJ/mol), and the change of entropy (ΔS^θ, J/mol/K), which could be calculated from the slope and the intercept of $\ln K_D$

against the 1/T plot (Figure 7). Table 3 shows the thermodynamic parameters obtained from MB-NPC adsorption. The ΔG^θ was negative, proving that the adsorption of MB on the NPC was spontaneous in the temperature set, and the positive values of ΔH^θ confirmed that the adsorption process was endothermic. Furthermore, ΔG^θ changed to more negative values with increasing temperature, indicating that the MB-NPC adsorption was more favorable at a higher temperature, which could also be revealed by the increased experimental adsorption, under a higher temperature. In addition, the values of ΔG^θ, within the range of −20–0 kJ/mol, indicated that the mechanism of the adsorption process was mainly physical adsorption. The ΔS^θ was positive, reflecting the randomness at the solid–liquid interface, during the increasing MB-NPC adsorption.

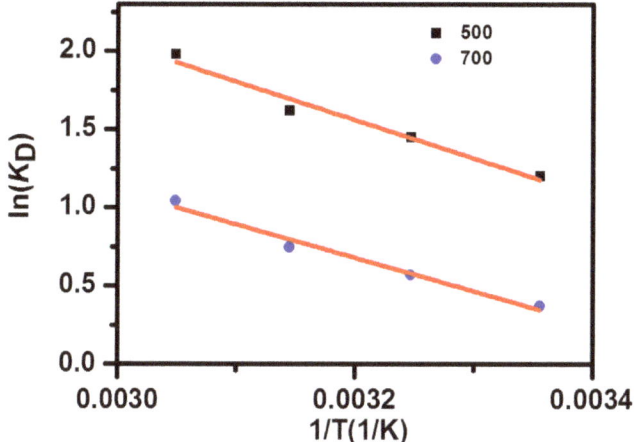

Figure 7. The plot of lnK_D versus 1/T for the adsorption of MB on the NPC samples.

Table 3. Thermodynamic parameters for the adsorption of MB on the NPC samples.

C_0 (mg/L)	T (K)	q_e (mg/g)	K_D	ΔG^θ (kJ/mol)	ΔH^θ (kJ/mol)	ΔS^θ (kJ/mol/K)
500	298	385	3.32	−2.98	20.28	0.078
	308	405	4.26	−3.71		
	318	417	5.05	−4.28		
	328	439	7.24	−5.40		
700	298	414	1.45	−0.92	17.69	0.062
	308	447	1.79	−1.46		
	318	475	2.11	−1.98		
	328	517	2.83	−2.84		

Different pH values of the solution were considered to be another factor in the MB-NPC adsorption. As the pH values of the wasted water were various in practice, it was necessary to evaluate the effect of the pH values on the dye removal technologies. The pH values could affect the adsorption capacity because they could remarkably change the surface charge of the adsorbent. They could also influence the electrostatic interactions and chemical reaction between adsorbates and the adsorbent on active sites [56]. In this study, the pH values were changed from 4 to 10 (Figure 8). The adsorption capacity showed a negligible change within the entire pH region, indicating that the adsorption capacity of the NPC with MB, was not influenced by this factor.

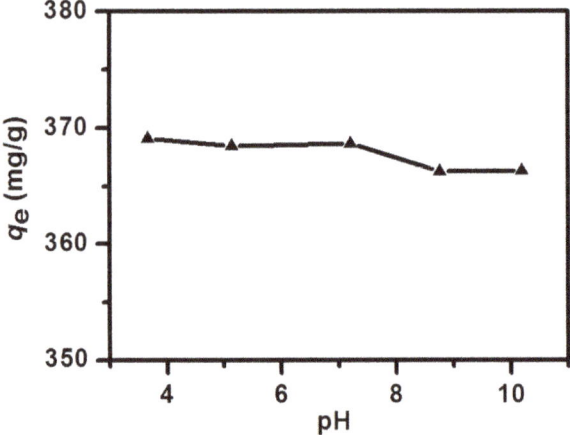

Figure 8. Influence of the solution's initial pH on the MB adsorption onto the NPC.

The electrochemical properties were studied in a full-cell setup, to evaluate the capacitive performance of NPC, in practice. Using 1 M Li_2SO_4 aqueous solution as the electrolyte, CV measurement could be tested in a wide voltage window from 0–1.2 V, at different scanning rates (Figure 9a). The similar shape for all curves suggested a negligible diffusion limitation in higher scanning rates. The CV curves showed a little distortion from rectangular, especially at a high scan rate, which might be due to the porosity saturation [57] and the poor mobility of highly solvated ion Li^+ and highly solvated anion SO_4^{2-} [58]. The specific capacitance derived from the CV, at a scanning rate of 1 mV/s, was calculated to be 226 F/g (Figure 9c). The large specific capacitance was benefits of the interconnected structure of the NPC. The interconnected structure could provide a good electron transfer, leading to a good electrical conductivity. In addition, this interconnected structure, formed from the aggregation of the ultra-small crystal, could supply the connected pores for a continuous and efficient ion diffusion. GCD was conducted to evaluate the stability in the potential window from 0 and 1.2 V, in a 1 M Li_2SO_4 aqueous solution (Figure 9b). NPC exhibits almost symmetrical triangular shapes within the wide potential window, implying a reversible capacitive performance, contributed by the electric double layer supercapacitor (EDLC), and excellent coulombic efficiency. To better analyze the rate performance of the electrode materials, the GCD curves of NPC, at various current densities from 1 to 20 A/g, were collected. The specific capacitances are 205, 179, 155, 143, and 128 F/g, at current densities of 1, 2, 5, 10, and 20 A/g, respectively (Figure 9c). The maximum energy density of NPC was 10.25 Wh/kg in aqueous electrolyte, among the highest values published. The Nyquist plot of the NPC revealed that the electrical conductivity was acquired from EIS measurements. In Figure 9d, from the intersection in the Z' axis, the interface contact resistance was as low as 0.12 Ω. The low charge transfer resistance could be revealed from the small diameter of a semicircle, which benefited from the interconnected structure. The almost vertical curve in the low-frequency region indicated a fast ion diffusion and an ideal capacitive behavior. Good cycling stability was another important factor for capacitors. The electrochemical stability of the NPC was further evaluated at a current density of 10 A/g. Figure S4 shows that there was still a 97.8% retention of capacitance, for the NPC electrodes, after 5000 cycles, confirming a long lifetime.

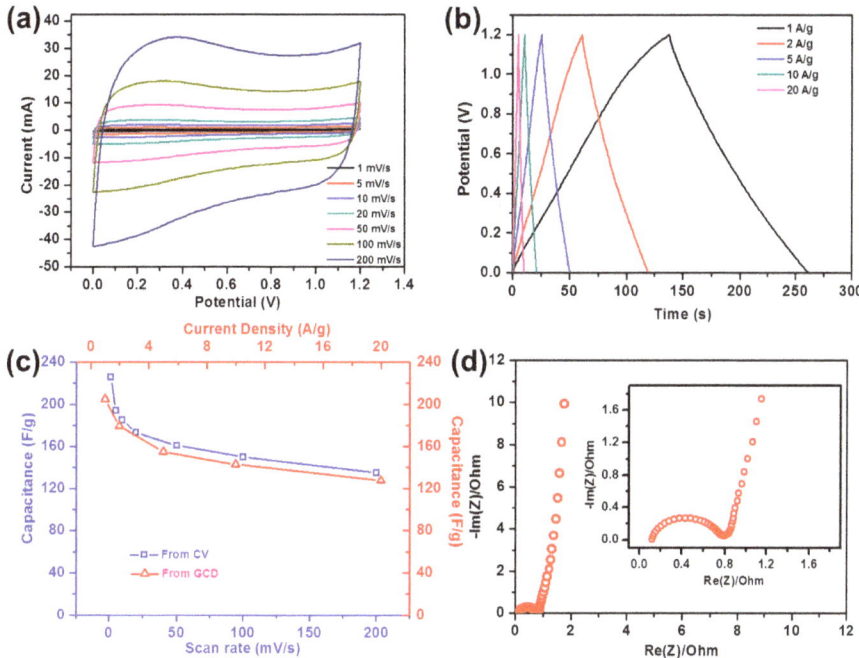

Figure 9. Cyclic voltammetry (CV) curves (**a**) and galvanostatic charging/discharging (GCD) curves (**b**) of NPC; Variations of the specific capacitances (**c**) of NPC from CV and GCD; Nyquist plots (**d**) of NPC (inset shows the Nyquist plots in a high-frequency region).

4. Conclusions

In summary, interconnected NPC was successfully prepared by direct carbonization of the ultra-small Al-MOC. With a porous structure, NPC had a high SSA (1593 m^2/g) and a large dye adsorption, with a maximum value of 415 mg/g, proceeding the physical adsorption. The equilibrium adsorption isotherm and the adsorption kinetics have been fully discussed. The adsorption process was not influenced by the different pH values. Moreover, it was a spontaneous and endothermic process, proved by the thermodynamic studies. NPC, with an excellent adsorption capacity of MB, was believed to have excellent capacitive characteristics in supercapacitors, due to its large ion accessible area, good electrical conductivity, and suitable PSD. NPC exhibited a high specific capacitance of 226 F/g, in a symmetrical supercapacitor, suggesting a large potential in supercapacitors.

Supplementary Materials: The following are available online at http://www.mdpi.com/2079-4991/9/4/601/s1, Figure S1. X-ray diffraction (XRD) pattern of Al-based metal-organic complex (Al-MOC) and the peak position of simulated MIL-53 (Al); Figure S2. Scanning Electron Microscopy (SEM) image of nanoporous carbon (NPC); Figure S3. (**a**) Transmission electron microscopy (TEM) image of NPC with mesopores pointed by arrows at the edge, (**b**,**c**) High-resolution TEM images of the thin edges of NPC with mesopores pointed by arrows; Figure S4. Specific capacitance retention for NPC at a current density of 10 A/g within 5000 cycles.

Author Contributions: Conceptualization, X.C.; Data curation, X.S. and S.Z.; Formal analysis, X.S.; Funding acquisition, X.C.; Supervision, E.M.; Writing—original draft, X.S.

Funding: This research was funded by the National Science Centre, Poland, within UMO-2017/25/B/ST8/02702.

Conflicts of Interest: The authors declare no conflict of interest.

References

1. Han, S.; Sohn, K.; Hyeon, T. Fabrication of New Nanoporous Carbons through Silica Templates and Their Application to the Adsorption of Bulky Dyes. *Chem. Mater.* **2000**, *12*, 3337–3341. [CrossRef]
2. Zhang, S.; Shi, X.; Moszyński, D.; Tang, T.; Chu, P.K.; Chen, X.; Mijowska, E. Hierarchical Porous Carbon Materials from Nanosized Metal–organic Complex for High-Performance Symmetrical Supercapacitor. *Electrochim. Acta* **2018**, *269*, 580–589. [CrossRef]
3. He, C.; Hu, X.J. Anionic Dye Adsorption on Chemically Modified Ordered Mesoporous Carbons. *Ind. Eng. Chem. Res.* **2011**, *50*, 14070–14083. [CrossRef]
4. Blackburn, R.S. Natural polysaccharides and their interactions with dye molecules: Applications in effluent treatment. *Environ. Sci. Technol.* **2004**, *38*, 4905–4909. [CrossRef] [PubMed]
5. Ramakrishna, K.R.; Viraraghavan, T. Dye removal using low cost adsorbents. *Water Sci. Technol.* **1997**, *36*, 189–196. [CrossRef]
6. Cazetta, A.L.; Vargas, A.M.M.; Nogami, E.M.; Kunita, M.H.; Guilherme, M.R.; Martins, A.C.; Silva, T.L.; Moraes, J.C.G.; Almeida, V.C. NaOH-activated carbon of high surface area produced from coconut shell: Kinetics and equilibrium studies from the methylene blue adsorption. *Chem. Eng. J.* **2011**, *174*, 117–125. [CrossRef]
7. Bedin, K.C.; Martins, A.C.; Cazetta, A.L.; Pezoti, O.; Almeida, V.C. KOH-activated carbon prepared from sucrose spherical carbon: Adsorption equilibrium, kinetic and thermodynamic studies for Methylene Blue removal. *Chem. Eng. J.* **2016**, *286*, 476–484. [CrossRef]
8. Zhang, Y.J.; Xing, Z.J.; Duan, Z.K.; Li, M.; Wang, Y. Effects of steam activation on the pore structure and surface chemistry of activated carbon derived from bamboo waste. *Appl. Surf. Sci.* **2014**, *315*, 279–286. [CrossRef]
9. Hameed, B.H.; Din, A.T.M.; Ahmad, A.L. Adsorption of methylene blue onto bamboo-based activated carbon: Kinetics and equilibrium studies. *J. Hazard. Mater.* **2007**, *141*, 819–825. [CrossRef]
10. Tan, I.A.W.; Ahmad, A.L.; Hameed, B.H. Adsorption of basic dye on high-surface-area activated carbon prepared from coconut husk: Equilibrium, kinetic and thermodynamic studies. *J. Hazard. Mater.* **2008**, *154*, 337–346. [CrossRef] [PubMed]
11. Arami, M.; Limaee, N.Y.; Mahmoodi, N.M.; Tabrizi, N.S. Removal of dyes from colored textile wastewater by orange peel adsorbent: Equilibrium and kinetic studies. *J. Colloid Interface Sci.* **2005**, *288*, 371–376. [CrossRef]
12. Namasivayam, C.; Muniasamy, N.; Gayatri, K.; Rani, M.; Ranganathan, K. Removal of dyes from aqueous solutions by cellulosic waste orange peel. *Bioresour. Technol.* **1996**, *57*, 37–43. [CrossRef]
13. Kuo, C.Y.; Wu, C.H.; Wu, J.Y. Adsorption of direct dyes from aqueous solutions by carbon nanotubes: Determination of equilibrium, kinetics and thermodynamics parameters. *J. Colloid Interface Sci.* **2008**, *327*, 308–315. [CrossRef] [PubMed]
14. Yao, Y.J.; He, B.; Xu, F.F.; Chen, X.F. Equilibrium and kinetic studies of methyl orange adsorption on multiwalled carbon nanotubes. *Chem. Eng. J.* **2011**, *170*, 82–89. [CrossRef]
15. Kang, D.J.; Yu, X.L.; Ge, M.F.; Xiao, F.; Xu, H. Novel Al-doped carbon nanotubes with adsorption and coagulation promotion for organic pollutant removal. *J. Environ. Sci.* **2017**, *54*, 1–12. [CrossRef] [PubMed]
16. Kumar, R.; Ansari, M.O.; Barakat, M.A. Adsorption of Brilliant Green by Surfactant Doped Polyaniline/MWCNTs Composite: Evaluation of the Kinetic, Thermodynamic, and Isotherm. *Ind. Eng. Chem. Res.* **2014**, *53*, 7167–7175. [CrossRef]
17. Rathinam, K.; Singh, S.P.; Li, Y.L.; Kasher, R.; Tour, J.M.; Arnusch, C.J. Polyimide derived laser-induced graphene as adsorbent for cationic and anionic dyes. *Carbon* **2017**, *124*, 515–524. [CrossRef]
18. Ramesha, G.K.; Kumara, A.V.; Muralidhara, H.B.; Sampath, S. Graphene and graphene oxide as effective adsorbents toward anionic and cationic dyes. *J. Colloid Interface Sci.* **2011**, *361*, 270–277. [CrossRef]
19. Abdi, J.; Vossoughi, M.; Mahmoodi, N.M.; Alemzadeh, I. Synthesis of metal–organic framework hybrid nanocomposites based on GO and CNT with high adsorption capacity for dye removal. *Chem. Eng. J.* **2017**, *326*, 1145–1158. [CrossRef]
20. Zhu, J.Y.; Wang, Y.M.; Liu, J.D.; Zhang, Y.T. Facile One-Pot Synthesis of Novel Spherical Zeolite-Reduced Graphene Oxide Composites for Cationic Dye Adsorption. *Ind. Eng. Chem. Res.* **2014**, *53*, 13711–13717. [CrossRef]

21. Qadri, S.; Ganoe, A.; Haik, Y. Removal and recovery of acridine orange from solutions by use of magnetic nanoparticles. *J. Hazard. Mater.* **2009**, *169*, 318–323. [CrossRef]
22. Asfaram, A.; Ghaedi, M.; Hajati, S.; Goudarzi, A.; Dil, E.A. Screening and optimization of highly effective ultrasound-assisted simultaneous adsorption of cationic dyes onto Mn-doped Fe_3O_4-nanoparticle-loaded activated carbon. *Ultrason. Sonochem.* **2017**, *34*, 1–12. [CrossRef] [PubMed]
23. El-Kady, M.F.; Strong, V.; Dubin, S.; Kaner, R.B. Laser Scribing of High-Performance and Flexible Graphene-Based Electrochemical Capacitors. *Science* **2012**, *335*, 1326–1330. [CrossRef] [PubMed]
24. Xu, Y.X.; Lin, Z.Y.; Zhong, X.; Huang, X.Q.; Weiss, N.O.; Huang, Y.; Duan, X.F. Holey graphene frameworks for highly efficient capacitive energy storage. *Nat. Commun.* **2014**, *5*, 4554. [CrossRef] [PubMed]
25. Winter, M.; Brodd, R.J. What are batteries, fuel cells, and supercapacitors? *Chem. Rev.* **2004**, *104*, 4245–4269. [CrossRef] [PubMed]
26. Balducci, A.; Dugas, R.; Taberna, P.L.; Simon, P.; Plee, D.; Mastragostino, M.; Passerini, S. High temperature carbon-carbon supercapacitor using ionic liquid as electrolyte. *J. Power Sources* **2007**, *165*, 922–927. [CrossRef]
27. Zhang, L.L.; Zhao, X.; Stoller, M.D.; Zhu, Y.W.; Ji, H.X.; Murali, S.; Wu, Y.P.; Perales, S.; Clevenger, B.; Ruoff, R.S. Highly Conductive and Porous Activated Reduced Graphene Oxide Films for High-Power Supercapacitors. *Nano Lett.* **2012**, *12*, 1806–1812. [CrossRef] [PubMed]
28. Chen, X.Y.; Chen, C.; Zhang, Z.J.; Xie, D.H.; Deng, X. Nitrogen-Doped Porous Carbon Prepared from Urea Formaldehyde Resins by Template Carbonization Method for Supercapacitors. *Ind. Eng. Chem. Res.* **2013**, *52*, 10181–10188. [CrossRef]
29. Zhang, Z.J.; Cui, P.; Chen, X.Y. Structure and Capacitive Performance of Porous Carbons Derived from Terephthalic Acid-Zinc Complex via a Template Carbonization Process. *Ind. Eng. Chem. Res.* **2013**, *52*, 16211–16219. [CrossRef]
30. Vinu, A.; Ariga, K.; Mori, T.; Nakanishi, T.; Hishita, S.; Golberg, D.; Bando, Y. Preparation and characterization of well-ordered hexagonal mesoporous carbon nitride. *Adv. Mater.* **2005**, *17*, 1648–1652. [CrossRef]
31. Ania, C.O.; Khomenko, V.; Raymundo-Pinero, E.; Parra, J.B.; Beguin, F. The large electrochemical capacitance of microporous doped carbon obtained by using a zeolite template. *Adv. Funct. Mater.* **2007**, *17*, 1828–1836. [CrossRef]
32. Lee, J.; Kim, J.; Hyeon, T. Recent progress in the synthesis of porous carbon materials. *Adv. Mater.* **2006**, *18*, 2073–2094. [CrossRef]
33. Xiao, J.-D.; Qiu, L.-G.; Jiang, X.; Zhu, Y.-J.; Ye, S.; Jiang, X. Magnetic porous carbons with high adsorption capacity synthesized by a microwave-enhanced high temperature ionothermal method from a Fe-based metal–organic framework. *Carbon* **2013**, *59*, 372–382. [CrossRef]
34. Torad, N.L.; Hu, M.; Ishihara, S.; Sukegawa, H.; Belik, A.A.; Imura, M.; Ariga, K.; Sakka, Y.; Yamauchi, Y. Direct Synthesis of MOF-Derived Nanoporous Carbon with Magnetic Co Nanoparticles toward Efficient Water Treatment. *Small* **2014**, *10*, 2096–2107. [CrossRef] [PubMed]
35. Jiao, C.; Wang, Y.; Li, M.; Wu, Q.; Wang, C.; Wang, Z. Synthesis of magnetic nanoporous carbon from metal–organic framework for the fast removal of organic dye from aqueous solution. *J. Magn. Magn. Mater.* **2016**, *407*, 24–30. [CrossRef]
36. Yang, S.J.; Kim, T.; Im, J.H.; Kim, Y.S.; Lee, K.; Jung, H.; Park, C.R. MOF-Derived Hierarchically Porous Carbon with Exceptional Porosity and Hydrogen Storage Capacity. *Chem. Mater.* **2012**, *24*, 464–470. [CrossRef]
37. Li, C.X.; Hu, C.G.; Zhao, Y.; Song, L.; Zhang, J.; Huang, R.D.; Qu, L.T. Decoration of graphene network with metal–organic frameworks for enhanced electrochemical capacitive behavior. *Carbon* **2014**, *78*, 231–242. [CrossRef]
38. Miao, L.; Zhu, D.Z.; Zhao, Y.H.; Liu, M.X.; Duan, H.; Xiong, W.; Zhu, Q.J.; Li, L.C.; Lv, Y.K.; Gan, L.H. Design of carbon materials with ultramicro-, supermicro- and mesopores using solvent- and self-template strategy for supercapacitors. *Microporous Mesoporous Mater.* **2017**, *253*, 1–9. [CrossRef]
39. Loiseau, T.; Serre, C.; Huguenard, C.; Fink, G.; Taulelle, F.; Henry, M.; Bataille, T.; Férey, G. A Rationale for the Large Breathing of the Porous Aluminum Terephthalate (MIL-53) Upon Hydration. *Chem. Eur. J.* **2004**, *10*, 1373–1382. [CrossRef]
40. Zhao, J.; Yang, L.; Li, F.; Yu, R.; Jin, C. Structural evolution in the graphitization process of activated carbon by high-pressure sintering. *Carbon* **2009**, *47*, 744–751. [CrossRef]
41. Dresselhaus, M.S.; Jorio, A.; Hofmann, M.; Dresselhaus, G.; Saito, R. Perspectives on Carbon Nanotubes and Graphene Raman Spectroscopy. *Nano Lett.* **2010**, *10*, 751–758. [CrossRef]

42. Langmuir, I. The adsorption of gases on plane surfaces of glass, mica and platinum. *J. Am. Chem. Soc.* **1918**, *40*, 1361–1403. [CrossRef]
43. Hall, K.R.; Eagleton, L.C.; Acrivos, A.; Vermeulen, T. Pore- and solid-diffusion kinetics in fixed-bed adsorption under constant-pattern conditions. *Ind. Eng. Chem. Fundam.* **1966**, *5*, 212–223. [CrossRef]
44. Freundlich, H. Über die Adsorption in Lösungen. *Z. Phys. Chem.* **1907**, *57*, 385–470. [CrossRef]
45. Gupta, V.K.; Ali, I.; Saini, V.K. Adsorption studies on the removal of Vertigo Blue 49 and Orange DNA13 from aqueous solutions using carbon slurry developed from a waste material. *J. Colloid Interface Sci.* **2007**, *315*, 87–93. [CrossRef] [PubMed]
46. Namasivayam, C.; Jeyakumar, R.; Yamuna, R.T. Dye removal from wastewater by adsorption on 'waste' Fe(III)/Cr(III) hydroxide. *Waste Manag.* **1994**, *14*, 643–648. [CrossRef]
47. Lagergren, S. Zur theorie der sogenannten adsorption gelöster stoffe. *Kungliga Svenska Vetenskapsakademiens* **1898**, *24*, 1–39.
48. Ho, Y.S.; McKay, G. Pseudo-second order model for sorption processes. *Process Biochem.* **1999**, *34*, 451–465. [CrossRef]
49. Aksakal, O.; Ucun, H. Equilibrium, kinetic and thermodynamic studies of the biosorption of textile dye (Reactive Red 195) onto *Pinus sylvestris* L. *J. Hazard. Mater.* **2010**, *181*, 666–672. [CrossRef]
50. Vargas, A.M.M.; Cazetta, A.L.; Martins, A.C.; Moraes, J.C.G.; Garcia, E.E.; Gauze, G.F.; Costa, W.F.; Almeida, V.C. Kinetic and equilibrium studies: Adsorption of food dyes Acid Yellow 6, Acid Yellow 23, and Acid Red 18 on activated carbon from flamboyant pods. *Chem. Eng. J.* **2012**, *181–182*, 243–250. [CrossRef]
51. Khan, T.A.; Chaudhry, S.A.; Ali, I. Equilibrium uptake, isotherm and kinetic studies of Cd(II) adsorption onto iron oxide activated red mud from aqueous solution. *J. Mol. Liq.* **2015**, *202*, 165–175. [CrossRef]
52. Nethaji, S.; Sivasamy, A. Adsorptive removal of an acid dye by lignocellulosic waste biomass activated carbon: Equilibrium and kinetic studies. *Chemosphere* **2011**, *82*, 1367–1372. [CrossRef] [PubMed]
53. Cheung, W.H.; Szeto, Y.S.; McKay, G. Intraparticle diffusion processes during acid dye adsorption onto chitosan. *Bioresour. Technol.* **2007**, *98*, 2897–2904. [CrossRef] [PubMed]
54. Weber, W.J.; Morris, J.C. Kinetics of Adsorption on Carbon from Solution. *J. Sanit. Eng. Div.* **1963**, *89*, 31–60.
55. Fan, J.; Zhang, J.; Zhang, C.; Ren, L.; Shi, Q. Adsorption of 2,4,6-trichlorophenol from aqueous solution onto activated carbon derived from loosestrife. *Desalination* **2011**, *267*, 139–146. [CrossRef]
56. Zhang, C.; Ye, F.; Shen, S.; Xiong, Y.; Su, L.; Zhao, S. From metal–organic frameworks to magnetic nanostructured porous carbon composites: Towards highly efficient dye removal and degradation. *RSC Adv.* **2015**, *5*, 8228–8235. [CrossRef]
57. Vaquero, S.; Díaz, R.; Anderson, M.; Palma, J.; Marcilla, R. Insights into the influence of pore size distribution and surface functionalities in the behaviour of carbon supercapacitors. *Electrochim. Acta* **2012**, *86*, 241–247. [CrossRef]
58. Fic, K.; Lota, G.; Meller, M.; Frackowiak, E. Novel insight into neutral medium as electrolyte for high-voltage supercapacitors. *Energy Environ. Sci.* **2012**, *5*, 5842–5850. [CrossRef]

 © 2019 by the authors. Licensee MDPI, Basel, Switzerland. This article is an open access article distributed under the terms and conditions of the Creative Commons Attribution (CC BY) license (http://creativecommons.org/licenses/by/4.0/).

Article

Interaction between Persistent Organic Pollutants and ZnO NPs in Synthetic and Natural Waters

Rizwan Khan [1], Muhammad Ali Inam [1], Sarfaraz Khan [2], Du Ri Park [1] and Ick Tae Yeom [1,*]

[1] Graduate School of Water Resources, Sungkyunkwan University (SKKU) 2066, Suwon 16419, Korea; rizwankhan@skku.edu (R.K.); aliinam@skku.edu (M.A.I.); enfl8709@skku.edu (D.R.P.)
[2] Key Laboratory of the Three Gorges Reservoir Region Eco-Environment, State Ministry of Education, Chongqing University, Chongqing 400045, China; Sfk.jadoon@yahoo.com
* Correspondence: yeom@skku.edu; Tel.: +82-31-299-6699

Received: 28 February 2019; Accepted: 15 March 2019; Published: 21 March 2019

Abstract: The use of zinc oxide nanoparticles (ZnO NPs) and polybrominated diphenyl ethers (PBDPEs) in different products and applications leads to the likelihood of their co-occurrence in the aquatic system, making it important to study the effect of PBDPEs on the fate and transport of ZnO NPs. In this study, we determine the influence of PBDPEs (BDPE-47 and BDPE-209) on the colloidal stability and physicochemical properties of ZnO NPs in different aqueous matrices. The results indicated the shift in ζ potential of ZnO NP from positive to negative in the presence of both PBDPEs in all tested waters; however, the effect on the NPs surface potential was specific to each water considered. The lower concentration of the PBDPEs (e.g., 0.5 mg/L) significantly reduced the ζ potential and hydrodynamic diameter (HDD) of ZnO NP, even in the presence of high content of dissolved organic matter (DOM) in both freshwater and industrial wastewater. Moreover, both BDPE-47 and BDPE-209 impede the agglomeration of ZnO NP in simple and natural media, even in the presence of monovalent and polyvalent cations. However, the effect of BDPE-47 on the ζ potential, HDD, and agglomeration of ZnO NP was more pronounced than that of BDPE-209 in all tested waters. The results of Fourier transform infrared (FT-IR) and X-ray Photon Spectroscopy (XPS) further confirm the adsorption of PBDPEs onto ZnO NP surface via aromatic ether groups and Br elements. The findings of this study will facilitate a better understanding of the interaction behavior between the ZnO NPs and PBDPEs, which can reduce the exposure risk of aquatic organisms to both pollutants.

Keywords: agglomeration; interaction; organic pollutants; stability; ZnO nanoparticles

1. Introduction

Engineered nanoparticles (ENPs) are used in various fields, because of their unique structural properties and high reactivity. Among them, zinc oxide (ZnO) NPs is the third extensively produced ENP, and is used in cosmetics, marine antifouling paints, and packaging. The industrial applications of ZnO include catalysts, photovoltaic devices, lasers, transducers, and sensors [1]. The global production of ZnO NPs was around 1600 tons in 2010, which is estimated to increase to 58,000 tons/year by 2020 [2]. Moreover, a recent study [3] reported the concentration of ZnO NPs in wastewater treatment plant biofilm to be (22–24) μg/Kg, which is anticipated to increase over time. As a consequence, a considerable fraction of ZnO NPs may enter into water, which may pose a threat to aquatic life and human health. Previous studies [4,5] indicated the adverse effects of ZnO NPs on different aquatic organisms, such as zebrafish, sea urchin, algae, bacteria, and plants. A recent study have reported the toxic effect of ZnO NPs on the organisms living on the ground, which include the accumulation of tiny NPs in mice liver causing endoplasmic reticulum (ER) stress response [6]. Accordingly, the effect of ZnO NPs on the reproduction of *Folsomia candida* and *Eisenia* fetida showed that the toxicity of

ZnO NPs highly influenced by soil pH, as the rate of dissolution increased in more acidic soils [7,8]. Moreover, the toxic effects of these NPs on human cells, including damage to DNA and cell membranes, have also been well reported [9]. Thus, understanding the fate and mobility of ZnO NPs in water is essential to assess their potential risk in the ecosystem.

The toxicity and colloidal stability of ZnO NPs depend upon several environmental factors, such as primary particle size, pH, ionic strength (IS), dissolved organic matter (DOM), and extracellular polymeric substances (EPS) [10–13]. The small size of NPs may cause significant variation in the aggregation kinetics under various water matrices. Moreover, they are considered stable in suspension with high capability to convey toxic substance, while large size NPs have a tendency to form large agglomerates which decrease the dissolution of NPs in solution [9]. Some recent studies [14,15] demonstrated that ZnO NPs are unstable in waters with high IS (e.g., seawater); however, high concentration of DOM enhances the colloidal stability. In addition, natural waters may also contain synthetic organic pollutants resulting from direct or indirect anthropogenic activities. Among many persistent organic pollutants (POPs), polybrominated diphenyl ethers (PBDPEs) are mostly used as flame retardants in paints, industrial applications, and other consumer products [16]. Therefore, there are growing concerns regarding the release of PBDPEs into the natural environment during the usage and disposal of these products. Earlier study [17] has shown the adsorption of organic pollutants onto the NPs surface in water. Several developed countries have restricted the use of PBDPEs; however, these pollutants are still found in products manufactured prior to the phase-out completion in these countries. Moreover, their products are frequently available in countries with unrestricted use of PBDPEs, thus leading to their risk in the eco-toxicological context [18]. A few researchers [19,20] have reported the concentration of PBDPEs in surface waters up to 1000 ng/L, while it has been also found in the atmosphere, sediments, and humans [21–23]. Moreover, the elevated concentration of organic pollutants such as PBDPEs may lead to their adsorption onto the NPs surface in an aquatic environment, thus enhancing the overall colloidal stability [24]. After an extensive literature search, we found that the interaction between POPs and ZnO NPs in water has barely been touched by the environmental scholars. Moreover, previous studies also seem insufficient regarding the effect of hydrophobic POPs on the environmental fate and mobility of ENPs in water.

Accordingly, the objective of the present study was to investigate the effect of two widely used PBDPEs, BDPE-47 and BDPE-209, on the colloidal stability of ZnO NP in aqueous matrices. Moreover, this work also hopes to assess the influence of POPs on the physicochemical properties of ZnO NPs in both synthetic and natural waters.

2. Materials and Methods

2.1. Materials and Reagents

The ZnO powder (CAS No:1314-13-2, purity > 97%, see Table S1 of the Supplementary Information (SI) for detailed properties), and two BDPE congeners 2,2′,4,4′-tetrabromodiphenyl ether (BDPE-47) and 2,2′,3,3′,4,4′,5,5′,6,6′-decabromodiphenyl ether (BDPE-209) with purity > 97% were purchased from Sigma Aldrich (St. Louis, MO, USA). Table 1 shows the physicochemical properties of both BDPEs. The potassium chloride (KCl), magnesium chloride ($MgCl_2$), aluminum sulfate $Al_2(SO_4)_3$, hydrochloric acid (HCl), and sodium hydroxide (NaOH) were obtained from Samchun (Samchun Pure Chemicals Co., Ltd., Pyeongteak-si, Korea). The Synergy water system (Milli-Q, Millipore, Burlington, MA, USA) was used to produce deionized (DI) water (18.2 $M\Omega$ cm^{-1} resistivity), and it was used in the preparation of the stock solutions.

2.2. Preparation of Stock Suspension

The stock solution of ZnO NPs was prepared by adding 100 mg of ZnO powder in 1 L of DI water. The detailed procedure can be found in our previous study [25]. Stock solutions (500 mg/L) of BDPE-47 and BDPE-209 were prepared in anhydrous ethanol and dimethyl sulfoxide (DMSO, purity > 99.5%), respectively, by continuous mixing with a magnetic stirrer for 2 h. The ZnO–PBDPEs suspensions were prepared with a lower ratio of 1:20 (solvent:water) to reduce the effect of solvent during experiments, and further diluted to achieve the desired concentration.

2.3. Collection of Natural Waters

To investigate the effect of PBDPEs on ZnO NPs stability in synthetic and natural waters, the freshwater (FW) was prepared in the laboratory, while the industrial wastewater (IWW) was obtained from metal processing industry (Onsan National Industrial Complex, Ulsan, Korea). Table S2 of the SI shows the major properties of both waters. The concentration of various ions present in IWW was measured with ion chromatography (861-Advanced Compact IC, Herisau, Switzerland) using standard methods [26]. The pH and total organic carbon (TOC) of collected samples were measured with pH meter (HACH-HQ40d portable multi meter, Loveland, CO, USA) and TOC analyzer (TOC-5000A, Shimadzu Corporation, Kyoto, Japan). In order to remove the particulates, both waters were filtered using 0.45 µm glass fiber filter.

Table 1. Important properties of the PBDPEs used in the present work.

Property	BDPE-209	BDPE-47
Solubility (µg/L)	20–30	1–15
Log k_{ow}	6.27–9.98	6.77–6.81
Molecular weight (g/mol)	959.2	485.79
Empirical formula	$C_{12}Br_{10}O$	$C_{12}H_6Br_4O$
Molecular structure		

Sources: [22,27,28].

2.4. Effect of PBDPEs on ζ Potential and Size of ZnO NPs

The ENPs stability and their interactions with other molecules in aqueous environment are mostly controlled by the absolute surface charges of the colloids. The controlled experiments were performed to investigate the influence of PBDPEs type (BDPE-47, BDPE-209) and concentration ((0–5) mg/L) on the ζ potential and hydrodynamic diameter (HDD) of ZnO NPs in DI water at pH 7. Moreover, additional experiments were conducted in synthetic and natural waters to quantify the effect of both PBDPEs on the ζ potential and HDD of ZnO NPs in these waters.

2.5. Aggregation Kinetics in Various Aqueous Matrix

The aggregation kinetics of ZnO NP was studied through the time-resolved dynamic light scattering (DLS) method. The predetermined amount of ZnO NPs suspension was added into the glass vials with known volume of either DI water or electrolyte solution (KCl, $MgCl_2$, and $Al_2(SO_4)_3$). Afterwards, the vials were instantly vortexed for 5 s, to ensure complete mixing. Subsequently, 1 mL of the suspension was transferred into a DTS0012 cuvette (Malvern) and then placed in the Zetasizer sample chamber for hydrodynamic size measurements. Following the same procedure, further experiments were conducted by adding known concentrations (0–10 mg/L) of BDPE-47 or BDPE-209, prior to the addition of ZnO NPs in solutions. In order to obtain sufficient signal for DLS analysis, the ZnO NPs concentration was set to 10 mg/L. Since the higher polydispersity index (PDI) limits the reliability of particle HDD values, the PDI values above 0.7 in datasets were not considered [29]. The DLS data were collected in triplicates at 30 s intervals for 30 min at 25 °C, while pH was adjusted to 7.0 using 0.1 M HCl or NaOH solution. The aggregation kinetics of ZnO NPs in synthetic and natural waters were determined using Equation (1) as described in earlier studies [30,31]:

$$k_a \propto \frac{1}{N_0}\left(\frac{dD_h(t)}{dt}\right)_{t \to 0} \to k_a \cdot N_0 = \left(\frac{dD_h(t)}{dt}\right)_{t \to 0} \quad (1)$$

where k_a is the aggregation rate constant, and N_0 is the initial particle number concentration. The early stage aggregation period was defined for D_h values equal to, or less than two times of initial D_h [30].

The attachment efficiency α was used to measure ZnO NP aggregation kinetics in various waters [32]. The α can be determined from the aggregation rate constant (k_a) normalized by the aggregation rate constant (k_a)$_{fast}$ in the diffusion-limited regime, as shown in Equation (2):

$$\alpha = \frac{1}{w} = \frac{k_a}{k_{a\,fast}} = \frac{\frac{1}{N_0}\left(\frac{dD_h(t)}{dt}\right)_{t \to 0}}{\frac{1}{(N_0)_{fast}}\left(\frac{dD_h(t)}{dt}\right)_{t \to 0,\,fast}} \quad (2)$$

where (k_a)$_{fast}$ shows the favorable suspension condition, where fast, diffusion-limited aggregation occurs [33].

Furthermore, the intersection of two lines extrapolated through the diffusion-limited and reaction-limited regimes yields the critical coagulation concentration (CCC) of KCl for ZnO NP [34]. The CCC can be defined as the minimum amount of an electrolyte required to destabilize NP suspension completely. The value of CCC provides essential information about NPs stability, and can thus be used to predict the fate and transport of NPs in natural waters [14,23].

2.6. Other Analytical Procedures

The ζ potential of ZnO NPs was measured with Zetasizer (Nano ZS90, Malvern Instruments, Worcestershire, UK). The instrument was equipped with a 633-nm red laser and was capable of analyzing particles with diameters ranging from 0.3 nm to 5.0 micron using DLS as the basic principle of operation, and ζ potential measurement through Doppler electrophoresis. Folded capillary cell (DTS1060) was used for ζ potential measurements, and disposable low volume polystyrene (DTS0012) cuvette was used for particle size measurement. The ζ potential (mV) was measured at 25 °C with 10 repeated measurements, where the refractive indices of ZnO and water were set to 2.00 and 1.33, respectively. The absorption spectra of ZnO NPs were determined using UV–Vis spectrophotometer (Optizen-2120, Mecasys, Daejeon, Korea) in the 250–800 nm wavelength range. The X-ray photoelectron spectroscopy (XPS), Raman spectroscopy, X-ray diffractometry (XRD), and Brunauer–Emmett–Teller (BET) surface area analysis of the ZnO NPs were done using XSAM HS (KRATOS), D max C III (Rigaku Corporation, Tokyo, Japan), Jobin Yvon microscope (Horiba, Bensheim, Germany), and ASAP 2020 (Micromeritics, Norcross, GA, USA) respectively. Fourier transform infrared (FT-IR) spectroscopy (JASCO, FT-IR-4700, Easton, PA, USA) of ZnO and PBDPEs before and after interaction were conducted to explore the possible attachment of functional groups during the interaction between the two pollutants. Moreover, XPS analysis of ZnO–PBDPEs complexes was conducted to confirm the presence of different elements in the structure of ZnO NP after interaction with PBDPEs.

3. Results

3.1. ZnO NP Characterization

Figure 1A shows the FT-IR spectroscopy of pristine ZnO powder, which reveals the broad band at 556 cm^{-1} corresponding to the stretching vibration of Zn–O bonds [25]. Moreover, the peaks observed at 1378, 1629, and 3402 cm^{-1} are attributed to the OH groups and displacement of weakly adsorbed water molecules on the NPs surface. Figure 1B shows the XPS spectra of pure ZnO, which revealed that the corresponding binding energies of Zn 2p$_{1/2}$ and 2p$_{3/2}$ were 1042 and 1021 eV, respectively [35].

In addition, the Raman scattering showed peaks at 386 and 441 cm^{-1}, which correspond to A1 (TO) and E1 (TO) mode (Figure 1C).

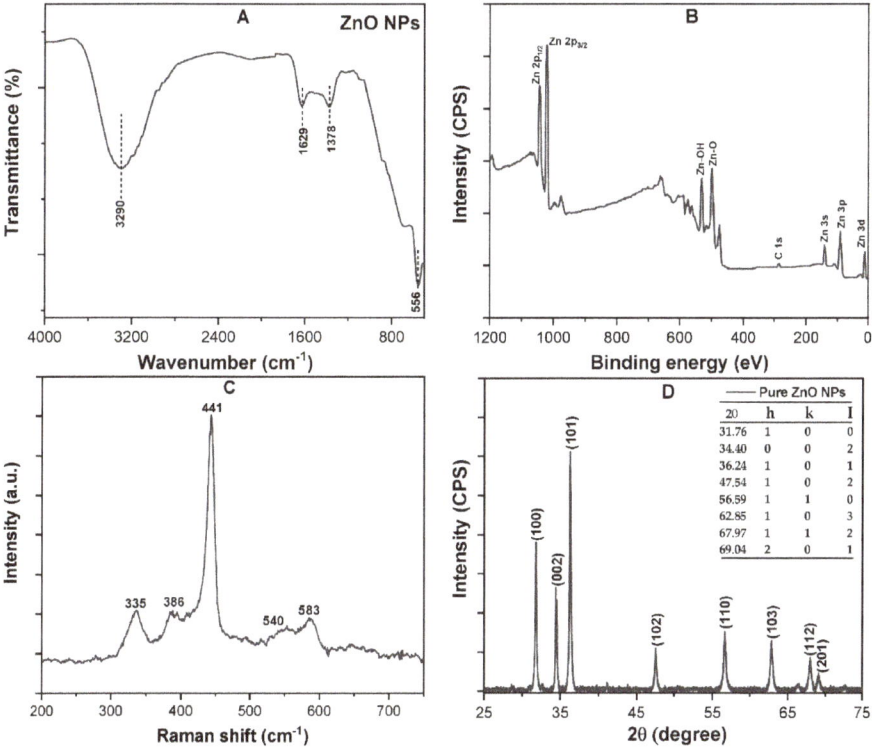

Figure 1. Characterization of the zinc oxide nanoparticles (ZnO NPs) powder (**A**) FT-IR; (**B**) X-ray photoelectron spectroscopy (XPS) survey spectrum; (**C**) Raman spectrum; (**D**) XRD spectra expressed in arbitrary units and counts per second.

The XRD spectrum pattern confirms the crystalline structure of the ZnO NPs (Figure 1D). In the spectrum, peaks observed at 31.76°, 34.37°, 36.24°, 47.54°, 56.58°, 62.85°, and 67.97°, were corresponding to (100), (002), (101), (102), (110), (103), and (112) planes of pure ZnO, respectively. Therefore, all detectable peaks can be indexed to the ZnO wurtzite structure (JCPDS: 00-036-1451) [36]. In addition, the crystallite size of ZnO NPs was determined using Scherer formula [37], and the average size of ZnO NPs has been calculated to be 45 ± 2 nm. The detailed description can be found in (Supplementary Materials, Section 1.1). Furthermore, Figure S1A of the SI shows the Brunauer–Emmett–Teller (BET) surface area of the ZnO NPs to be around 12.5 m^2/g.

3.2. Effects of PBDPEs on the ζ Potential of ZnO NPs

Figure 2 presents the effects of both PBDPEs on the ζ potential of ZnO NPs (10 mg/L) at pH 7 in the absence and presence of the electrolyte. As indicated in Figure S1B of the Supplementary Information, the ζ potential of ZnO NPs was +12.6 mV, while the isoelectric point (pH$_{iep}$) was observed to be approximately 9.2, which is consistent with previous studies [38,39]. The ζ potential of ZnO NPs became negative upon the addition of BDPE-47 and BDPE-209 (Figure 2A). It is noteworthy that the higher concentration of both PBDPEs resulted in a further decrease in ζ potential (more negative) of the ZnO NPs. For example, the significant decrease in surface charge of ZnO NPs in the presence of

5 mg/L of BDPE-47 and BDPE-209 was found to be −33.2 and −27.1 mV, respectively. However, the ζ potential of 5 mg/L BDPE-47 and BDPE-209 in DI water was observed to be −50.0 and −42.8 mV, respectively (Figure 2A).

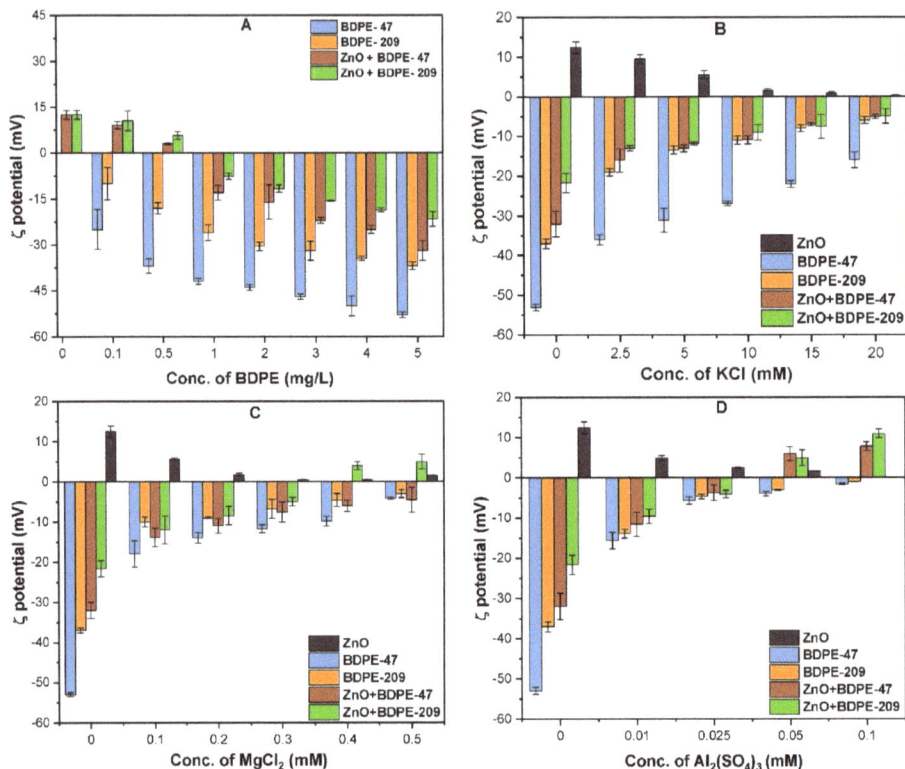

Figure 2. (**A**) ζ potential of ZnO NPs with polybrominated diphenyl ethers (PBDPEs) ((0–5) mg/L) in DI water. ζ potential of BDPEs and ZnO NPs coated with (5 mg/L) PBDPEs in the presence of (**B**) KCl; (**C**) MgCl$_2$, and (**D**) Al$_2$(SO$_4$)$_3$ electrolytes.

These results suggest that both PBDPEs interact with ZnO NPs and reverse their surface potential from positive to negative, thus enhancing the overall colloidal stability [28]. In addition, the adsorption of both PBDPEs on the ZnO NPs surface might decrease the van der Wall (vdW) forces, and increase steric hindrance among the NPs [23]. A more prominent effect of BDPE-47 on the ζ potential was observed as compared to BDPE-209, which might be ascribed to greater electrostatic repulsion and more steric stabilization.

The effects of various electrolytes ((0–20 mM) KCl, (0–0.5 mM) MgCl$_2$, and (0–0.1 mM) Al$_2$(SO$_4$)$_3$) on the ζ potential of ZnO NPs in the presence of both PBDPEs (5 mg/L) were investigated (Figure 2B–D). The presence of all electrolytes significantly increases the ζ potential of ZnO NPs in the presence of BDPE-47 and BDPE-209. For example, the ζ potential of NPs with BDPE-47 indicated significant shifts to be −5.0, −4.5, and +7.7 mV at the concentration of each electrolyte, i.e., KCl (20 mM), MgCl$_2$ (0.5 mM), and Al$_2$(SO$_4$)$_3$ (0.1 mM), respectively. The effect of trivalent cations on the ζ potential of ZnO NPs was more pronounced, as it required 0.1 mM Al$_2$(SO$_4$)$_3$ to reverse the surface charge from (−52.0 to +7.7) mV with BDPE-47. The possible explanation for such phenomena may be related to the effective charge screening, due to the adsorption of the cations around the EDL of the NPs [10,13]. Moreover, according to the Schulze–Hardy rule, higher valence cations, such as Mg^{2+} and Al^{3+}, have a greater ability for charge screening. These results are consistent with a previous studies [40,41], which reported that trivalent cations effectively compress the EDL of NPs in the solution. Furthermore, the ζ potential of BDPE-47 and BDPE-209 indicated a similar trend in the presence of each electrolyte (Figure 2B–D). Our results are in good agreement with earlier studies [14,36] that found charge screening and the neutralization effect from counterions played a substantial role, in comparison to the compressive effect of co-ions.

3.3. Effects of PBDPEs on the HDD of ZnO NPs

Figure 3 shows the effects of (5 mg/L) BDPE-47 and BDPE-209 on the HDD and size distribution of ZnO NPs suspension. Prior to HDD measurement, the influence of solvent on the ZnO NPs aggregation was determined, which showed that the aggregation kinetics remains the same in the absence and presence of solvents (Figure S1C of the SI). The ZnO NP suspension obtained in DI water showed intensity-weighted HDD of 226 nm, which was much larger than the vender-reported primary particle size (<50 nm). Such observation may be attributed to enhanced vdW forces among the NPs, thereby forming large NP aggregates in aqueous solution [14]. The HDD of ZnO NPs increased to 278 nm in the presence of BDPE-209; however, HDD remained the same (~235 nm) in the case of BDPE-47 (Figure 3A). Moreover, the size distribution of ZnO NPs in DI water and BDPE-47 showed that NPs typically exist in the range of 180–250 nm, while in the case of BDPE-209, the size distribution was found to be between 250 and 350 nm (Figure 3B). These results indicate that the size distribution of NPs in suspension highly depends upon the aggregation, as well as the magnitude, of the surface coating of PBDPEs [29]. It is noteworthy that HDD and the size distribution of ZnO NPs slightly increase in the presence of BDPE-209, which might be ascribed to its higher molecular weight, as compared to BDPE-47 (Table 1). In addition, the enhancement in size distribution of ZnO NPs in the presence of BDPE-209 as compared to BDPE-47 and DI water may also suggest the lower colloidal stability of ZnO NPs in suspension.

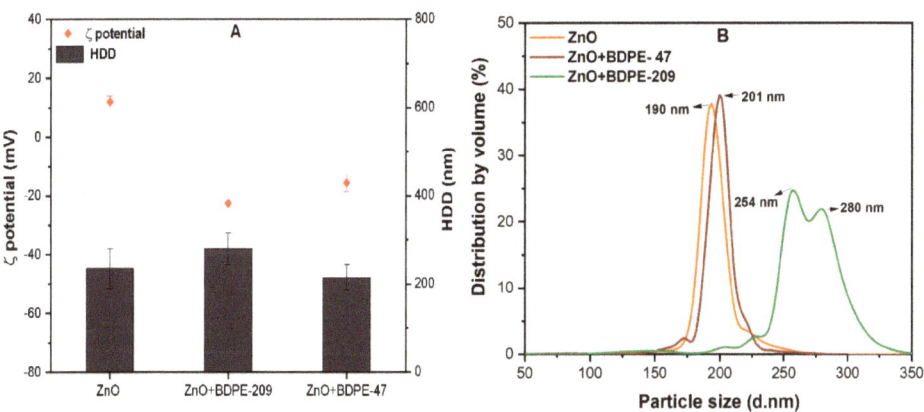

Figure 3. (**A**) Hydrodynamic diameter (HDD) of ZnO NPs (10 mg/L) with and without PBDPEs (5 mg/L) in DI water; (**B**) distribution by volume (%) of ZnO NPs in the presence of (5 mg/L) of BDPE-47 and BDPE-209 at pH 7.

3.4. Effects of PBDPEs on ζ Potential and HDD of ZnO NP in Natural Waters

The ζ potential of ZnO NP in FW and IWW was found to be 5.51 and −8.64 mV, respectively (Figure 4A). The interactions between NPs and organic matter present in IWW resulted in the charge reversal of ZnO NPs; moreover, the cations present in FW provide an effective EDL compression on the NPs surface [42]. The addition of low concentration (0.5 mg/L) of both PBDPEs in these waters reduces the ζ potential of ZnO NP with different degrees (Figure 4A). It can be observed that the ζ potential of

ZnO NP in FW reduces from 5.51 to −6.8 mV in the presence of 0.5 mg/L of BDPE-47, while it further decreases upon increasing the BDPE-47 concentration. For example, at higher concentration (10 mg/L) of BDPE-47 and BDPE-209 in FW, the reduction in surface charge of ZnO NPs was observed to be −18.5 and −22.0 mV, respectively. Moreover, the results of ζ potential of ZnO NP in IWW showed a similar trend, which was indicated (−18.0 and −17.4 mV) in the presence of 10 mg/L of BDPE-47 and BDPE-209, respectively. It is noteworthy that the effect of PBDPEs on the ζ potential of ZnO NP in FW was more pronounced than that of IWW.

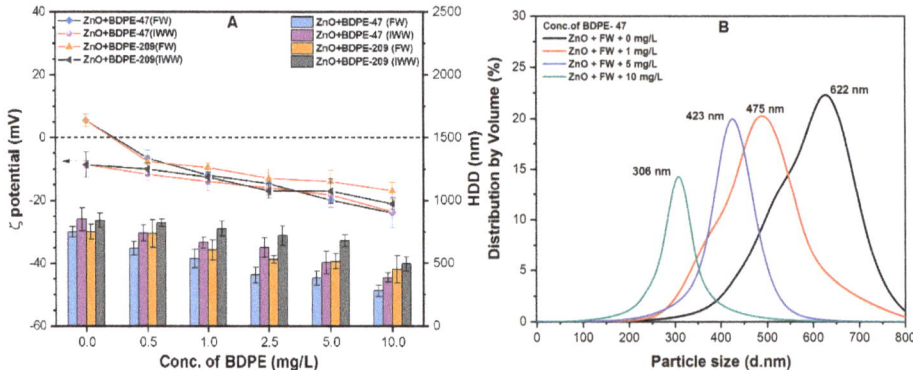

Figure 4. (**A**) The ζ potential and HDD of ZnO NPs coated with different concentrations (0–10 mg/L) of both BDPEs in freshwater (FW) and industrial wastewater (IWW); (**B**) size distribution of ZnO NPs with various BDPE-47 (0, 1, 5, and 10 mg/L) concentrations in FW.

Figure 4A shows significant increases in the HDD of ZnO NPs from 226 to 659 nm in FW and 769 nm in IWW. Moreover, in comparison to DI water, the particle size of ZnO NPs was found to be in the range of 400–850 nm in FW, and 350–900 nm in IWW (Figures S2A, S3A, and S4A of the Supplementary Information). This might be attributed to the higher conductivity in FW (150 µS/cm) and IWW (619 µS/cm), which might result in the formation of large agglomerates. The reduction in HDD and size distribution was observed upon the addition of both PBDPEs for all tested concentrations (Figure 4A,B and Figures S2–S4).

For example, the addition of 0.5 mg/L of BDPE-47 reduces the HDD of NPs to 306 nm (FW) and 493 nm (IWW), which further decrease upon increasing BDPE-47 concentration in these waters. The higher concentration (10 mg/L) of BDPE-47 and BDPE-209 presented low HDD values of 280 and 385 nm in FW, and 450 and 495 nm in IWW, respectively. This is consistent with our previous ζ potential observation, which indicated the significant influence of both PBDPEs, even at the high concentration of DOM present in natural waters (Table S2). It is interesting to note that the more drastic decrease in HDD of ZnO NPs was observed with the increasing concentration of BDPE-47 in comparison to BDPE-209. The shifts in size range further specify that the smaller size NPs may form due to the coating of PBDPEs onto ZnO NPs surface in natural waters, which may enhance the bioavailability and toxicity to the aquatic environment. Therefore, aggregation kinetic studies were conducted in the presence of various electrolytes to understand the fate and transport of the NPs in the heterogeneous aqueous environment.

3.5. Effects of PBDPEs on the Aggregation Kinetics of ZnO NP

Figure 5A shows the effects of BDPE-47 and BDPE-209 on the aggregation kinetics (30 min) of ZnO NPs in the presence of 20 mM KCl in both waters that were studied. It can be observed that the NPs suspension become unstable in the absence of PBDPEs, as indicated by the higher HDD value that reached up to 600 nm within the studied period (Figure S5A,B). The addition of

0.1 mg/L BDPE-47 slightly suppressed the aggregation, thereby reducing the HDD by up to ~500 nm (Figure S5A). Furthermore, the increase in BDPE-47 concentration to 5 mg/L remarkably suppressed the aggregation, as well as decreasing the HDD value (220 nm). In contrast, the effect of BDPE-209 on the aggregation of NPs was found to be insignificant where HDD of ZnO NPs observed ~600 nm, even at 5 mg/L of BDPE-209 (Figure S5B). Interestingly, the aggregation of NPs was fully suppressed at higher BDPE-209 concentration (10 mg/L), as shown in Figure S5C. These results suggest the higher stabilizing ability of BDPE-47 than BDPE-209, even at low concentration. This phenomenon is in good agreement with our previous observation that the higher adsorption of BDPE-47 molecules onto ZnO NPs contributes to higher surface charge than for BDPE-209.

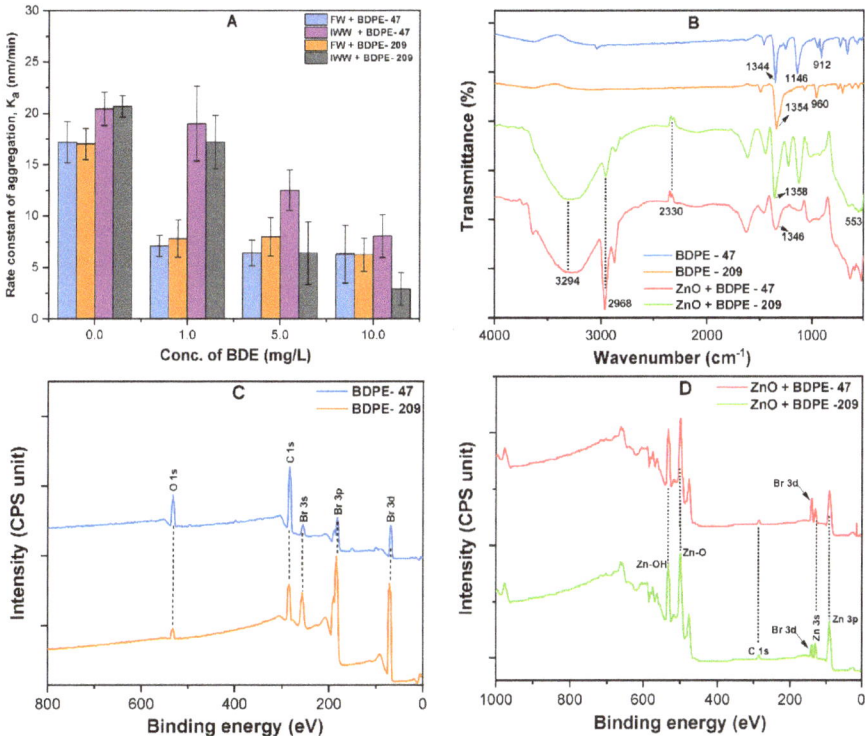

Figure 5. (**A**) Aggregation rate constant (k_a) of ZnO NPs at various concentrations (0–10 mg/L) of both BDPEs in FW and IWW; (**B**) FT-IR spectra of both pristine BDPEs and ZnO NPs coated with BDPE-47 and BDPE-209; (**C**) XPS survey spectra of both BDPEs; and (**D**) ZnO NPs coated with BDPE-47 and BDPE-209.

To further understand the influence of PBDPEs on the colloidal stability, the aggregation rates (k_a) were measured using the kinetic data. In the absence of PBDPEs, the k_a value reached to 13.75 nm/min, while it reduced to 0.87 nm/min at the concentration of 5 mg/L BDPE-47 (Figure S5D). In contrast, the effect of BDPE-209 on the stability of ZnO NPs was insignificant, even at higher concentration (5 mg/L). Moreover, a slight decrease in the aggregation rate of ZnO NPs was observed, where the k_a was found to be 10.38 and 9.96 nm/min upon the addition of 0.5 and 1 mg/L BDPE-209, respectively (Figure S5D). This observation indicated that the sorption of PBDPEs onto NPs surface mainly occurs via electrostatic forces, thereby affecting the pH and IS of the solution [15]. Another possible reason might be related to the competitive adsorption behavior between cations and PBDPEs onto the NPs surface, which increases with the hydrophobicity of the organic pollutant. It is therefore noteworthy that the influence of 20 mM KCl on the binding affinity of ZnO NPs was maximum in the presence of BDPE-209 (for concentration up to 5 mg/L), rather than for the less hydrophobic BDPE-47.

Figure 5A shows the results of the aggregation kinetics and rates (k_a) of ZnO NPs in the absence and presence of PBDPEs in natural waters. It can be observed that the addition of PBDPEs in natural waters resulted in the decrease in k_a values (Figure 5A). The aggregation rate of ZnO NPs in FW significantly decreased from 17.5 to 7.5 nm/min upon the addition of 1 mg/L BDPE-47; afterwards it slightly decreased to 6.8 nm/min at a higher concentration of BDPE-47 (10 mg/L). Similar trends have been observed in FW containing BDPE-209; however, both PBDPEs showed discrepant aggregation kinetics behavior in IWW (Figure S6). These results indicate that the lower concentration of PBDPEs may significantly influence the fate and mobility of ZnO NPs, even in waters containing high DOM concentration. The low solubility (1–30 µg/L) of organic pollutants in water might enhance the stability of ZnO NPs, thereby reducing the risk of releasing Zn^{2+} ions into natural water bodies. Moreover, the extremely high hydrophobicity of BDPE-209 (log K_{ow} = (6.27–9.97)) and BDPE-47 (log K_{ow} = (6.77–6.81)) might facilitate the adsorption of organic pollutant onto ZnO NPs surface in these waters [14,23]. These results strongly suggest that a low concentration of PBDPEs and other organic contaminants may adsorb on the NPs surface via hydrophobic ligands, thereby affecting the overall colloidal stability.

The interaction behavior of PBDPEs with ZnO NPs depends upon the surface charges, physicochemical interactions, and characteristics of PBDPEs. Moreover, vdW forces and hydrogen bonding might play a vital role during the interactions between NPs and PBDPEs [39]. Figure 5B shows the FT-IR spectra of ZnO NPs before and after interactions with BDPE-47 and BDPE-209. The peaks at ~3294, ~2968, and ~2330 cm^{-1} that appear in ZnO–PBDPEs complex correspond to the stretching vibrations of OH, and symmetric stretching vibration of aliphatic C–H and C=O, respectively [19,25]. In comparison to the pristine ZnO NPs, a new band was observed at ~1346–1358 cm^{-1}, which is attributed to the adsorption of PBDPEs onto ZnO NPs surface via aromatic ether (-O-) groups [27,42] Furthermore, Figure 5C,D, indicate the XPS analysis of PBDPEs before and after the interaction with ZnO NPs, which further confirm the attachment of Br (present in BDPE-47 and BDPE-209) onto the ZnO NPs surface [43,44] Therefore, it can be inferred that surface Zn-OH groups (donor) of the NP and ether (-O-) groups in PBDPEs (acceptor), or induced dipole, might be responsible for the strong interaction between PBDPEs and ZnO NPs.

3.6. Effects of PBDPEs on ZnO Colloidal Stability

The stability of ZnO NPs was further explored by measuring the CCC of KCl for ZnO NPs in the absence and presence of PBDPEs in aqueous media (Figure 6). The energy barrier among the particles is diminished at CCC and electrolyte concentrations above CCC, and hence promotes the diffusion-controlled aggregation. The commonly used DOM substitute humic acid (HA) was used to compare the NPs stability with PBDPEs. Different diffusion-limited agglomeration (DLA) and reaction-limited agglomeration (RLA) regimes were found in the absence and presence of both PBDPEs (Figure 6). The CCC of KCl for ZnO NPs in the absence of PBDPEs was found to be 1.3 mM. This has practical implications in natural waters with IS (1–15 mM), suggesting the rapid settling of ZnO NPs in those environments [10,14]. Moreover, increasing IS weakens the electrostatic repulsion forces and increase the attractive forces among the NPs, thereby enhance the sedimentation due to an increase in attachment efficiency [34,45]. Upon the addition of 0.5 mg/L BDPE-47 and BDPE-209, the CCC of ZnO NPs increase to 6.8 and 11.7 mM KCl, respectively, which indicate the enhancement in stability of ZnO NPs in the presence of PBDPEs (even at low concentrations). This may be attributable to the surface coating of organic pollutant onto NPs increasing the steric hindrance, thus reducing the compression-induced effect of electrolytes [46]. Compared with organic pollutants, the HA increases the CCC of KCl for ZnO NPs two-fold, i.e., 32 mM. These results are in good agreement with previous [13,36,39] that affirm that high molecular weight (HMW) compounds such as HA have higher affinity of sorption onto NPs, due to the presence of several aliphatic and aromatic groups. These results are also consistent with our previous DLS observation, where BDPE-209 shows an insignificant effect on the aggregation of NPs for concentration up to 5 mg/L (Figure S5). Thus, further experiments were conducted to quantify the effect of different concentration of BDPE-209 on the CCC of KCl for ZnO NPs (Figure 6B). The slight increase in CCC was observed to be 20 mM KCl at 3 mg/L BDPE-209 concentration; the CCC becomes 78 mM when the BDPE-209 concentration reaches 6 mg/L. Similar behavior has been observed in the kinetic aggregation study, where the stability of ZnO NPs was significantly increased above 5 mg/L BDPE-209 concentration (Figure S5C). In general, such observations suggest that PBDPEs might improve the colloidal stability of ZnO NPs in the heterogeneous aqueous environment.

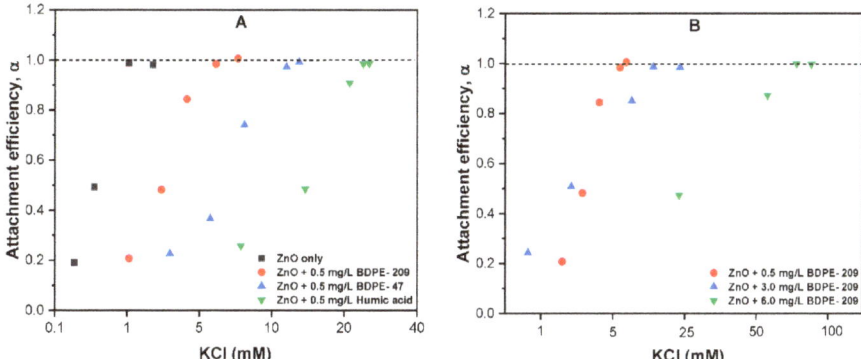

Figure 6. (**A**) Effects of 0.5 mg/L BDPE 47, BDPE 209, and humic acid (HA) on the critical coagulation concentration (CCC) of KCl, and (**B**) effects of different concentrations of BDPE-209 on the CCC of KCl for ZnO NPs (10 mg/L) at pH 7.

3.7. Effects of Electrolyte Type on Aggregation Kinetics

To further understand the effect on the aggregation behavior of ZnO NPs in the presence of various cations and PBDPEs, additional experiments were performed (Figure 7). The aggregation kinetics and rate (k_a) of ZnO NPs at 1 and 5 mM KCl were found to be 320 and 430 nm, and 7.2 and 12.9 nm/min, respectively (Figures 7A and S6). This might be related to the closeness to the value of CCC of ZnO NPs for KCl; thus, most of the NPs are unstable under these conditions. Similar observation was reported in an earlier study [23,39], where enhanced agglomeration was observed in the presence of K^+, as depicted by the larger HDD of NPs. In contrast, lower concentration of $MgCl_2$ (0.5 mM), $Al_2(SO_4)_3$ (0.15 mM) significantly destabilized the NPs in solution, as described by their higher HDD (1250, 1600 nm) and k_a (89.3, 92 nm/min) values (Figures 7A and S6B,C). These results are in good agreement with the recent study [10,38], that demonstrated the greater charge screening ability of polyvalent cations, thus effectively destabilizing the NPs in suspension.

The addition of PBDPEs suppresses the ZnO NPs agglomeration even in the presence of electrolytes, regardless of their type and concentration (Figure 7B,D). For example, the addition of 0.5 mg/L of BDPE-47 and BDPE-209 in ZnO NPs suspension containing 5 mM KCl decreases the HDD to 350 and 370 nm (Figure 7B). Moreover, the value of k_a of ZnO NPs decreases to 5 and 7 nm/min in the presence of both PBDPEs (Figure S7A). The greater decrease in k_a value in the presence of BDPE-47 may be attributable to the higher stabilizing ability, as observed in our previous section results. The addition of BDPE-209 and BDPE-47in ZnO NPs suspension containing divalent cations (0.5 mM $MgCl_2$) slightly decreased the HDD of 900 and 700 nm and k_a of 71 and 54 nm/min, respectively (Figures 7C and S7B). However, the effect of BDPE-47 and BDPE-209 in ZnO NPs suspension with trivalent cations (0.15 mM $Al_2(SO_4)_3$) on the colloidal stability of NPs was insignificant (Figures 7D and S7C). In general, the stabilization ability of PBDPEs showed a remarkable effect in the presence of monovalent cations, as compared to polyvalent cations. This might be related to the competitive adsorption, as well as the strong interactive behavior between the contaminants in the ternary environment. However, further study is needed to confirm this interpretation.

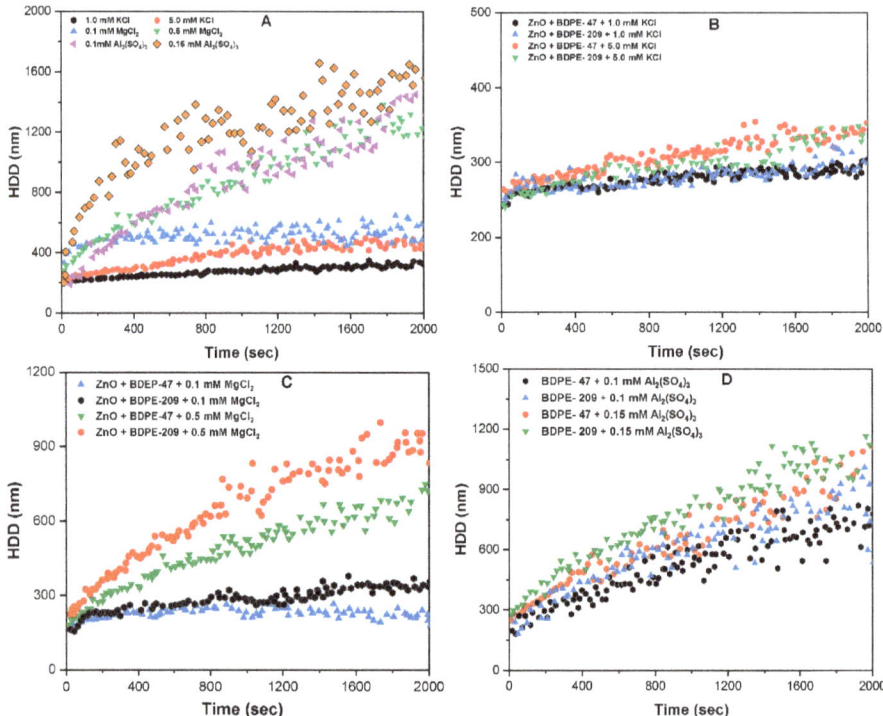

Figure 7. Influence of BDPE-47 and BDPE-209 on the aggregation kinetics of ZnO NPs (10 mg/L) in (**A**) KCl, MgCl$_2$, and Al$_2$(SO$_4$)$_3$ at different IS ((0–5) mM); (**B**) monovalent (0–5mM) KCl; (**C**) divalent ions ((0.1 and 0.5) mM of MgCl$_2$); and (**D**) trivalent ions ((0.1 and 0.15) mM Al$_2$(SO$_4$)$_3$) at pH 7.

3.8. Study Significance

This study is the first approach to provide some insight into the interaction behavior of ZnO NPs in synthetic and natural waters containing different PBDPEs. Several studies study [13,15,25,34], have described the influence of DOMs on the individual parameters of ENPs. Although the PBDPEs found in the natural environment are at very low concentration, due to their hydrophobic characteristics, they are mostly found in the dissolved state. Thus, the large surface area of ZnO NPs may provide active absorption site to these hydrophobic compounds, and could alter their fate and transport behavior in natural water bodies. The higher concentration of PBDPEs was used in the present short-term study to simulate the amount of POPs absorbed onto the ZnO NPs surface for a more extended period. Moreover, such concentration of PBDPEs (i.e., 0.5 mg/L) facilitates an understanding of the possible interaction mechanism of PBDPEs with ZnO NPs.

4. Conclusions

In the present work, we investigated the interaction behavior between ZnO NPs and the two common organic pollutants, BDPE-47 and BDPE-209, in synthetic and natural waters. The results showed that at similar concentrations, BDPE-47 improved the stability of ZnO NPs suspension more than did BDPE-209. Moreover, the BDPE-47 imparts a higher surface potential and more effective surface coating of the ZnO NP, than does BDPE-209. The presence of both PBDPEs suppresses the aggregation rate of ZnO due to the electrosteric hinderance effect, even in the presence of monovalent and polyvalent cations. The FT-IR analysis of ZnO–PBDPEs complexes indicated that aromatic ether groups of PBDPEs played an essential role in the interactions between the POPs and NPs. The results of XPS further confirm the attachment of Br onto the ZnO NPs surface. The presence of both PBDPEs in environmental waters (freshwater and industrial wastewater) results in a discrete adverse effect on the aggregate kinetics and rate of ZnO NPs. Further research and endeavors will focus on the impact of other factors, such as media pH, DOM type, etc., on the interaction behavior between PBDPEs and ZnO NPs in the aquatic system. The findings of this study provide new insights into the enhanced stability of PBDPEs-sorbed ZnO NPs in natural bodies, which may influence the fate and mobility of NPs, thereby increasing their exposure risk to aquatic life and humans.

Supplementary Materials: The following are available online at http://www.mdpi.com/2079-4991/9/3/472/s1, **Figure S1**. (A) Brunauer–Emmett–Teller (BET) analysis of zinc oxide nanoparticles (ZnO NPs) powder; (B) isoelectric point of ZnO NPs in DI water; (C) aggregation kinetics of ZnO NPs in the absence and presence of both solvents used for dissolving the polybrominated diphenyl ethers (PBDPEs); and (D) hydrodynamic size distribution of ZnO NPs in DI water after 30 min sonication. **Figure S2**. Volume distribution of ZnO NPs (10 mg/L) with different concentrations of BDPE-47 of (A) 0 mg/L; (B) 1 mg/L; (C) 5 mg/L; and (D) 10 mg/L in in industrial wastewater (IWW). **Figure S3**. Volume distribution of ZnO NPs (10 mg/L) with different concentration of BDPE-209 of (A) 0 mg/L; (B) 1 mg/L; (C) 5 mg/L; and (D) 10 mg/L in freshwater (FW). **Figure S4**. Volume distribution of ZnO NPs (10 mg/L) with different concentration of BDPE-209 showing (A) 0 mg/L; (B) 1 mg/L; (C) 5 mg/L; and (D) 10 mg/L in IWW. **Figure S5**. Effects of different concentrations (0–5 mg/L) of (A) BDPE-47 and (B) BDPE- 209 on the aggregation kinetics of ZnO NPs (10 mg/L) in the presence of 20 mM KCl at pH 7; (C) aggregation kinetics of ZnO NPs in DI water containing 0, 5, and 10 mg/L BDPE-209, while inset showing aggregation rates at same concentrations; and (D) aggregation rates of ZnO NPs at various concentrations (0–5 mg/L) of BDPE-47 and BDPE-209. **Figure S6**. Effects of (A) BDPE-47 and (B) BDPE-209 on the aggregation kinetics of ZnO NPs (10 mg/L) in FW, and (C) BDPE-47 and (D) BDPE-209 on the aggregation kinetics of ZnO NPs in IWW. **Figure S7**. Aggregation rates of ZnO NPs (10 mg/L) in the absence and presence (1 mg/L) of BDPE-47 and BDPE-209 with (A) monovalent (KCl); (B) divalent ($MgCl_2$); and (C) trivalent $Al_2(SO_4)_3$ cations at pH 7. **Table S1**. Detailed properties of Zinc oxide nanoparticles used in this study. **Table S2**. Properties of the synthetic and natural waters used in the current study.

Author Contributions: R.K. and I.T.Y. conceived and designed the study; R.K. and M.A.I. performed the experiment and analyzed the data; M.A.I., S.K., and D.R.P. provided critical feedback, and helped shape the research; and R.K. wrote the final version of the manuscript.

Funding: This work was supported by the BK21 plus program through the National Research Foundation of Korea (NRF), funded by the Ministry of Education of Korea (Grant No. 22A20152613545).

Conflicts of Interest: The authors declare no conflict of interest.

References

1. Wang, Z.L. Zinc oxide nanostructures: Growth, properties and applications. *J. Phys. Condens. Matter* **2004**, *16*, R829. [CrossRef]
2. Gottschalk, F.; Sun, T.; Nowack, B. Environmental concentrations of engineered nanomaterials: Review of modeling and analytical studies. *Environ. Pollut.* **2013**, *181*, 287–300. [CrossRef] [PubMed]
3. Lowry, G.V.; Gregory, K.B.; Apte, S.C.; Lead, J.R. Transformations of nanomaterials in the environment. *Environ. Sci. Technol.* **2012**, *46*, 6893–6899. [CrossRef] [PubMed]
4. Fairbairn, E.A.; Keller, A.A.; Mädler, L.; Zhou, D.; Pokhrel, S.; Cherr, G.N. Metal oxide nanomaterials in seawater: Linking physicochemical characteristics with biological response in sea urchin development. *J. Hazard. Mater.* **2011**, *192*, 1565–1571. [CrossRef] [PubMed]
5. Bondarenko, O.; Juganson, K.; Ivask, A.; Kasemets, K.; Mortimer, M.; Kahru, A. Toxicity of Ag, CuO and ZnO nanoparticles to selected environmentally relevant test organisms and mammalian cells in vitro: A critical review. *Arch. Toxicol.* **2013**, *87*, 1181–1200. [CrossRef]
6. Kuang, H.; Yang, P.; Yang, L.; Aguilar, Z.P.; Xu, H. Size dependent effect of ZnO nanoparticles on endoplasmic reticulum stress signaling pathway in murine liver. *J. Hazard. Mater.* **2016**, *317*, 119–126. [CrossRef]
7. Waalewijn-Kool, P.L.; Ortiz, M.D.; Van Straalen, N.M.; van Gestel, C.A.M. Sorption, dissolution and pH determine the long-term equilibration and toxicity of coated and uncoated ZnO nanoparticles in soil. *Environ. Pollut.* **2013**, *178*, 59–64. [CrossRef]

8. Heggelund, L.R.; Diez-Ortiz, M.; Lofts, S.; Lahive, E.; Jurkschat, K.; Wojnarowicz, J.; Cedergreen, N.; Spurgeon, D.; Svendsen, C. Soil pH effects on the comparative toxicity of dissolved zinc, non-nano and nano ZnO to the earthworm Eisenia fetida. *Nanotoxicology* **2014**, *8*, 559–572. [CrossRef]
9. Teow, Y.; Asharani, P.V.; Hande, M.P.; Valiyaveettil, S. Health impact and safety of engineered nanomaterials. *Chem. Commun.* **2011**, *47*, 7025–7038. [CrossRef]
10. Lopes, S.; Ribeiro, F.; Wojnarowicz, J.; Łojkowski, W.; Jurkschat, K.; Crossley, A.; Soares, A.M.V.M.; Loureiro, S. Zinc oxide nanoparticles toxicity to Daphnia magna: Size-dependent effects and dissolution. *Environ. Toxicol. Chem.* **2014**, *33*, 190–198. [CrossRef]
11. Liu, W.S.; Peng, Y.H.; Shiung, C.E.; Shih, Y.H. The effect of cations on the aggregation of commercial ZnO nanoparticle suspension. *J. Nanopart. Res.* **2012**, *14*. [CrossRef]
12. Mudunkotuwa, I.A.; Rupasinghe, T.; Wu, C.-M.; Grassian, V.H. Dissolution of ZnO Nanoparticles at Circumneutral pH: A Study of Size Effects in the Presence and Absence of Citric Acid. *Langmuir* **2012**, *28*, 396–403. [CrossRef]
13. Philippe, A.; Schaumann, G.E. Interactions of dissolved organic matter with natural and engineered inorganic colloids: A review. *Environ. Sci. Technol.* **2014**, *48*, 8946–8962. [CrossRef]
14. Majedi, S.M.; Kelly, B.C.; Lee, H.K. Combined effects of water temperature and chemistry on the environmental fate and behavior of nanosized zinc oxide. *Sci. Total Environ.* **2014**, *496*, 585–593. [CrossRef]
15. Keller, A.A.; Wang, H.; Zhou, D.; Lenihan, H.S.; Cherr, G.; Cardinale, B.J.; Miller, R.; Zhaoxia, J.I. Stability and aggregation of metal oxide nanoparticles in natural aqueous matrices. *Environ. Sci. Technol.* **2010**, *44*, 1962–1967. [CrossRef]
16. Zhu, L.Y.; Hites, R.A. Temporal trends and spatial distributions of brominated flame retardants in archived fishes from the Great Lakes. *Environ. Sci. Technol.* **2004**, *38*, 2779–2784. [CrossRef] [PubMed]
17. Branchi, I.; Capone, F.; Alleva, E.; Costa, L.G. Polybrominated diphenyl ethers: Neurobehavioral effects following developmental exposure. *Neurotoxicology* **2003**, *24*, 449–462. [CrossRef]
18. Hale, R.C.; Alaee, M.; Manchester-Neesvig, J.B.; Stapleton, H.M.; Ikonomou, M.G. Polybrominated diphenyl ether flame retardants in the North American environment. *Environ. Int.* **2003**, *29*, 771–779. [CrossRef]
19. Johnson-Restrepo, B.; Kannan, K. An assessment of sources and pathways of human exposure to polybrominated diphenyl ethers in the United States. *Chemosphere* **2009**, *76*, 542–548. [CrossRef]
20. Peng, X.; Tang, C.; Yu, Y.; Tan, J.; Huang, Q.; Wu, J.; Chen, S.; Mai, B. Concentrations, transport, fate, and releases of polybrominated diphenyl ethers in sewage treatment plants in the Pearl River Delta, South China. *Environ. Int.* **2009**, *35*, 303–309. [CrossRef]
21. Moon, H.-B.; Kannan, K.; Lee, S.-J.; Choi, M. Atmospheric deposition of polybrominated diphenyl ethers (PBDEs) in coastal areas in Korea. *Chemosphere* **2007**, *66*, 585–593. [CrossRef] [PubMed]
22. U.S. Environmental Protection Agency. *An Exposure Assessment of Polybrominated Diphenyl Ethers*; U.S. Environmental Protection Agency: Washington, DC, USA, 2010.
23. Covaci, A.; Voorspoels, S.; Roosens, L.; Jacobs, W.; Blust, R.; Neels, H. Polybrominated diphenyl ethers (PBDEs) and polychlorinated biphenyls (PCBs) in human liver and adipose tissue samples from Belgium. *Chemosphere* **2008**, *73*, 170–175. [CrossRef]
24. Wang, X.; Adeleye, A.S.; Wang, H.; Zhang, M.; Liu, M.; Wang, Y.; Li, Y.; Keller, A.A. Interactions between polybrominated diphenyl ethers (PBDEs) and TiO2 nanoparticle in artificial and natural waters. *Water Res.* **2018**, *146*, 98–108. [CrossRef]
25. Khan, R.; Inam, M.; Iqbal, M.; Shoaib, M.; Park, D.; Lee, K.; Shin, S.; Khan, S.; Yeom, I. Removal of ZnO Nanoparticles from Natural Waters by Coagulation-Flocculation Process: Influence of Surfactant Type on Aggregation, Dissolution and Colloidal Stability. *Sustainability* **2019**, *11*, 17. [CrossRef]
26. Rice, E.W.; Baird, R.B.; Eaton, A.D. (Eds.) *Standard Methods for the Examination of Water and Wastewater*; American Public Health Association: Washington, DC, USA, 1995.
27. Hua, I.; Kang, N.; Jafvert, C.T.; Fábrega-Duque, J.R. Heterogeneous photochemical reactions of decabromodiphenyl ether. *Environ. Toxicol. Chem.* **2003**, *22*, 798–804. [CrossRef]
28. Tittlemier, S.A.; Halldorson, T.; Stern, G.A.; Tomy, G.T. Vapor pressures, aqueous solubilities, and Henry's law constants of some brominated flame retardants. *Environ. Toxicol. Chem.* **2002**, *21*, 1804–1810. [CrossRef]
29. Kroll, A.; Behra, R.; Kaegi, R.; Sigg, L. Extracellular polymeric substances (EPS) of freshwater biofilms stabilize and modify CeO2 and Ag nanoparticles. *PLoS ONE* **2014**, *9*, e110709. [CrossRef]

30. Adeleye, A.S.; Conway, J.R.; Perez, T.; Rutten, P.; Keller, A.A. Influence of extracellular polymeric substances on the long-term fate, dissolution, and speciation of copper-based nanoparticles. *Environ. Sci. Technol.* **2014**, *48*, 12561–12568. [CrossRef] [PubMed]
31. Adeleye, A.S.; Keller, A.A. Long-term colloidal stability and metal leaching of single wall carbon nanotubes: Effect of temperature and extracellular polymeric substances. *Water Res.* **2014**, *49*, 236–250. [CrossRef]
32. Bouchard, D.; Ma, X.; Isaacson, C. Colloidal properties of aqueous fullerenes: Isoelectric points and aggregation kinetics of C60 and C60 derivatives. *Environ. Sci. Technol.* **2009**, *43*, 6597–6603. [CrossRef]
33. Chen, K.L.; Elimelech, M. Aggregation and deposition kinetics of fullerene (C60) nanoparticles. *Langmuir* **2006**, *22*, 10994–11001. [CrossRef] [PubMed]
34. Chowdhury, I.; Duch, M.C.; Mansukhani, N.D.; Hersam, M.C.; Bouchard, D. Colloidal properties and stability of graphene oxide nanomaterials in the aquatic environment. *Environ. Sci. Technol.* **2013**, *47*, 6288–6296. [CrossRef] [PubMed]
35. Wagner, C.D.; Gale, L.H.; Raymond, R.H. Two-dimensional chemical state plots: A standardized data set for use in identifying chemical states by x-ray photoelectron spectroscopy. *Anal. Chem.* **1979**, *51*, 466–482. [CrossRef]
36. Khan, R.; Inam, M.; Park, D.; Zam Zam, S.; Shin, S.; Khan, S.; Akram, M.; Yeom, I. Influence of Organic Ligands on the Colloidal Stability and Removal of ZnO Nanoparticles from Synthetic Waters by Coagulation. *Processes* **2018**, *6*, 170. [CrossRef]
37. Patterson, A.L. The Scherrer formula for X-ray particle size determination. *Phys. Rev.* **1939**, *56*, 978. [CrossRef]
38. Bian, S.W.; Mudunkotuwa, I.A.; Rupasinghe, T.; Grassian, V.H. Aggregation and dissolution of 4 nm ZnO nanoparticles in aqueous environments: Influence of pH, ionic strength, size, and adsorption of humic acid. *Langmuir* **2011**, *27*, 6059–6068. [CrossRef] [PubMed]
39. Peng, Y.H.; Tsai, Y.C.; Hsiung, C.E.; Lin, Y.H.; Shih, Y.H. Influence of water chemistry on the environmental behaviors of commercial ZnO nanoparticles in various water and wastewater samples. *J. Hazard. Mater.* **2017**, *322*, 348–356. [CrossRef]
40. Zhou, X.H.; Huang, B.C.; Zhou, T.; Liu, Y.C.; Shi, H.C. Aggregation behavior of engineered nanoparticles and their impact on activated sludge in wastewater treatment. *Chemosphere* **2015**, *119*, 568–576. [CrossRef]
41. Khan, R.; Inam, M.A.; Zam, S.Z.; Park, D.R.; Yeom, I.T. Assessment of key environmental factors influencing the sedimentation and aggregation behavior of zinc oxide nanoparticles in aquatic environment. *Water* **2018**, *10*, 660. [CrossRef]
42. Vargas, R.; Núñez, O. Hydrogen bond interactions at the TiO_2 surface: Their contribution to the pH dependent photo-catalytic degradation of p-nitrophenol. *J. Mol. Catal. A Chem.* **2009**, *300*, 65–71. [CrossRef]
43. Peng, Y.-H.H.; Chen, M.K.; Shih, Y.H. Adsorption and sequential degradation of polybrominated diphenyl ethers with zerovalent iron. *J. Hazard. Mater.* **2013**, *260*, 844–850. [CrossRef] [PubMed]
44. Lambert, J.B.; Shurvell, H.F.; Lightner, D.A.; Cooks, R.G. *Introduction to organic spectroscopy*; Macmillan Publishing Company: London, UK, 1987; ISBN 0023673001.
45. Yang, K.; Chen, B.; Zhu, X.; Xing, B. Aggregation, adsorption, and morphological transformation of graphene oxide in aqueous solutions containing different metal cations. *Environ. Sci. Technol.* **2016**, *50*, 11066–11075. [CrossRef] [PubMed]
46. Miao, L.; Wang, C.; Hou, J.; Wang, P.; Ao, Y.; Li, Y.; Lv, B.; Yang, Y.; You, G.; Xu, Y. Effect of alginate on the aggregation kinetics of copper oxide nanoparticles (CuO NPs): Bridging interaction and hetero-aggregation induced by Ca^{2+}. *Environ. Sci. Pollut. Res.* **2016**, *23*, 11611–11619. [CrossRef] [PubMed]

© 2019 by the authors. Licensee MDPI, Basel, Switzerland. This article is an open access article distributed under the terms and conditions of the Creative Commons Attribution (CC BY) license (http://creativecommons.org/licenses/by/4.0/).

Article

Facile Preparation of Rod-like MnO Nanomixtures via Hydrothermal Approach and Highly Efficient Removal of Methylene Blue for Wastewater Treatment

Yuelong Xu [1,2,3,4], Bin Ren [3,4], Ran Wang [2], Lihui Zhang [3,4], Tifeng Jiao [1,2,*] and Zhenfa Liu [3,4,*]

1. State Key Laboratory of Metastable Materials Science and Technology, Yanshan University, Qinhuangdao 066004, China; xudalong.cool@163.com
2. Hebei Key Laboratory of Applied Chemistry, School of Environmental and Chemical Engineering, Yanshan University, Qinhuangdao 066004, China; wr1422520780@163.com
3. Institute of Energy Resources, Hebei Academy of Sciences, Shijiazhuang 050081, China; RENBINTS@126.com (B.R.); zlhkxy@126.com (L.Z.)
4. Hebei Engineering Research Center for Water Saving in Industry, Shijiazhuang 050081, China
* Correspondence: tfjiao@ysu.edu.cn (T.J.); lzf63@sohu.com (Z.L.); Tel.: +86-335-8056854 (T.J.)

Received: 12 December 2018; Accepted: 13 December 2018; Published: 22 December 2018

Abstract: In the present study, nanoscale rod-shaped manganese oxide (MnO) mixtures were successfully prepared from graphitic carbon nitride (C_3N_4) and potassium permanganate ($KMnO_4$) through a hydrothermal method. The as-prepared MnO nanomixtures exhibited high activity in the adsorption and degradation of methylene blue (MB). The as-synthesized products were characterized by scanning electron microscopy (SEM), transmission electron microscopy (TEM), surface area analysis, X-ray diffraction (XRD), and X-ray photoelectron spectroscopy (XPS). Furthermore, the effects of the dose of MnO nanomixtures, pH of the solution, initial concentration of MB, and the temperature of MB removal in dye adsorption and degradation experiments was investigated. The degradation mechanism of MB upon treatment with MnO nanomixtures and H_2O_2 was studied and discussed. The results showed that a maximum adsorption capacity of 154 mg g^{-1} was obtained for a 60 mg L^{-1} MB solution at pH 9.0 and 25 °C, and the highest MB degradation ratio reached 99.8% under the following optimum conditions: 50 mL of MB solution (20 mg L^{-1}) at room temperature and pH ≈ 8.0 with 7 mg of C, N-doped MnO and 0.5 mL of H_2O_2.

Keywords: hydrothermal method; manganese oxide; adsorption; degradation; nanomixtures

1. Introduction

Water pollution is currently among the major environmental challenges and has attracted increasing research attention. The wide use of dyes has resulted in organic pollution in water, and dyes are considered a severe threat to ecosystems [1–6]. As untreated dyes are very active and stable, adsorption followed by oxidative degradation has emerged as a practical and effective technique to accelerate the treatment of dye effluent pollution. Thus, the following technological systems have been developed for the removal of dyes from water: physical adsorption [7], biodegradation [8,9] and chemical reaction and adsorption [10]. In recent years, photocatalytic decomposition [11–13] and chemical oxidation reduction have become highly efficient techniques for the degradation of methylene blue (MB) in water.

Over the last decades, nanomixtures, mostly nanorods/nanotubes-like structured, have been widely used for contaminant adsorption/removal [14–17]. Cavallaro et al. [15] investigated comprehensively the effect of anionic surfactants (sodium dodecanoate and sodium dodecylsulfate) on pristine halloysite nanotubes (HNT), which was beneficial for the solubilization and delivery of hydrophobic compounds from such hybrid materials. Recently, the oxidation degradation of dyes in water using environmentally benign oxidants has attracted considerable attention [18–21]. On this basis, some nontoxic and low-cost metal oxides have been widely used as catalysts for the oxidation of organic compounds [22–25]. Huang et al. [26] reported the application of Prussian blue (PB)-modified γ-Fe_2O_3 magnetic nanoparticles (PBMNPs) in the degradation of MB. The PBMNPs were used as peroxidase-like catalysts with H_2O_2 as the oxidant to completely degrade MB. The optimal conditions were as follows: pH range of 3 to 10, degradation temperature of 25 °C and degradation time of 120 min. However, the preparation process for the PBMNPs was very complicated and involved the use of toxic chemicals. Wolski et al. [27] investigated the effects of ZnO, Nb_2O_5 and $ZnNb_2O_6$ on the degradation of dyes, and MB could be completely degraded under optimal conditions. Nevertheless, the as-reported metal oxides (Nb_2O_5 and $ZnNb_2O_6$) were highly toxic and expensive.

In recent years, the synergistic application of metal oxides and H_2O_2 as peroxidase-like catalysts and an oxidant, respectively, in the degradation of dyes has been reported. Metal oxides can catalyze the generation of active oxygen (such as hydroxyl radicals (HO$^\bullet$), peroxides (HO$_2^-$) and superoxide anions (HO$_2^\bullet$)) upon H_2O_2 treatment, and this active oxygen can catalyze the degradation of dyes in water [28]. Saha et al. reported a novel method to prepare nanodimensional copper ferrite which exhibited high activity in the degradation of dyes in water with H_2O_2 as an oxidant [29]. The researchers used ethylenediaminetetraacetic acid and citric acid as the complexing agent and the fuel, respectively, in a modified complexometric method to prepare $CuFe_2O_4$, which had the capability to degrade 96% of the total MB. Because of its size-, structure- and morphology-dependent characteristics, and the variety of unique physical, chemical and functional properties, hausmannite (MnO) has been widely investigated in the fields of materials science, chemistry and physics. Zhang et al. prepared MnO nanocrystals of various sizes and shapes by soft-template self-assembly and studied the synthetic conditions and degradation mechanism of MB with H_2O_2 treatment [30]. In their report, cetyltrimethylammonium bromide (CTAB), polyvinyl pyrrolidine (PVP) and P123 were used as structure-directing agents; manganese sulfate was used as the source of manganese; and the size and shape of MnO could be controlled by varying the growth time, reaction temperature, surfactant, and manganese source. The as-prepared MnO showed a very high capacity for (above 99.7%) MB degradation.

Recently, Because of its excellent chemical and thermal stabilities and nontoxicity, graphitic carbon nitride (g-C_3N_4) [31–35], a novel 2D material, which was prepared through simple and green pyrolysis of melamine, has been used in many applications, such as energy conversion, biomedical applications and hydrogen production. According to the literature, g-C_3N_4 can absorb aromatic pollutants via the conjugated π region, which makes g-C_3N_4 a potential effective adsorbent. In this paper, the preparation of MnO Nanomixtures through a hydrothermal method with C_3N_4 as the source of carbon and nitrogen and potassium permanganate ($KMnO_4$) as the source of manganese was investigated. The effects of the hydrothermal reaction time, molar ratio of C_3N_4 to $KMnO_4$, and hydrothermal temperature on the adsorption capacity for MB were studied. In addition, the adsorption and degradation properties of the as-prepared product were systematically studied, and thermodynamic and kinetic analyses of the adsorption–degradation process were performed through experiments.

2. Materials and Methods

2.1. Materials

All reagents were purchased from Shanghai Aladdin Bio-Chem Technology Co., Ltd, Shanghai, China. All reagents were of analytical reagent (AR) grade and were used as received without further treatment.

2.2. Synthesis of C_3N_4

C_3N_4 was prepared by heating melamine (10 g) at 650 °C for 4 h in an air atmosphere. After the heat treatment, a light yellow solid was obtained.

2.3. Synthesis of MnO Nanomixtures

MnO nanomixtures were prepared via a hydrothermal method with C_3N_4 as the source of carbon and nitrogen and potassium permanganate ($KMnO_4$) as the source of manganese. Typically, certain amounts of C_3N_4 power and $KMnO_4$ were put into a 100 mL hydrothermal reactor. The molar ratios of C_3N_4 to $KMnO_4$ were 2.0, 4.0 and 6.0, and the mass concentration of the reactants (C_3N_4 + $KMnO_4$) in the solution was 12%. The hydrothermal temperature was set as 180 °C, and the hydrothermal reaction times were 24 h and 30 h. The as-prepared MnO nanomixtures with a hydrothermal reaction time of 30 h were denoted MnO-X (X = 2, 4, and 6), where X represents the reactants molar ratio of C_3N_4 to $KMnO_4$. The sample prepared with a molar ratio of 4.0 and a hydrothermal reaction time of 24 h was denoted as MnO-24.

2.4. MB Adsorption and Degradation Experiments

In the adsorption experiments, 50 mL of 10–60 mg L^{-1} MB aqueous solutions containing 5 mg of the MnO nanomixtures adsorbent were stirred at different temperatures (293.15–333.15 K) and different pH values (3.0–11.0) for MB adsorption. After an adsorption time of 20–300 min, the adsorbent solution was centrifuged, and the supernatant was examined by a UV-Vis spectrophotometer (TU-1900, Beijing Persee Instruments Co. Ltd., Beijing, China) to determine the MB concentration. The maximum wavelength of MB absorption was observed at λ = 665 nm.

The reusability of the MnO nanomixtures adsorbent was also investigated via 10 consecutive adsorption/desorption cycles. Briefly, the MnO nanomixtures with MB adsorbed were stirred in 50 mL of HCl solution (0.1 M) for 120 min, and then, the adsorbent was washed three times with distilled water. The adsorbed MB was desorbed from the MnO nanomixtures adsorbent, and the recovered MnO nanomixtures adsorbent was used to adsorb MB in another cycle. This cycle of adsorption and desorption was performed 10 times. The amount of MB adsorbed (q_t) was calculated according to Equation (1):

$$q_t = \frac{(C_0 - C_t)V}{W} \quad (1)$$

where C_0 is the initial concentration of MB (mg L^{-1}), C_t is the concentration of MB at contact time t (mg L^{-1}), V is the volume of the MB solution (L), and W is the weight of the adsorbent (g).

The MB degradation process was carried out in a 100 mL beaker containing 50 mL of a MB dye solution (20 mg L^{-1} or 40 mg L^{-1}), 0.5 mL of 30% H_2O_2, and 7 mg of MnO nanomixtures. The degradation time was varied from 0 h to 24 h, and the MB concentration was monitored by a UV-Vis spectrophotometer.

2.5. Characterization

MnO nanomixtures were characterized by X-ray diffraction (XRD, SMART LAB, Rigaku, Akishima, Japan) with CuKa radiation (λ = 1.54 Å), scanning electron microscopy (SEM, Field Emission Gun FEI QUANTA FEG 250, FEI Corporate, Hillsboro, OR, USA), transmission electron microscopy (TEM, HT7700, High-Technologies Corp., Ibaraki, Japan) and X-ray photoelectron spectroscopy (XPS, ESCALAB 250Xi XPS, Thermo Fisher Scientific, San Jose, CA, USA). The Brunauer−Emmett−Teller (BET) method was utilized to calculate the specific surface areas (ASAP2420 surface area analyzer, Micromeritics, Norcross, GA, USA). The pore volume and pore size were calculated from the adsorption–desorption isotherms using the Barrett−Joyner−Halenda (BJH) model. The total pore volume (V_{total}) was estimated from the amount adsorbed at a relative pressure (P/P_0) of 0.998.

2.6. Kinetic, Adsorption and Degradation Isotherm Models

The kinetics of the adsorption process were studied through kinetic models in our work. The pseudo-first-order kinetic model (2) and pseudo-second-order kinetic model (3) were adopted to fit the experimental data.

$$\ln(q_e - q_t) = \ln q_e - k_1 t \tag{2}$$

$$\frac{t}{q_t} = \frac{1}{k_2 q_e^2} + \frac{t}{q_e} \tag{3}$$

In these equations, q_e represents the equilibrium absorption capacity (mg g^{-1}), q_t represents the absorption amount (mg g^{-1}) at an absorption time of t (min), and k_1 and k_2 are the pseudo-first-order rate constant (min^{-1}) and the pseudo-second-order rate constant (g mg^{-1}·min^{-1}), respectively.

The Langmuir isotherm model (4) was adopted to investigate the surface properties, adsorbate affinity and adsorption capacity of MnO nanomixtures.

$$\frac{C_e}{q_e} = \frac{1}{q_m b} + \frac{C_e}{q_m} \tag{4}$$

In this equation, q_e (mg g^{-1}) is the equilibrium adsorption capacity, q_m (mg g^{-1}) is the maximum adsorption capacity (corresponding to complete monolayer coverage), C_e (mg L^{-1}) is the adsorbate concentration at the adsorption equilibrium, and b (L mg^{-1}) is a constant. The kinetics of the degradation process was also investigated via the pseudo-first-order kinetic model (2).

2.7. Thermodynamic Evaluation of the Adsorption Process

The thermodynamics of the adsorption process were obtained from Equations (5)–(7).

$$K_c = \frac{q_e}{C_e} \tag{5}$$

$$\Delta G^0 = -RT \ln K_c \tag{6}$$

$$\ln K_c = \frac{\Delta S^0}{R} - \frac{\Delta H^0}{RT} \tag{7}$$

In these equations, ΔG^0 is the standard Gibbs free energy change, ΔH^0 is the standard enthalpy change, ΔS^0 is the standard entropy change, q_e is the equilibrium adsorption capacity, C_e (mg L^{-1}) is the adsorbate concentration at the adsorption equilibrium, K_c is the distribution coefficient, R is the molar gas constant (8.314 J mol^{-1} K^{-1}), and T is the adsorption temperature (K).

3. Results and Discussion

The XRD patterns of the as-prepared MnO nanomixtures samples are shown in Figure 1. As presented in Figure 1, the peaks of (111), (200), (220), (311) and (222) were attributed to MnO [36], which indicated that MnO nanomixtures were successfully prepared via a novel hydrothermal self-assembly method. We also investigated the effect of the hydrothermal reaction time on the formation of MnO nanomixtures. We found that other manganese oxides were produced when the hydrothermal reaction time was less than 30 h. In the experiment, manganese oxide was the only product when the hydrothermal reaction time exceeded 30 h.

The nitrogen adsorption–desorption isotherms are shown in Figure 2a, and the pore size distribution curves are shown in Figure 2b. As seen in Figure 2a, all the curves corresponded to type-IV isotherms, and hysteresis loops could be clearly observed, illustrating the presence of a pore structure. The high P/P$_0$ of the hysteresis loops indicated a large pore size distribution, which was in accordance with the pore size distribution curves. As shown in Figure 2b, the as-prepared MnO nanomixtures samples exhibited a micro-mesoporous structure. The surface properties, consisting of the specific surface area (S$_{BET}$), micropore surface area (S$_{micro}$), average pore diameter (D$_{average}$) and total pore volume (V$_{total}$), are listed in Table 1. MnO-4 showed the largest surface area and total pore volume, which were beneficial for adsorption. As presented in Table 1, the molar ratio of C$_3$N$_4$ to KMnO$_4$ and the hydrothermal reaction time exerted obvious effects on the textural properties, in which shorter hydrothermal reaction times and higher or lower molar ratios affected the hydrothermal self-assembly process.

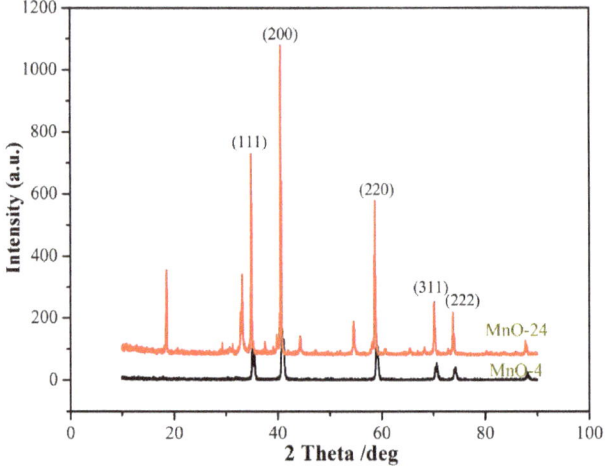

Figure 1. XRD patterns of the as-prepared MnO nanomixtures samples.

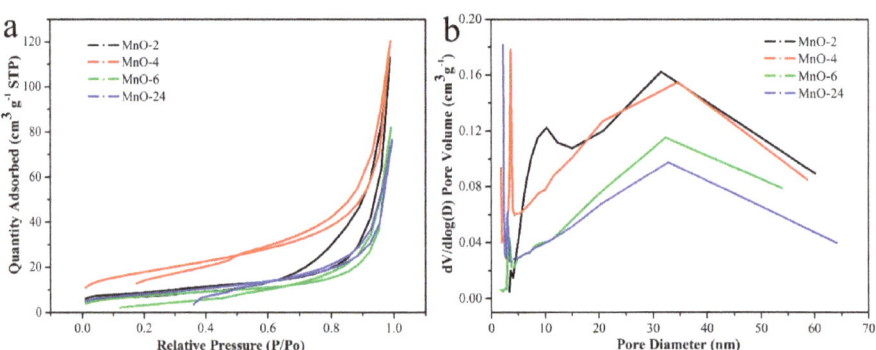

Figure 2. Nitrogen adsorption–desorption isotherms (a) and pore size distributions (b).

Table 1. Surface characterization of different samples.

Entry	S$_{BET}$ (m^2/g)	S$_{micro}$ (m^2/g)	Daverage (nm)	V$_{total}$ (cm^3/g)
MnO-2	30.6	6.3	22.81	0.175
MnO-4	38.7	8.5	23.09	0.193
MnO-6	35.5	1.0	21.46	0.127
MnO-24	33.9	1.6	15.97	0.168

SEM images of the as-prepared MnO nanomixtures samples and TEM images of MnO-4 are shown in Figures 3 and 4. The nanoscale rod-shape of C, N-doped MnO can be clearly seen in Figure 3; this product was formed via the polymerization of C$_3$N$_4$ and oxidation by KMnO$_4$. As shown in Figure 3d, the amount of rod-shaped MnO nanomixtures particles in MnO-24 was less than that in the other samples, which was caused by the shorter hydrothermal reaction time. When the molar ratio of C_3N_4 to $KMnO_4$ was more than 4.0, many linked spherical particles were formed, as shown in Figure 3c; there particles formed through the polymerization of excess C_3N_4 in the hydrothermal process. As presented in Figure 4a,b, nanoscale rod-shaped MnO nanomixtures particles were clearly observed. The lattice fringe spacing was determined from Figure 4c and was attributed to the presence of manganese.

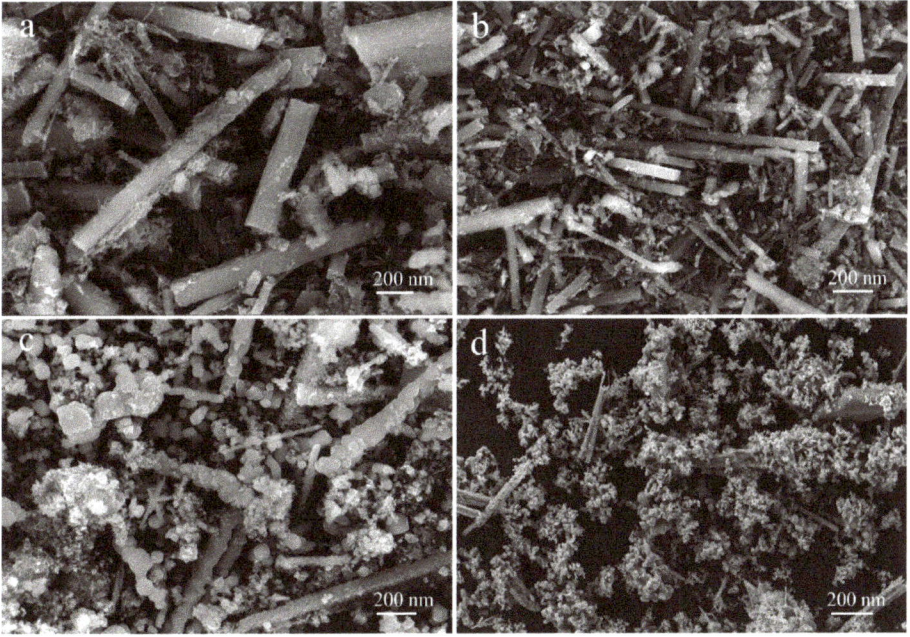

Figure 3. Images of MnO-2 (**a**), MnO-4 (**b**), MnO-6 (**c**) and MnO-24 (**d**).

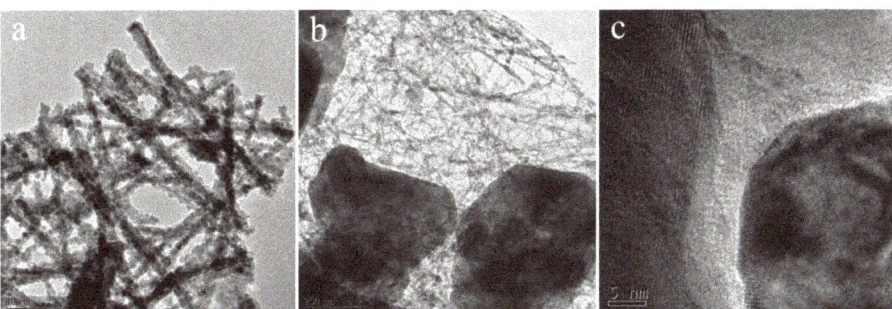

Figure 4. TEM images (**a**,**b**) and high resolution image (**c**) of MnO-4.

XPS was performed to analyze the chemical nature of MnO-4; the results are shown in Figure 5. Figure 5a reveals the presence of C, K, O, N and Mn, which corresponded to peaks at 285 eV, 300 eV, 535 eV, 410 eV and 650 eV, respectively. As presented in Figure 5b, five peaks were observed (284.6 eV, 285.3 eV, 285.9 eV, 287.4 eV and 289.0 eV), and these peaks were attributed to C–N–C, C–C, C–O, C=O, and O–C=O groups. This result indicated that C_3N_4 was oxidized by $KMnO_4$ in the hydrothermal self-assembly process. The peaks shown in Figure 5c corresponded to C=O (531.0 eV), COOH (532.0 eV) and C–O–C (535.0 eV). As shown in Figure 5d, two peaks [7] were observed at 400.3 eV and 398.8 eV, which were assigned to N–C$_3$ and C–N–C, respectively. The presence of N–C$_3$ was beneficial for MB adsorption [37]. The peaks at 641.8 eV and 653.4 eV corresponded to Mn 2p, which indicated the presence of manganese.

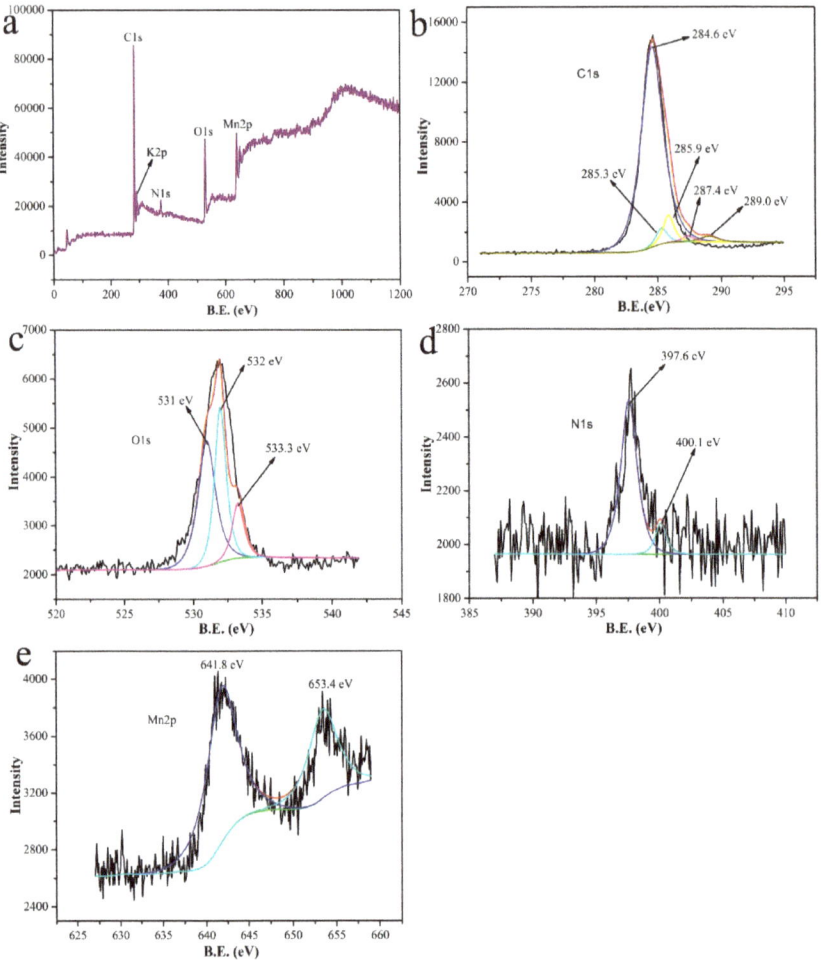

Figure 5. XPS spectra of MnO-4: (**a**) full-scan spectrum, (**b**) C 1s spectrum, (**c**) O 1s spectrum, (**d**) N 1s spectrum, and (**e**) Mn 2p spectrum.

The effect of the different samples on the MB adsorption amount was investigated, and the result is shown in Figure 6. As seen in Figure 6, MnO-4 and MnO-6 exhibited larger adsorption amounts than MnO-2, which was attributed to the higher reactant molar ratio of C$_3$N$_4$ to KMnO$_4$. C$_3$N$_4$ introduced a π-conjugation system in MnO nanomixtures during the hydrothermal process, which could improve the adsorption capacity. Meanwhile, a moderate dosage of KMnO$_4$ could improve the surface area to increase the adsorption of MB. From the comparison of the adsorption amounts of MnO-4 and MnO-24, the hydrothermal reaction time exerted an effect on the adsorption capacity, in which MnO-4 had a higher adsorption capacity of up to 137 mg g^{-1} in a 20 mg L^{-1} MB solution at 20 °C. The zeta potentials of MnO-2, MnO-4, MnO-6 and MnO-24 in water were as follows: −29.8 mV, −42.3 mV,

−37.1 mV, and −34.7 mV, respectively. MB is a cationic dye; thus, the lower the zeta potential is, the better the adsorption.

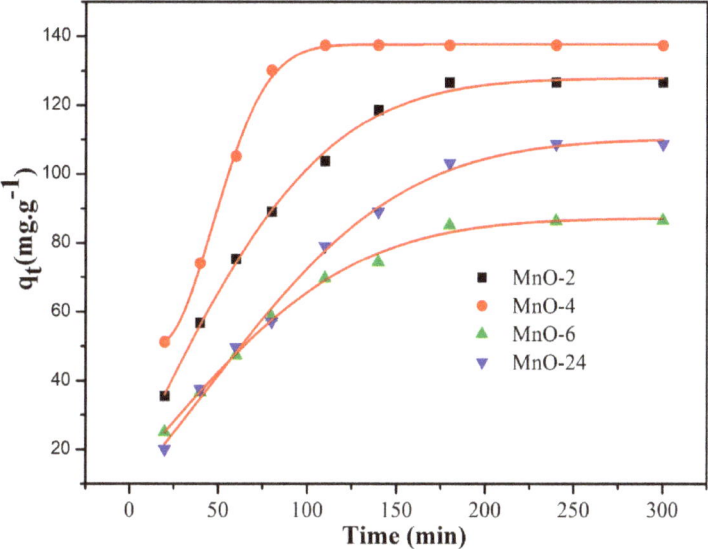

Figure 6. MB adsorption curves of the as-prepared samples.

The effect of the MB concentration on the adsorption capacity is shown in Figure 7a, in which the adsorption capacity was observed to increase with the MB concentration. The higher the MB concentration, the shorter the adsorption equilibrium time was. The MB adsorption efficiency was up to 96% for an MB concentration of 10 mg L^{-1} at 150 min. As seen in Figure 7b, an equilibrium plateau was reached, which indicated that MnO-4 acted as a monolayer adsorbent in MB absorption. The Langmuir model was adopted to investigate the adsorption process on the MnO-4 surface, and the results are shown in Figure 7c. The correlation coefficient (R^2) of the fitted curve was 0.996, which indicated that adsorption occurred through a Langmuir process, meaning that it was a monolayer process. This analysis result was in accordance with the results of Figure 7b.

The effect of the MB solution pH on the adsorption capacity was studied, and the results are presented in Figure 8, in which the maximum adsorption capacity was achieved with a strong basic MB solution and the adsorption capacity increased with the solution pH. This result was attributed to the electrostatic interaction between the MB molecules and MnO nanomixtures. In the previous discussion, the zeta potentials exerted an effect on the adsorption capacity, as MB is a cationic dye. In an acidic solution, the zeta potentials of MnO nanomixtures were positive, which inhibited MB adsorption. In contrast, at lower pH values, the zeta potentials were negative and lower. Therefore, MnO-4 had a high adsorption capacity in a basic MB solution. Meanwhile, the nitrogen doping of MnO could improve the alkalinity of the solution, which was beneficial for MB adsorption.

The pseudo-first-order and pseudo-second-order kinetic models were used to analyze the kinetics of the adsorption process. The theoretical adsorption capacity of MnO-4 calculated from the pseudo-first-order model was 194 mg g^{-1}, and that calculated from the pseudo-second-order model was 164 mg g^{-1} (Table 2), which fit well with the experimental data (154 mg g^{-1}). As shown in Figure 9, the R^2 values obtained from the pseudo-second-order model were better than the R^2 values obtained from the pseudo-first-order model. In conclusion, the pseudo-second-order model was more suitable for investigation of the MB adsorption process.

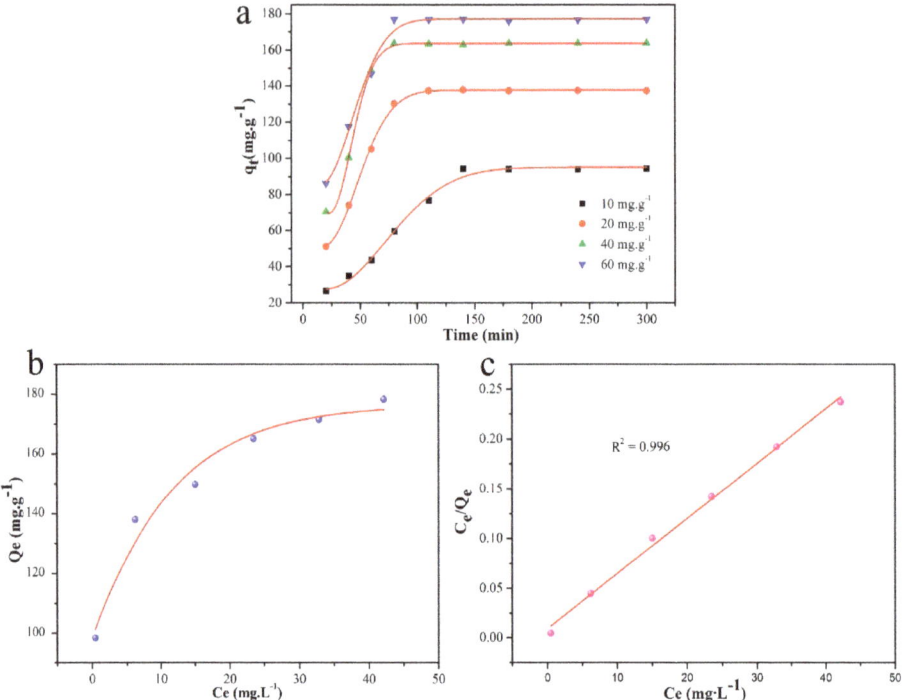

Figure 7. (**a**) Adsorption curves under different concentrations of MB; (**b**) adsorption isotherm of an MB solution in MnO-4; (**c**) Langmuir isotherm plot for MB adsorption in MnO-4.

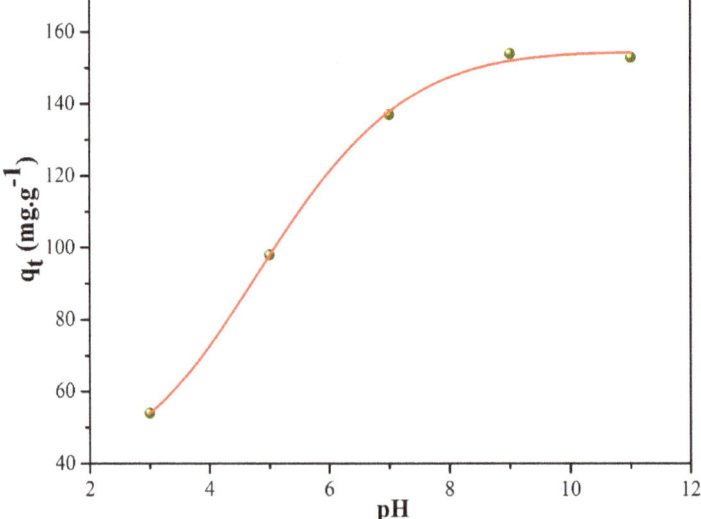

Figure 8. MB adsorption capacity of MnO-4 at different pH values.

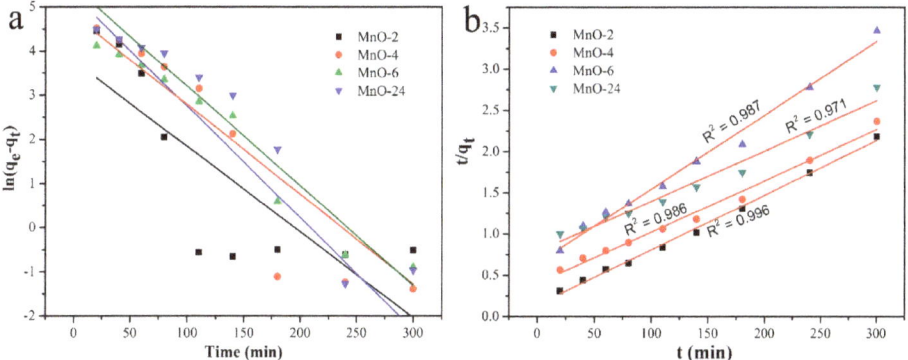

Figure 9. Pseudo-first-order kinetic model plot (**a**) and pseudo-second-order kinetic model plot (**b**).

Table 2. Parameters of pseudo-first-order kinetic model and pseudo-second-order kinetic model for the adsorption of MB in MnO nanomixtures.

Entry	Pseudo-First-Order Kinetic Model		Pseudo-Second-Order Kinetic Model	
	K_1	q_e (mg g^{-1})	K_2	q_e (mg g^{-1})
MnO-2	0.019	43.70	0.00031	150.38
MnO-4	0.025	194.03	0.000097	164.12
MnO-6	0.021	122.85	0.00012	111.48
MnO-24	0.022	234.40	0.000047	160.67

MB adsorption experiments were performed at different temperatures, and the results are shown in Figure 10a. At the same time, the plot of ln Kc versus 1/T for MnO-4 is demonstrated in Figure 10b. As presented in Figure 10a, a higher adsorption capacity was obtained at a higher temperature, which indicated that a high temperature was beneficial for MB adsorption. The ΔG^0, ΔH^0 and ΔS^0 values of MB adsorption on MnO-4 were calculated from Equations (6) and (7) [38] to be -7.4 kJ mol^{-1}, 21.5 kJ mol^{-1} and 97.0 J mol^{-1}, respectively. The value of ΔG^0 was negative, which demonstrated that spontaneous MB adsorption occurred on the MnO-4 surface. In addition, the value of ΔS^0 was positive, which was attributed to an increase in the chaos at the adsorbent/solution interface during MB adsorption in MnO-4. In addition, the value of ΔH^0 was below 40 kJ mol^{-1}, as demonstrated by the physisorption of MB in MnO-4, and the positive value indicated that the process was endothermic, which was in accordance with the experimental results.

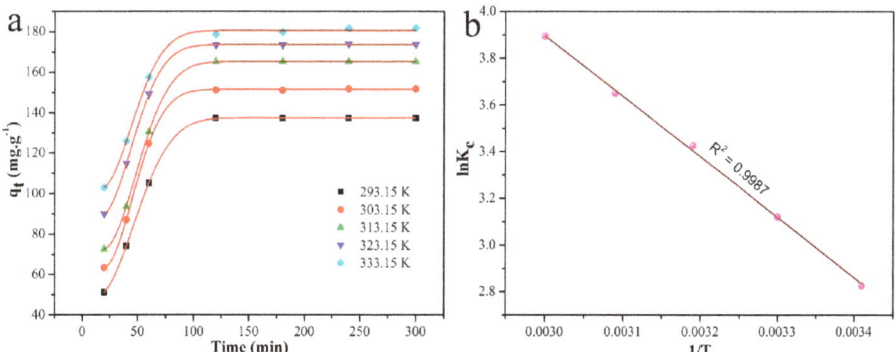

Figure 10. Adsorption curves measured at different temperatures (**a**) and the plot of ln Kc versus 1/T for MnO-4 (**b**).

Repeated experiments were conducted to investigate the reusability of MnO-4 for MB adsorption, and the results are shown in Figure 11. The adsorption capacity was 137 mg g^{-1} in the first cycle, and 96% of the adsorption capacity, corresponding to 132 mg g^{-1}, was retained in the last cycle. Therefore, this reusability indicated that MnO-4 was a good adsorbent for MB. Meanwhile, the obtained MnO-4 exhibited excellent adsorption capacity, which could be roughly compared with other reported absorbents shown in Table 3.

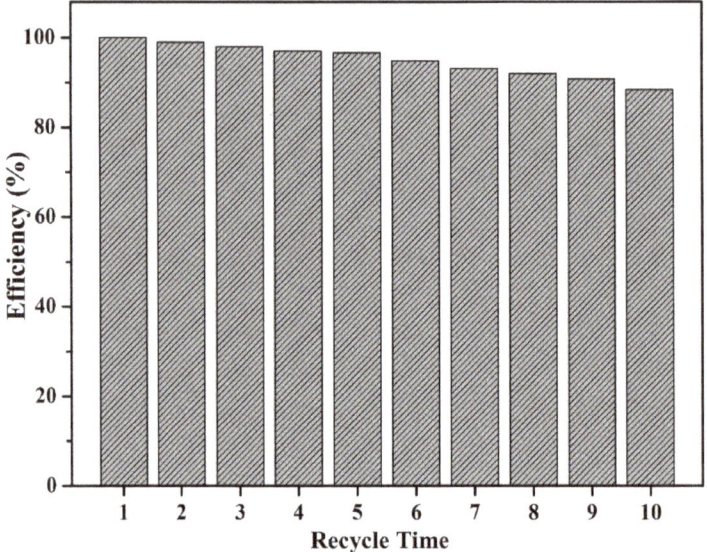

Figure 11. MB adsorption capacities of MnO-4 in 10 adsorption cycles.

Table 3. Comparison of the adsorption capacities of different absorbents from previous reports with that of C, N-MnO-4.

Adsorbent	mg g^{-1}	Reference
Wheat shells	21.5	[39]
Chitosan-modified zeolite	37	[40]
Fe$_3$O$_4$@Ag/SiO$_2$ nanospheres	128.5	[41]
α-Fe$_2$O$_3$@carboxyl-functionalized yeast composite	49.5	[42]
N, O-codoped porous carbon	100.2	[43]
Kaolin	52.7	[44]
C, N-doped MnO	154	Present work

The degradation efficiency of MB in MnO nanomixtures was investigated in this work. As shown in Figure 12a, MnO nanomixtures exhibited high degradation efficiency under different MB concentrations (99.8%, ≈142 mg g^{-1} at a MB concentration of 20 mg L^{-1}). As presented in Figure 12b,c, the MB solution exhibited a sharp absorption band at 656 nm in the UV-Vis spectrum, and this absorption band obviously decreased with increasing degradation time. The degradation kinetics were well fitted by the pseudo-first-order model shown in Figure 12d, and the theoretical De (the degradation amount at the degradation equilibrium) value was 146 mg g^{-1}, which was in good agreement with the experimental data. This analysis result indicated that the pseudo-first-order model could effectively describe the MB degradation process in MnO nanomixtures [45–52].

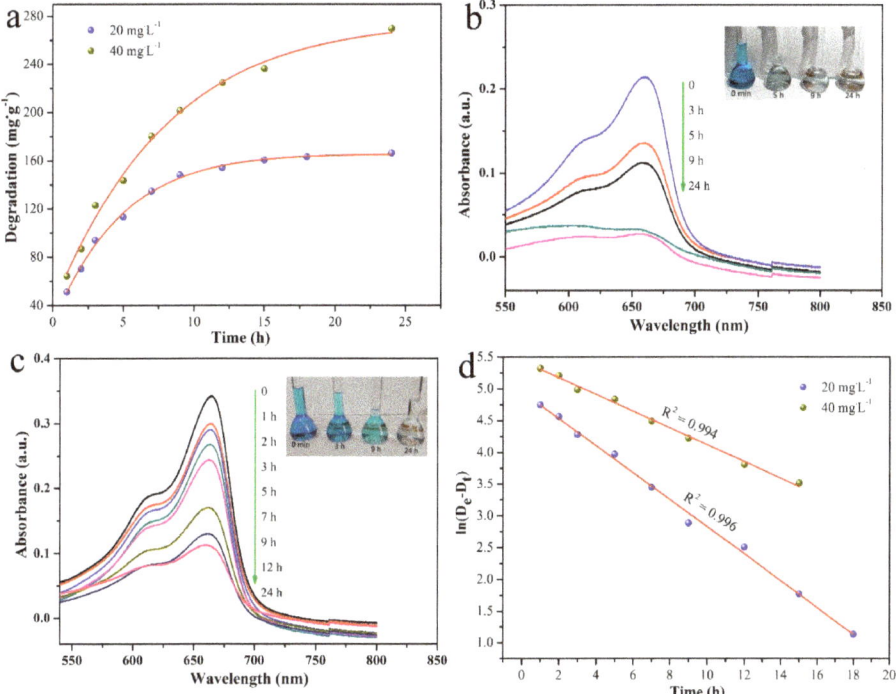

Figure 12. MB degradation curves under different MB concentrations (**a**); UV-Vis spectra of 20 mg L^{-1} MB after various degradation times (**b**); UV-Vis spectra of 40 mg L^{-1} MB after various degradation times (**c**); and pseudo-first-order kinetic model plot of the degradation process (**d**).

The degradation mechanism of MB in MnO nanomixtures was proposed (Figure 13). Active superoxide anions and/or peroxide species could form in the H$_2$O$_2$-MnO system according to previous reports [16,53], and these species could oxidize MB. As shown in Figure 13, H$_2$O$_2$ was used as an oxidant to form various superoxide anions and peroxide species, and C; N-doped MnO was used as a catalyst to catalyze the decomposition of H$_2$O$_2$. Mn(III)/Mn(II) played an important role in the MB degradation process and contributed to ideal MB degradation in C, N-doped MnO. Present obtained MnO nanomixtures demonstrated potential applications in self-assembled materials design and composites for wide applications [54–65].

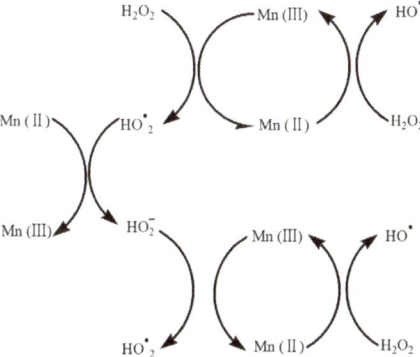

Figure 13. Degradation mechanism of MB in MnO nanomixtures.

4. Conclusions

In summary, novel nanoscale rod-shaped MnO nanomixtures were successfully prepared via a hydrothermal self-assembly method with C_3N_4 as the source of carbon and nitrogen and potassium permanganate ($KMnO_4$) as the source of manganese. The as-prepared materials exhibited good MB adsorption and degradation with H_2O_2 as the oxidant. The maximum adsorption capacity was 154 mg g^{-1}, and the optimum degradation efficiency was 99.8%. The adsorption process was very well fitted by the pseudo-second-order model, and the degradation process was very well fitted by the pseudo-first-order model. MB adsorption occurred through physicorption, and MB degradation was caused by a chemical reaction. Meanwhile, MnO nanomixtures exhibited excellent reusability. The as-prepared MnO nanomixtures are potential and effective materials for extensive pollutant removal.

Author Contributions: T.J. and Z.L. conceived and designed the experiments; Y.X., B.R., R.W., and L.Z. performed the experiments; T.J. and Y.X. analyzed the data; Y.X. wrote the paper.

Funding: This research was funded by Foundation of Key R&D Program of Hebei Province (No. 18393616D), science and technology projects of Hebei Academy of Sciences (No. 18707), Natural Science Fund and Key Basic Research Project (No. 18964005D), National Natural Science Foundation of China (Nos. 21872119, 21473153), Support Program for the Top Young Talents of Hebei Province, China Postdoctoral Science Foundation (No. 2015M580214), Research Program of the College Science & Technology of Hebei Province (No. ZD2018091), and Scientific and Technological Research and Development Program of Qinhuangdao City (No. 201701B004).

Conflicts of Interest: The authors declare no conflict of interest.

References

1. Marrakchi, F.; Ahmed, M.J.; Khanday, W.A.; Asif, M.; Hameed, B.H. Mesoporous-activated carbon prepared from chitosan flakes via single-step sodium hydroxide activation for the adsorption of methylene blue. *Int. J. Biol. Macromol.* **2017**, *98*, 233–239. [CrossRef] [PubMed]
2. Guo, R.; Jiao, T.F.; Li, R.F.; Chen, Y.; Guo, W.C.; Zhang, L.X.; Zhou, J.X.; Zhang, Q.R.; Peng, Q.M. Sandwiched Fe_3O_4/carboxylate graphene oxide nanostructures constructed by layer-by-layer assembly for highly efficient and magnetically recyclable dye removal. *ACS Sustain. Chem. Eng.* **2018**, *6*, 1279–1288. [CrossRef]
3. Vaz, M.G.; Pereira, A.G.B.; Fajardo, A.R.; Azevedo, A.C.N.; Rodrigues, F.H.A. Methylene Blue Adsorption on Chitosan-g-Poly(Acrylic Acid)/Rice Husk Ash Superabsorbent Composite: Kinetics, Equilibrium, and Thermodynamics. *Water Air Soil Pollut.* **2017**, *228*, 14. [CrossRef]
4. Liu, Y.; Hou, C.; Jiao, T.; Song, J.; Zhang, X.; Xing, R.; Zhou, J.; Zhang, L.; Peng, Q. Self-assembled AgNP-containing nanocomposites constructed by electrospinning as efficient dye photocatalyst materials for wastewater treatment. *Nanomaterials* **2018**, *8*, 35. [CrossRef] [PubMed]
5. Ullah, R.; Dutta, J. Photocatalytic degradation of organic dyes with manganese-doped ZnO nanoparticles. *J. Hazard. Mater.* **2008**, *156*, 194–200. [CrossRef]
6. Zhao, X.N.; Jiao, T.F.; Ma, X.L.; Huang, H.; Hu, J.; Qu, Y.; Zhou, J.X.; Zhang, L.X.; Peng, Q.M. Facile fabrication of hierarchical diamond-based AuNPs-modified nanomixtures via layer-by-layer assembly with enhanced catalytic capacities. *J. Taiwan Inst. Chem. Eng.* **2017**, *80*, 614–623. [CrossRef]
7. Atchudan, R.; Edison, T.N.J.I.; Perumal, S.; Karthik, N.; Karthikeyan, D.; Shanmugam, M.; RokLee, Y. Concurrent synthesis of nitrogen-doped carbon dots for cell imaging and ZnO@nitrogen-doped carbon sheets for photocatalytic degradation of methylene blue. *J. Photochem. Photobiol. A Chem.* **2018**, *350*, 75–85. [CrossRef]
8. Tanhaei, M.; Mahjoub, A.R.; Safarifard, V. Sonochemical synthesis of amide-functionalized metal-organic framework/graphene oxide nanocomposite for the adsorption of methylene blue from aqueous solution. *Ultrason. Sonochem.* **2018**, *41*, 189–195. [CrossRef]
9. More, A.T.; Vira, A.; Fogel, S. Biodegradation of trans-1, 2-dichloroethylene by methane-utilizing bacteria in an aquifer simulator. *Environ. Sci. Technol.* **1989**, *23*, 403–406. [CrossRef]
10. Slokar, Y.M.; Marechal, A.M.L. Methods of decoloration of textile wastewaters. *Dyes Pigment.* **1998**, *37*, 335–356. [CrossRef]

11. Tang, L.; Jia, C.T. Fabrication of compressible and recyclable macroscopic g-C_3N_4/GO aerogel hybrids for visible-light harvesting: A promising strategy for water remediation. *Appl. Catal. B Environ.* **2017**, *219*, 241–248. [CrossRef]
12. Xu, J.; Wang, Z.P. Enhanced visible-light-driven photocatalytic disinfection performance and organic pollutant degradation activity of porous g-C_3N_4 nanosheets. *ACS Appl. Mater. Interfaces* **2017**, *9*, 27727–27735. [CrossRef]
13. Houas, A.; Lachheb, H.; Ksibi, M.; Elaloui, E.; Guillard, C.; Herrmann, J.M. Photocatalytic degradation pathway of methylene blue in water. *Appl. Catal. B Environ.* **2001**, *31*, 145–157. [CrossRef]
14. Cavallaro, G.; Gianguzza, A.; Lazzara, G.; Milioto, S.; Piazzese, D. Alginate gel beads filled with halloysite nanotubes. *Appl. Clay Sci.* **2013**, *72*, 132–137. [CrossRef]
15. Cavallaro, G.; Grillo, I.; Gradzielski, M.; Lazzara, G. Structure of Hybrid Materials Based on Halloysite Nanotubes Filled with Anionic Surfactants. *J. Phys. Chem. C* **2016**, *120*, 13492–13502. [CrossRef]
16. Li, T.T.; Wang, Z.H.; Liu, C.C.; Tang, C.M.; Wang, X.K.; Ding, G.S.; Ding, Y.C.; Yang, L.X. TiO_2 Nanotubes/Ag/MoS_2 Meshy Photoelectrode with Excellent Photoelectrocatalytic Degradation Activity for Tetracycline Hydrochloride. *Nanomaterials* **2018**, *8*, 666. [CrossRef] [PubMed]
17. Riahi-Madvaar, R.; Taher, M.A.; Fazelirad, H. Synthesis and characterization of magnetic halloysite-iron oxide nanocomposite and its application for naphthol green B removal. *Appl. Clay Sci.* **2017**, *137*, 101–106. [CrossRef]
18. Minero, C.; Lucchiari, M.; Vione, D.; Maurino, V. Fe(III)-Enhanced Sonochemical Degradation of Methylene Blue In Aqueous Solution. *Environ. Sci. Technol.* **2005**, *39*, 8936–8942. [CrossRef]
19. Jiang, F.; Yan, T.T.; Chen, H.; Sun, A.; Xu, C.M.; Wang, X. A g-C_3N_4–CdS composite catalyst with high visible-light-driven catalytic activity and photostability for methylene blue degradation. *Appl. Surf. Sci.* **2014**, *295*, 164–172. [CrossRef]
20. Xing, S.T.; Zhou, Z.C.; Ma, Z.C.; Wu, Y.S. Characterization and reactivity of Fe_3O_4/$FeMnOx$ core/shell nanoparticles for methylene blue discoloration with H_2O_2. *Appl. Catal. B Environ.* **2011**, *107*, 386–392. [CrossRef]
21. Li, Y.Q.; Qu, J.Y.; Gao, F.; Lv, S.Y.; Shi, L.; He, C.X.; Sun, J.C. In situ fabrication of Mn_3O_4 decorated graphene oxide as a synergistic catalyst for degradation of methylene blue. *Appl. Catal. B Environ.* **2015**, *162*, 268–274. [CrossRef]
22. Randorn, C.; Wongnawa, S.; Boonsin, P. Bleaching of Methylene Blue by Hydrated Titanium Dioxide. *Sci. Asia* **2004**, *30*, 149–156. [CrossRef]
23. Wang, Z.H.; Zhao, H.Q.; Qi, H.B.; Liu, X.Y.; Liu, Y. Free radical behaviours during methylene blue degradation in the Fe^{2+}/H_2O_2 system. *Environ. Technol.* **2017**. [CrossRef] [PubMed]
24. Bhattacharyya, G.K.; Sharma, A. Kinetics and thermodynamics of Methylene Blue adsorption on Neem (*Azadirachta indica*) leaf powder. *Dyes Pigment.* **2005**, *65*, 51–59. [CrossRef]
25. de Brito Benetoli, L.O.; Cadorin, B.M.; Baldissarelli, V.Z.; Geremias, R.; de Souza, I.G.; Debacher, N.A. Pyrite-enhanced methylene blue degradation in non-thermal plasma water treatment reactor. *J. Hazard. Mater.* **2012**, *237*, 55–62. [CrossRef] [PubMed]
26. Wang, H.; Huang, Y.M. Prussian-blue-modified iron oxide magnetic nanoparticles as effective peroxidase-like catalysts to degrade methylene blue with H_2O_2. *J. Hazard. Mater.* **2011**, *191*, 163–169. [CrossRef] [PubMed]
27. Wolski, L.; Ziolek, M. Insight into pathways of methylene blue degradation with H_2O_2 over mono and bimetallic Nb, Zn oxides. *Appl. Catal. B Environ.* **2018**, *224*, 634–647. [CrossRef]
28. Prathap, M.U.A.; Kaur, B.; Srivastava, R. Hydrothermal synthesis of CuO micro-/nanostructures and their applications in the oxidative degradation of methylene blue and non-enzymatic sensing of glucose/H_2O_2. *J. Colloid Interface Sci.* **2012**, *370*, 144–154. [CrossRef]
29. Saha, M.; Gayen, A.; Mukherjee, S. Microstructure, morphology, and methylene blue degradation over nano-$CuFe_2O_4$ synthesized by a modified complexometric method. *J. Aust. Ceram. Soc.* **2018**, *54*, 513–522. [CrossRef]
30. Zhang, P.Q.; Zhan, Y.G.; Cai, B.X.; Hao, C.C.; Wang, J.; Liu, C.X.; Meng, Z.J.; Yin, Z.L.; Chen, Q.Y. Shape-Controlled Synthesis of Mn_3O_4 Nanocrystals and Their Catalysis of the Degradation of Methylene Blue. *Nano Res.* **2010**, *3*, 235–243. [CrossRef]
31. Cai, X.G.; He, J.Y. A 2D-g-C_3N_4 nanosheet as an eco-friendly adsorbent for various environmental pollutants in water. *Chemosphere* **2017**, *171*, 192–201. [CrossRef] [PubMed]

32. Xin, G.; Meng, Y.L. Pyrolysis Synthesized g-C$_3$N$_4$ for Photocatalytic Degradation of Methylene Blue. *J. Chem.* **2013**, *2013*, 187912. [CrossRef]
33. Li, K.Y.; Fang, Z.L.; Xiong, S.; Luo, J. Novel graphitic-C$_3$N$_4$ nanosheets: Enhanced visible light photocatalytic activity and photoelectrochemical detection of methylene blue dye. *Mater. Technol.* **2017**, *32*, 391–398. [CrossRef]
34. Chang, F.; Xie, Y.C.; Li, C.L.; Chen, J.; Luo, J.R.; Hu, X.F.; Shen, J.W. A facile modification of g-C$_3$N$_4$ with enhanced photocatalytic activity for degradation of methylene blue. *Appl. Surf. Sci.* **2013**, *280*, 967–974. [CrossRef]
35. Mao, Y.; Wu, M.Z.; Li, G.; Dai, P.; Yu, X.X.; Bai, Z.M.; Chen, P. Photocatalytic degradation of methylene blue over boron-doped g-C$_3$N$_4$ together with nitrogenvacancies under visible light irradiation. *React. Kinet. Mech. Catal.* **2018**, *125*, 1179–1190. [CrossRef]
36. Chu, Y.T.; Guo, L.Y.; Xi, B.J.; Feng, Z.Y.; Wu, F.F.; Lin, Y.; Liu, J.C.; Sun, D.; Feng, J.K.; Qian, Y.T.; et al. Embedding MnO@Mn$_3$O$_4$ Nanoparticles in an N-Doped-Carbon Framework Derived from Mn-Organic Clusters for Effcient Lithium Storage. *Adv. Mater.* **2018**, *30*, 1704244. [CrossRef] [PubMed]
37. Dong, G.H.; Zhao, K. Carbon self-doping induced high electronic conductivity and photoreactivity of g-C$_3$N$_4$. *Chem. Commun.* **2012**, *48*, 6178–6180. [CrossRef] [PubMed]
38. Ren, B.; Xu, Y.L.; Zhang, L.H.; Liu, Z.F. Carbon-doped graphitic carbon nitride as environment-benign adsorbent for methylene blue adsorption: Kinetics, isotherm and thermodynamics study. *J. Taiwan Inst. Chem. Eng.* **2018**, *88*, 114–120. [CrossRef]
39. Bulut, Y.; Aydın, H. A kinetics and thermodynamics study of methylene blue adsorption on wheat shells. *Desalination* **2006**, *194*, 259–267. [CrossRef]
40. Xie, J.; Li, C.J. Chitosan modified zeolite as a versatile adsorbent for the removal of different pollutants from water. *Fuel* **2013**, *103*, 480–485. [CrossRef]
41. Saini, J.; Garg, V.K.; Gupta, R.K. Removal of Methylene Blue from aqueous solution by Fe$_3$O$_4$@Ag/SiO$_2$ nanospheres: Synthesis, characterization and adsorption performance. *J. Mol. Liq.* **2018**, *250*, 413–422. [CrossRef]
42. Wang, Y.; Zhang, W.; Qin, M.; Zhao, M.J.; Zhang, Y.S. Green one-pot preparation of α-Fe$_2$O$_3$@carboxyl-functionalized yeast composite with high adsorption and catalysis properties for removal of methylene blue. *Surf. Interface Anal.* **2017**, *50*, 311–320. [CrossRef]
43. Chen, B.L.; Yang, Z.X.; Ma, G.P.; Kong, D.L.; Xiong, W.; Wang, J.B.; Zhu, Y.Q.; Xia, Y.D. Heteroatom-doped porous carbons with enhanced carbon dioxide uptake and excellent methylene blue adsorption capacities. *Microporous Mesoporous Mater.* **2018**, *257*, 1–8. [CrossRef]
44. Mounia, L.; Belkhir, L.I.; Bollinger, J.C.; Bouzaza, A.; Assadi, A.; Tirri, A.; Dahmoune, F.; Madani, K.; Remini, H. Removal of Methylene Blue from aqueous solutions by adsorption on Kaolin: Kinetic and equilibrium studies. *Appl. Clay Sci.* **2018**, *153*, 38–45. [CrossRef]
45. Xing, R.; Wang, W.; Jiao, T.; Ma, K.; Zhang, Q.; Hong, W.; Qiu, H.; Zhou, J.; Zhang, L.; Peng, Q. Bioinspired polydopamine sheathed nanofibers containing carboxylate graphene oxide nanosheet for high-efficient dyes scavenger. *ACS Sustain. Chem. Eng.* **2017**, *5*, 4948–4956. [CrossRef]
46. Zhao, X.; Ma, K.; Jiao, T.; Xing, R.; Ma, X.; Hu, J.; Huang, H.; Zhang, L.; Yan, X. Fabrication of hierarchical layer-by-layer assembled diamond based core-shell nanocomposites as highly efficient dye absorbents for wastewater treatment. *Sci. Rep.* **2017**, *7*, 44076. [CrossRef] [PubMed]
47. Guo, H.; Jiao, T.; Zhang, Q.; Guo, W.; Peng, Q.; Yan, X. Preparation of graphene oxide-based hydrogels as efficient dye adsorbents for wastewater treatment. *Nanoscale Res. Lett.* **2015**, *10*, 272. [CrossRef] [PubMed]
48. Li, K.; Jiao, T.; Xing, R.; Zou, G.; Zhou, J.; Zhang, L.; Peng, Q. Fabrication of tunable hierarchical MXene@AuNPs nanocomposites constructed by self-reduction reactions with enhanced catalytic performances. *Sci. China Mater.* **2018**, *61*, 728–736. [CrossRef]
49. Wang, C.; Sun, S.; Zhang, L.; Yin, J.; Jiao, T.; Zhang, L.; Xu, Y.; Zhou, J.; Peng, Q. Facile preparation and catalytic performance characterization of AuNPs-loaded hierarchical electrospun composite fibers by solvent vapor annealing treatment. *Colloid Surf. A Physicochem. Eng. Asp.* **2019**, *561*, 283–291. [CrossRef]
50. Chen, K.; Li, J.; Zhang, L.; Xing, R.; Jiao, T.; Gao, F.; Peng, Q. Facile synthesis of self-assembled carbon nanotubes/dye composite films for sensitive electrochemical determination of Cd(II) ions. *Nanotechnology* **2018**, *29*, 445603. [CrossRef]

51. Huang, X.; Jiao, T.; Liu, Q.; Zhang, L.; Zhou, J.; Li, B.; Peng, Q. Hierarchical electrospun nanofibers treated by solvent vapor annealing as air filtration mat for high-efficiency PM2.5 capture. *Sci. China Mater.* **2018**. [CrossRef]
52. Li, N.; Tang, S.; Rao, Y.; Qi, J.; Wang, P.; Jiang, Y.; Huang, H.; Gu, J.; Yuan, D. Improved dye removal and simultaneous electricity production in a photocatalytic fuel cell coupling with persulfate. *Electrochim. Acta* **2018**, *270*, 330–338. [CrossRef]
53. Jiang, J.Z.; Zou, J.; Zhu, L.H.; Huang, L.; Jiang, H.P.; Zhang, Y.X. Degradation of Methylene Blue with H_2O_2 Activated by Peroxidase-Like Fe_3O_4 Magnetic Nanoparticles. *J. Nanosci. Nanotechnol.* **2011**, *11*, 4793–4799. [CrossRef] [PubMed]
54. Zhou, J.; Gao, F.; Jiao, T.; Xing, R.; Zhang, L.; Zhang, Q.; Peng, Q. Selective Cu(II) ion removal from wastewater via surface charged self-assembled polystyrene-Schiff base nanocomposites. *Colloid Surf. A Physicochem. Eng. Asp.* **2018**, *545*, 60–67. [CrossRef]
55. Tang, S.; Li, X.; Zhang, C.; Liu, Y.; Zhang, W.; Yuan, D. Strengthening decomposition of oxytetracycline in DBD plasma coupling with Fe-Mn oxides loaded granular activated carbon. *Plasma Sci. Technol.* **2018**. [CrossRef]
56. Luo, X.; Ma, K.; Jiao, T.; Xing, R.; Zhang, L.; Zhou, J.; Li, B. Graphene oxide-polymer composite Langmuir films constructed by interfacial thiol-ene photopolymerization. *Nanoscale Res. Lett.* **2017**, *12*, 99. [CrossRef] [PubMed]
57. Huo, S.; Duan, P.; Jiao, T.; Peng, Q.; Liu, M. Self-assembled luminescent quantum dots to generate full-color and white circularly polarized light. *Angew. Chem. Int. Ed.* **2017**, *56*, 12174–12178. [CrossRef]
58. Song, J.; Xing, R.; Jiao, T.; Peng, Q.; Yuan, C.; Möhwald, H.; Yan, X. Crystalline dipeptide nanobelts based on solid-solid phase transformation self-assembly and their polarization imaging of cells. *ACS Appl. Mater. Interfaces* **2018**, *10*, 2368–2376. [CrossRef]
59. Zhang, R.Y.; Xing, R.R.; Jiao, T.F.; Ma, K.; Chen, C.J.; Ma, G.H.; Yan, X.H. Synergistic in vivo photodynamic and photothermal antitumor therapy based on collagen-gold hybrid hydrogels with inclusion of photosensitive drugs Colloids and Surfaces A: Physicochemical and Engineering Aspects. *ACS Appl. Mater. Interfaces* **2016**, *8*, 13262–13269. [CrossRef]
60. Xing, R.; Jiao, T.; Liu, Y.; Ma, K.; Zou, Q.; Ma, G.; Yan, X. Co-assembly of graphene oxide and albumin/photosensitizer nanohybrids towards enhanced photodynamic therapy. *Polymers* **2016**, *8*, 181. [CrossRef]
61. Xing, R.; Liu, K.; Jiao, T.; Zhang, N.; Ma, K.; Zhang, R.; Zou, Q.; Ma, G.; Yan, X. An injectable self-assembling collagen-gold hybrid hydrogel for combinatorial antitumor photothermal/photodynamic therapy. *Adv. Mater.* **2016**, *28*, 3669–3676. [CrossRef]
62. Zhou, J.; Liu, Y.; Jiao, T.; Xing, R.; Yang, Z.; Fan, J.; Liu, J.; Li, B.; Peng, Q. Preparation and enhanced structural integrity of electrospun poly(ε-caprolactone)-based fibers by freezing amorphous chains through thiol-ene click reaction. *Colloid Surf. A Physicochem. Eng. Asp.* **2018**, *538*, 7–13. [CrossRef]
63. Liu, K.; Yuan, C.Q.; Zou, Q.L.; Xie, Z.C.; Yan, X.H. Self-assembled Zinc/cystine-based chloroplast mimics capable of photoenzymatic reactions for sustainable fuel synthesis. *Angew. Chem. Int. Ed.* **2017**, *56*, 7876–7880. [CrossRef] [PubMed]
64. Liu, K.; Xing, R.R.; Li, Y.X.; Zou, Q.L.; Möhwald, H.; Yan, X.H. Mimicking primitive photobacteria: Sustainable hydrogen evolution based on peptide-porphyrin co-assemblies with self-mineralized reaction center. *Angew. Chem. Int. Ed.* **2016**, *55*, 12503–12507. [CrossRef] [PubMed]
65. Liu, K.; Xing, R.R.; Chen, C.J.; Shen, G.Z.; Yan, L.Y.; Zou, Q.L.; Ma, G.H.; Möhwald, H.; Yan, X.H. Peptide-induced hierarchical long-range order and photocatalytic activity of porphyrin assemblies. *Angew. Chem. Int. Ed.* **2015**, *54*, 500–505. [CrossRef]

© 2018 by the authors. Licensee MDPI, Basel, Switzerland. This article is an open access article distributed under the terms and conditions of the Creative Commons Attribution (CC BY) license (http://creativecommons.org/licenses/by/4.0/).

Article

Silver Nanomaterial-Immobilized Desalination Systems for Efficient Removal of Radioactive Iodine Species in Water

Ha Eun Shim [1,2,†], Jung Eun Yang [3,†], Sun-Wook Jeong [4], Chang Heon Lee [1], Lee Song [1], Sajid Mushtaq [1,5], Dae Seong Choi [1], Yong Jun Choi [4,*] and Jongho Jeon [1,5,*]

1. Advanced Radiation Technology Institute, Korea Atomic Energy Research Institute, Jeongeup 56212, Korea; she0805@kaeri.re.kr (H.E.S.); chlee9406@kaeri.re.kr (C.H.L.); songlee@kaeri.re.kr (L.S.); sajidqau101@yahoo.com (S.M.); dschoi@kaeri.re.kr (D.S.C.)
2. Department of Chemistry, Kyungpook National University, Daegu 41566, Korea
3. Department of Chemical and Biomolecular Engineering, Korea Advanced Institute of Science and Technology, Daejeon 34141, Korea; jung-e@kaist.ac.kr
4. School of Environmental Engineering, University of Seoul, Seoul 02504, Korea; jeongsunwook@gmail.com
5. Radiation Biotechnology and Applied Radioisotope Science, University of Science and Technology, Daejeon 34113, Korea
* Correspondence: yongjun2165@uos.ac.kr (Y.J.C.); jeonj@kaeri.re.kr (J.J.); Tel.: +82-2-6490-2873 (Y.J.C.); +82-63-570-3374 (J.J.)
† These authors contributed equally to this work.

Received: 6 August 2018; Accepted: 24 August 2018; Published: 26 August 2018

Abstract: Increasing concerns regarding the adverse effects of radioactive iodine waste have inspired the development of a highly efficient and sustainable desalination process for the treatment of radioactive iodine-contaminated water. Because of the high affinity of silver towards iodine species, silver nanoparticles immobilized on a cellulose acetate membrane (Ag-CAM) and biogenic silver nanoparticles containing the radiation-resistant bacterium *Deinococcus radiodurans* (Ag-DR) were developed and investigated for desalination performance in removing radioactive iodines from water. A simple filtration of radioactive iodine using Ag-CAM under continuous in-flow conditions (approximately 1.5 mL/s) provided an excellent removal efficiency (>99%) as well as iodide anion-selectivity. In the bioremediation study, the radioactive iodine was rapidly captured by Ag-DR in the presence of high concentration of competing anions in a short time. The results from both procedures can be visualized by using single-photon emission computed tomography (SPECT) scanning. This work presents a promising desalination method for the removal of radioactive iodine and a practical application model for remediating radioelement-contaminated waters.

Keywords: bioremediation; desalination; membrane; nanocomposite; radioactive iodine; silver nanomaterials

1. Introduction

In recent decades, radioactive isotopes have widely been used for industrial and medical applications that have introduced drastic quantities of radioactive toxic pollutants to the environment [1,2]. Among many radioactive isotopes, large amounts of after-use radioactive iodine species (iodines) have been discarded following applications in radiation therapies and biomedical studies [3–8]. Moreover, the recalcitrant characteristics of exposed radioactive materials contribute to serious adverse effects such as acute diseases, metabolic imbalances, and genetic mutations [9–14]. Thus, the development of sustainable treatment methods for the removal of radioactive iodines is necessary for public health and environmental safety.

Many engineered nanomaterials have been used as adsorbents in nuclear waste treatment because of their large surface area. In addition, these materials possess high reactivity towards specific radioactive elements without additional chelating molecules [15]. Among them, silver metal-based materials have widely been used in the desalination of radioactive iodine wastes because of the high affinity of silver towards iodine species [16–30]. In a typical desalination procedure, these adsorbents should immerse in contaminated water for the removal of radioactive iodine, and thus radioactive elements-containing solid wastes generated by this process need to be separated from water [31]. Therefore, time-consuming processes such as centrifugation was required to harvest unsettled silver nanoparticles (AgNPs) and reprocess radioelement-contaminated solid adsorbents. Moreover, nano- or microscale silver materials are aggregated easily under high salt concentrations, causing physicochemical property losses [32,33].

Together, these observations inspired the design of a more efficient and stable silver particle-based desalination method for the removal of radioactive iodine waste in two ways by the incarceration of silver-based adsorbents. First, AgNPs were immobilized on a cellulose acetate membrane (CAM) and evaluated for desalination performance in a continuous-flow system. If radioisotopes are existed in homogeneous aqueous media, the engineered membrane can easily be applied to the separation step. However, when significant slurry or insoluble materials are contained in liquid contaminants, membrane-based equipment is unsuitable for purification procedures. In such cases, the bioremediation could be considered for an alternative process, as this method has some advantages over the membrane-based method, including: (1) possibility for on-site remediation; (2) removal of complexed radioelement contaminants by simple genetic engineering; and (3) easily scaled remediation processing. Therefore, as a second method, we report biogenic AgNP-containing radiation-resistant bacterial cells as efficient adsorbent carriers for use in a novel bioremediation platform.

2. Materials and Methods

2.1. General Methods

Silver nitrate, sodium borohydride, trisodium citrate, and sodium iodide (non-radioactive) were purchased from Sigma-Aldrich Korea (Yongin, Republic of Korea). CAMs (pore size = 0.20 μm, diameter = 25 mm) were purchased from Advantec MFS. [^{125}I]NaI was supplied by New Korea Industrial Co., Ltd. (Daejeon, Republic of Korea). The radioactivity of ^{125}I was measured by using a radioactivity dose calibrator (Capintec, Inc., Florham Park, NJ, USA) and a radio-thin-layer chromatography (TLC) scanner (Bioscan, AR-2000, Poway, CA, USA). The amount of radioactivity was determined by using a γ-counter (PerkinElmer, 2480 Automatic γ-counter, Waltham, MA, USA). Single-photon emission computed tomography/computed tomography (SPECT/CT) images were obtained by using an Inveon SPECT/CT system (Siemens, Erlangen, Germany). Silver nanomaterials on the CAMs and the surfaces of bacteria were observed using a field-emission scanning electron microscope (FE-SEM, Inspect F50, FEI, Mahwah, NJ, USA) under high-performance conditions with accelerating voltages reaching 15 kV. The elemental composition of the silver nanomaterials was analyzed by SEM-energy-dispersive X-ray (EDX) (EDAX Apollo XL, AMETEK, Mahwah, NJ, USA) analysis with accelerating voltages reaching 20 kV. EDX spectra were recorded in area scanning mode by focusing the electron beam onto a region of the sample surface.

2.2. Synthesis of Citrate-Stabilized AgNPs

AgNPs with the average diameter of 30 nm were synthesized as described in the previous report [34]. In brief, solutions of sodium borohydride (NaBH$_4$, 2.08 mM) and trisodium citrate (TSC, Na$_3$C$_6$H$_5$O$_7$, 2.08 mM) were mixed and the resulting solution was heated to 60 °C in the dark for 30 min. After heating, 2 mL of silver nitrate (AgNO$_3$, 1.17 mM) was added dropwise and then the temperature was raised to 90 °C. When the temperature reached 90 °C, the pH was adjusted to 10.5 by adding 0.1 M NaOH solution. The resulting solution was then heated for 20 min. After the suspension was cooled to room temperature, the concentration of AgNPs was measured by using ultraviolet-visible (UV/vis) spectrometry (concentration = 7.0×10^{-10} M with an extinction coefficient of 1.45×10^{10} at the peak of 406 nm with a 1-cm path length). The solution was stored at 4 °C. The mean diameter of citrate-stabilized AgNPs was approximately 30 nm [35].

2.3. Preparation of the Ag-CAM

The commercially available CAM filter (pore size = 0.2 μm, diameter = 25 mm) was washed with 10 mL deionized water using a syringe. Next, 5 mL of citrate-stabilized AgNPs (7.0×10^{-10} M), was passed through the filter using a syringe at the rate of 1 mL s^{-1}. The membrane filter was then washed with pure water three times, yielding a yellowish-brown colored filter. This Ag-CAM filter was kept under ambient conditions until testing in the desalination experiment. To analyze the surface of Ag-CAM by SEM-EDX, the composite membrane was isolated from the filter unit.

2.4. Desalination of Radioactive Iodine Using Ag-CAM Filter Unit under Continuous In-Flow Conditions

To evaluate the efficiency of the Ag-CAM filter unit under continuous-flow conditions, [^{125}I]NaI (3.7 MBq) was diluted with 50 mL aqueous media (pure water, 1.0 M NaCl, 1× phosphate-buffered saline (1× PBS), 10 mM NaI). Each aqueous radioactive iodine solution was then passed through an Ag-CAM at an in-flow rate of approximately 1.5 mL s^{-1} by using a syringe pump or by hand using lead gloves. The amount of residual radioactivity in the filtrate was measured by using a γ-counter.

The removal efficiency (%) was defined by the following equation to assess the adsorption capability of the Ag-CAM towards radioactive iodine:

$$\text{Removal efficiency (\%)} = (C_0 - C_e)/C_0 \times 100 \tag{1}$$

where C_0 and C_e represent the initial (before filtration) and equilibrium (after filtration) concentrations of radioactive iodine, respectively.

The distribution coefficient (K_d) was determined using the following equation:

$$K_d = (C_0 - C_e)/C_e \times \frac{V}{M} \tag{2}$$

where C_0 and C_e represent the initial and final concentrations of radioactive iodine, respectively; V denotes the volume of radioactive iodine solution (50 mL); and M is the mass of the adsorbent (0.1 mg AgNPs).

2.5. Preparation of Silver Nanomaterial-Containing Deinococcus radiodurans R1

Biogenic AgNPs were obtained using *Deinococcus radiodurans* cells as previously described [36,37]. In brief, *D. radiodurans* strain ATCC13939 was inoculated in a tryptone glucose yeast extract (TGY) liquid medium until the sample reached the optical density at 600 nm (OD$_{600}$) of 1.0. After cultivation, silver nitrate was added to the cell cultures adjusted to the final concentration of 2.5 mM and incubated for 24 h at 30 °C. The cultures were centrifuged at 4000 rpm for 30 min, and the resulting pellets were washed three times with deionized water. The pellets were then re-suspended in 5 mL deionized water and used for further analysis.

The absorption spectrum of AgNP-embedded *D. radiodurans* (Ag-DR) was monitored by a UV/vis spectrophotometer (Epoch Microplate Spectrophotometer, BioTek Instruments, Daejeon, Republic of Korea) from 400 to 800 nm. The dynamic light scattering (DLS) analysis of the biogenic AgNPs was performed as described previously [34]. For analysis by SEM-EDX, the samples were fixed with 2.5% glutaraldehyde solution and then dehydrated with 30%, 50%, 70%, 80%, 90%, 95%, and 100% ethanol (EtOH). After dehydration, samples were freeze-dried overnight. The prepared samples were subjected to FE-SEM after platinum coating using ion sputtering for 1 min and the morphology and existence of AgNPs were observed.

2.6. Remediation Procedure of Radioactive Iodine Using Ag-DR

[^{125}I]NaI solution (3.7 MBq) was added to the Ag-DR suspension in aqueous media (water or 1× PBS). The mixture was shaken on an orbital shaker. At each time point of 1, 5, and 15 min, an aliquot (0.5 µL) was withdrawn from the Ag-DR solution and spotted onto a silica-coated thin-layer chromatography (TLC) plate. The TLC plate was then developed using acetone as the mobile phase. After the solvent traveled to the top of the plate, the TLC plate was cut in half. The radioactivity of each piece was measured by using a γ-counter. The retention factor (R_f) values of free ^{125}I$^-$ in solution and ^{125}I$^-$ in Ag-DR were 0.8 and 0, respectively. Therefore, the removal efficiency (%) was defined by the following equation to evaluate the desalination performance of Ag-DR:

$$\text{Removal efficiency (\%)} = C_0/(C_0 + C_I) \times 100 \tag{3}$$

where C_0 and C_I represent the amounts of radioactivity at the bottom ($R_f = 0$) and higher position ($R_f = 0.8$), respectively.

2.7. SPECT/CT Imaging of Radioactive Iodine Captured by Ag-DR

Radioactive iodine ([^{125}I]NaI, 3.7 MBq) was diluted with 50 mL of pure water. An aqueous radioactive iodine solution was then passed through the CAM or Ag-CAM. After the filtration procedure, the radioactivity in the membrane filter was imaged by SPECT/CT scanning.

Radioactive iodine ([^{125}I]NaI, 3.7 MBq) was added to both Ag-DR and a wild *D. radiodurans* sample; each solution was shaken at room temperature for 15 min. Some of the cells were then transferred to a 1.5-mL tube and centrifuged for spinning down of the cells. Molecular imaging was performed by SPECT/CT.

3. Results and Discussion

3.1. Preparation of Ag-CAM

The main strategy for the desalination of radioactive iodines using silver nanomaterials is described in Figure 1. To examine the removal efficiency of radioactive iodine under continuous-flow conditions, the Ag-CAM filter is fabricated (Figure 2A,B). The immobilization of AgNPs on the CAM yields a homogeneous yellowish-brown color. The AgNPs incorporated in the filter unit were sustained stably without aggregate formation or elution from the membrane by continual washing with high-concentration salt solutions such as 1.0 M NaCl. Notably, Ag-CAM could be stored for several weeks without loss of stability or desalination performance. The hydroxyl and carbonyl groups in the polymeric cellulose acetate apparently efficiently stabilize the novel metal nanoparticles on the membrane [38,39]. SEM analysis of the surfaces of the Ag-CAM and CAM show that the nanomaterials are incorporated stably on the cellulose nanofibrils (Figure 2C,D). Elemental analysis of the membrane using EDX spectroscopy showed a set of peaks representing silver, along with carbon and oxygen atoms from the carbohydrate units in the cellulose polymer. These analyses verify the successful preparation of the composite membrane. The adsorption capacity of iodide anions on the Ag-CAM filter with a surface area of 4.91 cm^2 was measured as approximately 31 mg of I$^-$/g of AgNPs.

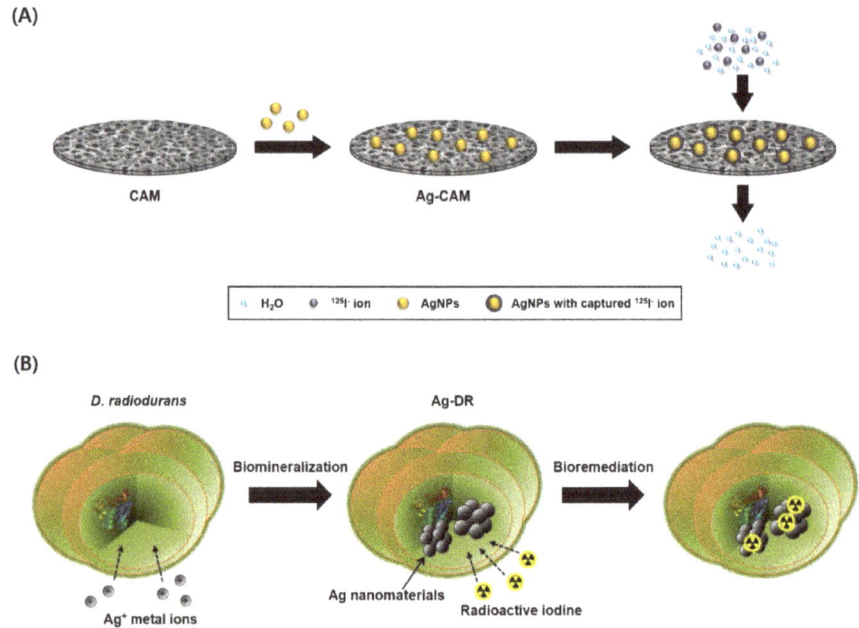

Figure 1. (**A**) Desalination of radioactive iodine by using Ag-CAM; and (**B**) bioremediation procedure of radioactive iodine anions using Ag-DR.

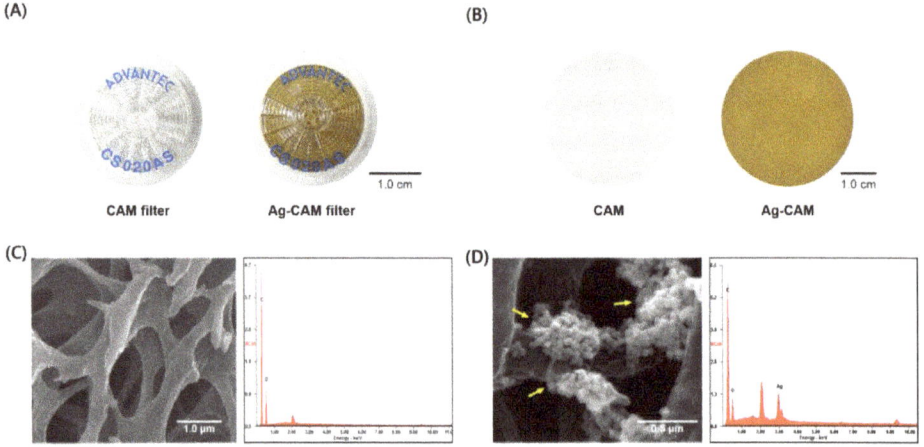

Figure 2. Characterization of Ag-CAM: (**A**) Photographic images of the CAM filter (left) and Ag-CAM (right) prepared using a syringe filter; (**B**) photographic images of the CAM (left) and Ag-CAM (right) prepared using a vacuum filter holder; (**C**) SEM–EDX analysis of the CAM (40,000×); and (**D**) Ag-CAM (100,000×). Yellow arrows in the images indicate AgNPs on cellulose fibers.

3.2. Desalination of Radioactive Iodine Using Ag-CAM

The removal efficiency of Ag-CAM was investigated in the continuous-flow system. The filtration process was performed as shown in Figure 1A. The radioactive iodine solutions (typical concentration of 1.0 nM, 3.7 MBq/50 mL) were passed through the Ag-CAM at an in-flow rate of 1.5 mL s^{-1}; the amount of radioactivity in the filtrate was then measured using a γ-counter. After a single filtration step, the concentration of ^{125}I$^-$ in pure water is dramatically decreased from 1 to 0.004 nM (Figure 3A). The removal efficiency reaches 99.6% and the distribution coefficient (K_d) exceeds 10^6 mL g^{-1}. However, the unmodified CAM does not remove radioactive iodine under the same operation. The Ag-CAM maintained excellent removal efficiency under the presence of high concentrations of competing anions (1.0 M NaCl and 1× PBS). An excellent removal efficiency (99.5%) is observed in 1.0 M NaCl solution, in which the ratio of Cl$^-$ to ^{125}I$^-$ anions ([Cl$^-$]:[^{125}I$^-$]) reached 10^9:1. However, in the presence of excess non-radioactive iodine, most radioactive iodine passed though the column, because the AgNPs immobilized on the CAM are covered with ^{127}I$^-$ anions. The desalination results are further visualized by SPECT/CT scanning. After filtering the [^{125}I]NaI solution, the radioactive species, initially detected in the aqueous solution, is efficiently captured by the nanocomposite membrane (Figure 3B). However, the CAM filter unit captures no iodide anions via the same operation (Figure 3C). Thus, these images also confirm the Ag-CAM-mediated removal of radioactive iodine in a continuous-flow aqueous system. The nanocomposite membrane used in this study offer a simpler and more practical method to efficiently capture radioactive species from various aqueous solutions. By a simple filtration process, the amount of radioactive iodine in water is reduced significantly, with a removal efficiency of ≥99.5% in the presence of competing anions. In addition, the desalination of 50 mL of aqueous solution can be accomplished in 1 min. These results compare favorably with our previous report, which applied gold nanoparticles as adsorbents for radioactive iodine [40]. In the previous study, approximately 1.6 mg of gold was necessary to prepare the desalination membrane filter [41]; the present method uses approximately 0.1 mg AgNPs to achieve high desalination performance. Because silver is much cheaper than gold, the proposed Ag-CAM is more practical for the treatment of radioactive iodine wastes.

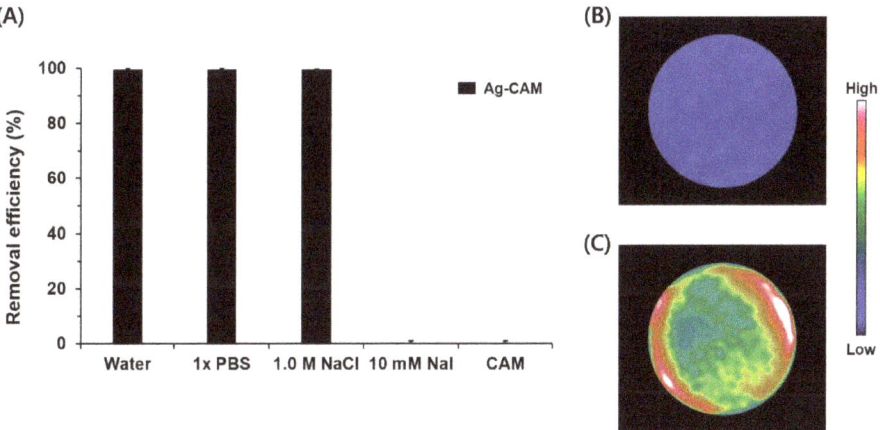

Figure 3. (**A**) Desalination of radioactive iodine using Ag-CAM in several aqueous solutions; (**B**) SPECT image of post-filtration non-modified CAM; and (**C**) SPECT image of post-filtration Ag-CAM.

3.3. Remediation of Radioactive Iodine Using Ag-DR

In general, the most important feature of a water treatment system is its direct applicability to a polluted environment. Although silver metal-based adsorbents show high removal efficiencies, they have physical and spatial limitations for on-site remediation [36]. Meanwhile, bioremediation has certain advantages over physicochemical methods, including cost-effectiveness, eco-friendliness, and practicality [42]. Thus, bioremediation is considered as an alternative to previous physicochemical treatment processes. However, no reports on the application of biogenic silver nanoparticles on the desalination of radioactive iodine wastes exist. This may be because of the high radiation dose emitted by radioactive iodine. Thus, treatment of radioactive iodine wastes using AgNPs immobilized in the radiation-resistant extremophile *D. radiodurans* was investigated (Figure 1B).

During the culturing of *D. radiodurans* with 2.5 mM AgNO$_3$, the color of the culture broth gradually changes from light orange to pale gray with increasing incubation time (Figure 4A). The UV/vis spectrum of the culture broth shows an absorption peak at 400–450 nm, indicating the formation of silver nanomaterials in the *D. radiodurans* strain (Figure 4B). The dynamic light scattering (DLS) analysis of nanomaterials isolated from *D. radiodurans* showed silver nanomaterials of approximately 30 nm in size. In addition, SEM-EDX analysis also clearly displays the silver nanomaterials synthesized by the bacterial cells (Figure 4C).

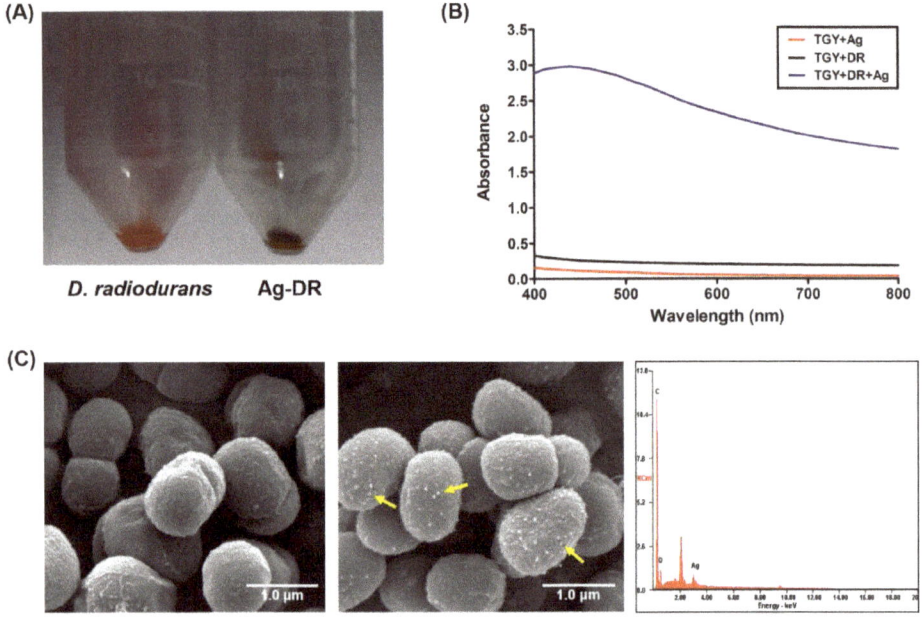

Figure 4. Characterization of Ag-DR: (**A**) Photographic image of *D. radiodurans* (left) and Ag-DR (right); (**B**) UV/vis spectra of *D. radiodurans* and Ag-DR; and (**C**) SEM image of *D. radiodurans* (left, 50,000×), Ag-DR (center, 50,000×), and EDX analysis of Ag-DR (right). Yellow arrows in the images indicate AgNPs on *D. radiodurans*.

Next, to investigate the desalination efficiency, radioactive iodine (3.7 MBq [^{125}I]NaI) was added to Ag-DR and wild *D. radiodurans* (~10^9 cells). The amount of radioactivity captured by the Ag-DR was then analyzed by a γ-counter and each experiment was performed in triplicate. As shown in Figure 5A, Ag-DR shows rapid uptake kinetics at the beginning of incubation; >97% radioactivity is captured by Ag-DR in 1 min. After overnight incubation (18 h), almost all iodine anions are stably retained in the Ag-DR, suggesting that Ag-DR provides efficient and sustainable remediation process. Meanwhile, *D. radiodurans* with no silver adsorbents show only non-specific retention of radioactive iodine. Although the remediation kinetics are slowed (84% in 15 min) in the presence of high concentrations of competing ions (1× PBS), the removal efficiency of 95% is obtained over a prolonged incubation time (18 h). Furthermore, the kinetics of the removal of radioactive iodine mostly depends on the amount of Ag-DR as can be seen in Figure 5B. However, a desalination efficiency of

~90% is obtained with a smaller amount of Ag-DR (0.1×) in a short time. We also investigated the removal capability of freeze-dried Ag-DR, considering the portability of the bacteria. Suspensions of 5, 10, 50, and 100 mg freeze-dried Ag-DR were mixed with equal amounts of radioactive iodine solution. As shown in Figure 5C, faster adsorption kinetics occurs with higher bacterial cell levels. After 15 min, satisfactory removal efficiency (>97%) is obtained with 5 mg freeze-dried Ag-DR.

To further confirm the bioremediation process, a molecular imaging study was performed using SPECT imaging. Radioactive iodine (3.7 MBq) was added to both Ag-DR and wild *D. radiodurans* and incubated for 15 min at room temperature. As shown in Figure 5D, the SPECT imaging analysis shows a strong radioactive signal at the bottom of the Ag-DR sample, exactly overlapping with photograph of the Ag-DR. However, *D. radiodurans* alone does not capture $^{125}I^-$ ions; thus, most of the radioactivity retains in the supernatant. These observations clearly demonstrate that the AgNP-containing bacterial cells successfully capture the radioactive $^{125}I^-$ ions. Although the Ag-DR shows good desalination performances, more successful examples of microbial bioremediation processes can be expected through guided strain development integrated with bioprocess engineering and systematic biotechnology.

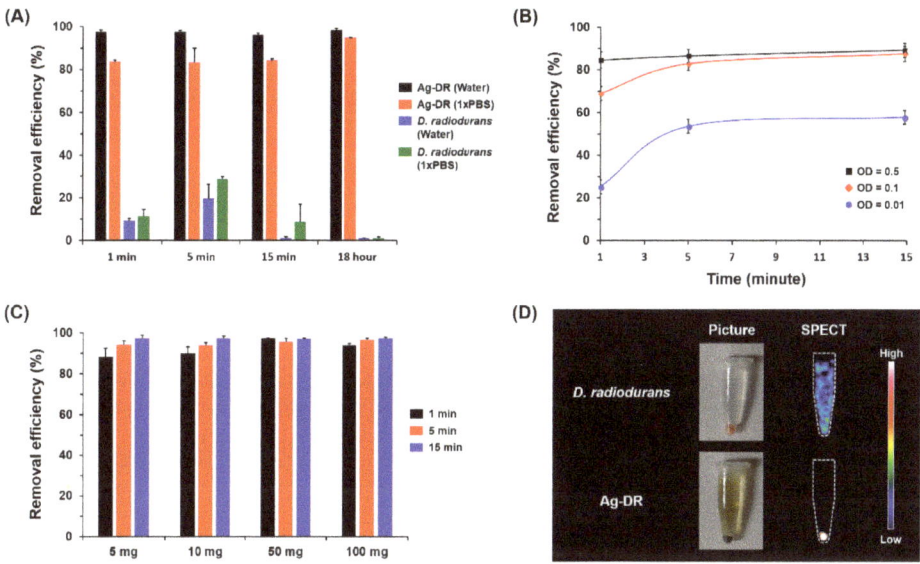

Figure 5. (**A**) Removal efficiency of Ag-DR in water and 1× PBS; (**B**) uptake kinetics for removal of radioiodine using smaller concentrations of Ag-DR in water for 15 min (OD = optical density at 600 nm); (**C**) removal efficiency of freeze-dried Ag-DR in water for 15 min; and **D**) photographic and SPECT/CT images of Ag-DR and *D. radiodurans* after $^{125}I^-$ incubation.

4. Conclusions

In the present study, we developed a desalination process for the removal of radioactive iodine from water. First, an Ag-CAM filter was constructed and tested regarding its removal efficiency of radioactive iodide ions ($^{125}I^-$). It showed the removal efficiency and distribution coefficient (K_d) of >99.6% and 10^6 mL g^{-1}, respectively. Next, biogenic AgNP-containing *D. radiodurans* was developed as a platform strain for on-site bioremediation. Approximately 3.7 MBq of radioactive iodine was successfully captured by the biogenic AgNPs immobilized in the cells within 15 min. Consequently, it is expected that the immobilized silver nanomaterials-based desalination method will provide a promising system worth to investigate large-scale nuclear waste management.

Author Contributions: J.J. and Y.J.C. conceived and designed the experiments. H.E.S., S.-W.J., C.H.L., L.S., S.M., and D.S.C. performed the desalination experiments. S.-W.J. and J.E.Y. performed the analysis of nano-composite materials. J.J., Y.J.C., and H.-E.S. wrote the manuscript. All authors approved the final version of manuscript.

Funding: This work was supported by the National Research Foundation of Korea (NRF) grant funded by the Korea government (MSIT) (Grant number: 2017M2A2A6A01070858).

Acknowledgments: We would like to thank to S. H. Kim for assistance in SEM-EDX analysis.

Conflicts of Interest: The authors declare no conflict of interest.

References

1. Khayet, M.; Matsuura, T. Radioactive decontamination of water. *Desalination* **2013**, *321*, 1–2. [CrossRef]
2. Abdel Rahman, R.O.; Ibrahium, H.A.; Hung, Y.-T. Liquid radioactive wastes treatment: A review. *Water* **2011**, *3*, 551–565. [CrossRef]
3. Bonnema, S.J.; Hegedüs, L. Radioiodine therapy in benign thyroid diseases effects, side effects, and factors affecting therapeutic outcome. *Endocr. Rev.* **2012**, *33*, 920–980. [CrossRef] [PubMed]
4. Prpic, M.; Dabelic, N.; Stanicic, J.; Jukic, T.; Milosevic, M.; Kusic, Z. Adjuvant thyroid remnant ablation in patients with differentiated thyroid carcinoma confined to the thyroid: A comparison of ablation success with different activities of radioiodine (I-131). *Ann. Nucl. Med.* **2012**, *26*, 744–751. [CrossRef] [PubMed]
5. Sabra, M.M.; Grewal, R.K.; Ghossein, R.A.; Tuttle, R.M. Higher administered activities of radioactive iodine are associated with less structural persistent response in older, but not younger, papillary thyroid cancer patients with lateral neck lymph node metastases. *Thyroid* **2014**, *24*, 1088–1095. [CrossRef] [PubMed]
6. Ravichandran, R. Management of radioactive wastes in a hospital environment. In *Modelling Trends in Solid and Hazardous Waste Management*; Sengupta, D., Agrahari, S., Eds.; Springer: Singapore, 2017; pp. 1–14.
7. Ravichandran, R.; Binukumar, J.P.; Sreeram, R.; Arunkumar, L.S. An overview of radioactive wastes disposal procedures of a nuclear medicine department. *J. Med. Phys.* **2011**, *36*, 95–99. [CrossRef] [PubMed]
8. Martin, J.E.; Fenner, F.D. Radioactivity in municipal sewage and sludge. *Public Health Rep.* **1997**, *112*, 308–316. [PubMed]
9. Grossman, C.M.; Nussbaum, R.H.; Nussbaum, F.D. Thyrotoxicosis among Hanford, Washington, downwinders: A community-based health survey. *Arch. Environ. Health Int. J.* **2002**, *57*, 9–15. [CrossRef] [PubMed]
10. Hou, X.; Povinec, P.P.; Zhang, L.; Shi, K.; Biddulph, D.; Chang, C.-C.; Fan, Y.; Golser, R.; Hou, Y.; Ješkovský, M.; et al. Iodine-129 in seawater offshore Fukushima: Distribution, inorganic speciation, sources, and budget. *Environ. Sci. Technol.* **2013**, *47*, 3091–3098. [CrossRef] [PubMed]
11. Thomas, G.D.; Smith, S.M.; Turcotte, J.A. Using public relations strategies to prompt populations at risk to seek health information: The Hanford community health project. *Health Promot. Pract.* **2009**, *10*, 92–101. [CrossRef] [PubMed]
12. Grossman, C.M.; Morton, W.E.; Nussbaum, R.H. Hypothyroidism and spontaneous abortions among Hanford, Washington, downwinders. *Arch. Environ. Health Int. J.* **1996**, *51*, 175–176. [CrossRef] [PubMed]
13. Goldsmith, J.R.; Grossman, C.M.; Morton, W.E.; Nussbaum, R.H.; Kordysh, E.A.; Quastel, M.R.; Sobel, R.B.; Nussbaum, F.D. Juvenile hypothyroidism among two populations exposed to radioiodine. *Environ. Health Perspect.* **1999**, *107*, 303–308. [CrossRef] [PubMed]
14. Grossman, C.M.; Nussbaum, R.H.; Nussbaum, F.D. Cancers among residents downwind of Hanford, Washington, plutonium production site. *Arch. Environ. Health Int. J.* **2003**, *58*, 267–274. [CrossRef] [PubMed]
15. Yuan, Y.; Wang, H.; Hou, S.; Xia, D. Chapter 18. Applications of Nanomaterials in Nuclear Waste Management. In *Multifunctional Nanocomposites for Energy and Environmental Applications*; Wiley-VCH: Hoboken, NJ, USA, 2018; pp. 543–566.
16. Mu, W.; Yu, Q.; Li, X.; Wei, H.; Jian, Y. Adsorption of radioactive iodine on surfactant-modified sodium niobate. *RSC Adv.* **2016**, *6*, 81719–81725. [CrossRef]
17. Yang, D.; Liu, H.; Liu, L.; Sarina, S.; Zheng, Z.; Zhu, H. Silver oxide nanocrystals anchored on titanate nanotubes and nanofibers: Promising candidates for entrapment of radioactive iodine anions. *Nanoscale* **2013**, *5*, 11011–11018. [CrossRef] [PubMed]

18. Yang, D.; Sarina, S.; Zhu, H.; Liu, H.; Zheng, Z.; Xie, M.; Smith, S.V.; Komarneni, S. Capture of radioactive cesium and iodide ions from water by using titanate nanofibers and nanotubes. *Angew. Chem. Int. Ed.* **2011**, *50*, 10594–10598. [CrossRef] [PubMed]
19. Li, B.; Dong, X.; Wang, H.; Ma, D.; Tan, K.; Jensen, S.; Deibert, B.J.; Butler, J.; Cure, J.; Shi, Z.; et al. Capture of organic iodides from nuclear waste by metal-organic framework-based molecular traps. *Nat. Commun.* **2017**, *8*, 485–493. [CrossRef] [PubMed]
20. Chapman, K.W.; Chupas, P.J.; Nenoff, T.M. Radioactive iodine capture in silver-containing mordenites through nanoscale silver iodide formation. *J. Am. Chem. Soc.* **2010**, *132*, 8897–8899. [CrossRef] [PubMed]
21. Sarina, S.; Bo, A.; Liu, D.; Liu, H.; Yang, D.; Zhuo, C.; Maes, N.; Komarneni, S.; Zhu, H. Separate or simultaneous removal of radioactive cations and anions from water by layered sodium vanadate-based sorbents. *Chem. Mater.* **2014**, *26*, 4788–4795. [CrossRef]
22. Mu, W.; Li, X.; Liu, G.; Yu, Q.; Xie, X.; Wei, H.; Jian, Y. Safe disposal of radioactive iodide ions from solutions by Ag_2O grafted sodium niobate nanofibers. *Dalton Trans.* **2016**, *45*, 753–759. [CrossRef] [PubMed]
23. Bo, A.; Sarina, S.; Liu, H.; Zheng, Z.; Xiao, Q.; Gu, Y.; Ayoko, G.A.; Zhu, H. Efficient removal of cationic and anionic radioactive pollutants from water using hydrotalcite-based getters. *ACS Appl. Mater. Interfaces* **2016**, *8*, 16503–16510. [CrossRef] [PubMed]
24. Yang, J.H.; Park, H.-S.; Cho, Y.-Z. Al_2O_3-containing silver phosphate glasses as hosting matrices for radioactive iodine. *J. Nucl. Sci. Technol.* **2017**, *54*, 1330–1337. [CrossRef]
25. Liu, S.; Wang, N.; Zhang, Y.; Li, Y.; Han, Z.; Na, P. Efficient removal of radioactive iodide ions from water by three-dimensional Ag_2O–Ag/TiO_2 composites under visible light irradiation. *J. Hazard. Mater.* **2015**, *284*, 171–181. [CrossRef] [PubMed]
26. Bo, A.; Sarina, S.; Zheng, Z.; Yang, D.; Liu, H.; Zhu, H. Removal of radioactive iodine from water using Ag_2O grafted titanate nanolamina as efficient adsorbent. *J. Hazard. Mater.* **2013**, *246–247*, 199–205. [CrossRef] [PubMed]
27. Chen, Y.-Y.; Yu, S.-H.; Yao, Q.-Z.; Fu, S.-Q.; Zhuo, G.-T. One-step synthesis of $Ag_2O@Mg(OH)_2$ nanocomposite as an efficient scavenger for iodine and uranium. *J. Colloid. Interface Sci.* **2018**, *510*, 280–291. [CrossRef] [PubMed]
28. Riley, B.J.; Vienna, J.D.; Strachan, D.M.; McCloy, J.S.; Jerden, J.L., Jr. Materials and processes for the effective capture and immobilization of radioiodine: A review. *J. Nucl. Mater.* **2016**, *470*, 307–326. [CrossRef]
29. Kim, T.; Lee, S.-K.; Lee, S.; Lee, J.S.; Kim, S.W. Development of silver nanoparticle-doped adsorbents for the separation and recovery of radioactive iodine from alkaline solutions. *Appl. Radiat. Isot.* **2017**, *129*, 215–221. [CrossRef] [PubMed]
30. Conde-González, J.E.; Peña-Méndez, E.M.; Rybáková, S.; Pasán, J.; Ruiz-Pérez, C.; Havel, J. Adsorption of silver nanoparticles from aqueous solution on copper-based metal organic frameworks (HKUST-1). *Chemosphere* **2016**, *150*, 659–666. [CrossRef] [PubMed]
31. Othman, S.H.; Sohsah, M.A.; Ghoneim, M.M.; Sokkar, H.H.; Badawy, S.M.; El-Anadouli, B.E. Adsorption of hazardous ions from radioactive waste on chelating cloth filter. *Radiat. Phys. Chem.* **2006**, *75*, 278–285. [CrossRef]
32. Badawy, A.M.E.; Luxton, T.P.; Silva, R.G.; Scheckel, K.G.; Suidan, M.T.; Tolaymat, T.M. Impact of environmental conditions (pH, ionic strength, and electrolyte type) on the surface charge and aggregation of silver nanoparticles suspensions. *Environ. Sci. Technol.* **2010**, *44*, 1260–1266. [CrossRef] [PubMed]
33. Hotze, E.M.; Phenrat, T.; Lowry, G.V. Nanoparticle aggregation: Challenges to understanding transport and reactivity in the environment. *J. Environ. Qual.* **2010**, *39*, 1909–1924. [CrossRef] [PubMed]
34. Agnihotri, S.; Mukherji, S.; Mukherji, S. Size-controlled silver nanoparticles synthesized over the range 5–100 nm using the same protocol and their antibacterial efficacy. *RSC Adv.* **2014**, *4*, 3974–3983. [CrossRef]
35. Paramelle, D.; Sadovoy, A.; Gorelik, S.; Free, P.; Hobley, J.; Fernig, D.G. A rapid method to estimate the concentration of citrate capped silver nanoparticles from UV-visible light spectra. *Analyst* **2014**, *139*, 4855–4861. [CrossRef] [PubMed]
36. Li, X.; Xu, H.; Chen, Z.-S.; Chen, G. Biosynthesis of nanoparticles by microorganisms and their applications. *J. Nanomater.* **2011**, *11*, 1–16. [CrossRef]
37. Kulkarni, R.R.; Shaiwale, N.S.; Deobagkar, D.N.; Deobagkar, D.D. Synthesis and extracellular accumulation of silver nanoparticles by employing radiation-resistant *Deinococcus radiodurans*, their characterization, and determination of bioactivity. *Int. J. Nanomed.* **2015**, *10*, 963–974.

38. Amini, E.; Azadfallah, M.; Layeghi, M.; Talaei-Hassanloui, R. Silver-nanoparticle-impregnated cellulose nanofiber coating for packaging paper. *Cellolose* **2016**, *23*, 557–570. [CrossRef]
39. Kaushil, M.; Moores, A. Review: Nanocelluloses as versatile supports for metal nanoparticles and their applications in catalysis. *Green Chem.* **2016**, *18*, 622–637. [CrossRef]
40. Choi, M.H.; Shim, H.E.; Yun, S.-J.; Park, S.H.; Choi, D.S.; Jang, B.-S.; Choi, Y.J.; Jeon, J. Gold-nanoparticle-immobilized desalting columns for highly efficient and specific removal of radioactive iodine in aqueous media. *ACS Appl. Mater. Interfaces* **2016**, *8*, 29227–29231. [CrossRef] [PubMed]
41. Mushtaq, S.; Yun, S.-J.; Yang, J.E.; Jeong, S.-W.; Shim, H.E.; Choi, M.H.; Park, S.H.; Choi, Y.J.; Jeon, J. Efficient and selective removal of radioactive iodine anions using engineered nanocomposite membranes. *Environ. Sci. Nano* **2017**, *4*, 2157–2163. [CrossRef]
42. Choi, M.H.; Jeong, S.-W.; Shim, H.E.; Yun, S.-J.; Mushtaq, S.; Choi, D.S.; Jang, B.-S.; Yang, J.E.; Choi, Y.J.; Jeon, J. Efficient bioremediation of radioactive iodine using biogenic gold nanomaterial-containing radiation-resistant bacterium, *Deinococcus radiodurans* R1. *Chem. Commun.* **2017**, *53*, 3937–3940. [CrossRef] [PubMed]

© 2018 by the authors. Licensee MDPI, Basel, Switzerland. This article is an open access article distributed under the terms and conditions of the Creative Commons Attribution (CC BY) license (http://creativecommons.org/licenses/by/4.0/).

Article

Fabrication and Highly Efficient Dye Removal Characterization of Beta-Cyclodextrin-Based Composite Polymer Fibers by Electrospinning

Rong Guo [1,2,3,†], Ran Wang [3,†], Juanjuan Yin [3], Tifeng Jiao [1,3,*], Haiming Huang [2], Xinmei Zhao [3], Lexin Zhang [3], Qing Li [3,*], Jingxin Zhou [3] and Qiuming Peng [1]

1. State Key Laboratory of Metastable Materials Science and Technology, Yanshan University, Qinhuangdao 066004, China; guorong@stumail.ysu.edu.cn (R.G.); pengqiuming@ysu.edu.cn (Q.P.)
2. School of Environment and Civil Engineering, Dongguan University of Technology, Dongguan 523808, China; huanghaiming52hu@163.com
3. Hebei Key Laboratory of Applied Chemistry, School of Environmental and Chemical Engineering, Yanshan University, Qinhuangdao 066004, China; wr1422520780@163.com (R.W.); jjy1729@163.com (J.Y.); zhaoxinmei2008@yeah.net (X.Z.); zhanglexin@ysu.edu.cn (L.Z.); zhoujingxin@ysu.edu.cn (J.Z.)
* Correspondence: tfjiao@ysu.edu.cn (T.J.); liqing_buct@163.com (Q.L.); Tel.: +86-335-805-6854 (T.J.)
† These authors contributed equally to this work.

Received: 30 November 2018; Accepted: 17 January 2019; Published: 20 January 2019

Abstract: Dye wastewater is one of the most important problems to be faced and solved in wastewater treatment. However, the treatment cannot be single and simple adsorption due to the complexity of dye species. In this work, we prepared novel composite fiber adsorbent materials consisting of ε-polycaprolactone (PCL) and beta-cyclodextrin-based polymer (PCD) by electrospinning. The morphological and spectral characterization demonstrated the successful preparation of a series of composite fibers with different mass ratios. The obtained fiber materials have demonstrated remarkable selective adsorption for MB and 4-aminoazobenzene solutions. The addition of a PCD component in composite fibers enhanced the mechanical strength of membranes and changed the adsorption uptake due to the cavity molecular structure via host–guest interaction. The dye removal efficiency could reach 24.1 mg/g towards 4-aminoazobenzene. Due to the admirable stability and selectivity adsorption process, the present prepared beta-cyclodextrin-based composite fibers have demonstrated potential large-scale applications in dye uptake and wastewater treatment.

Keywords: beta-cyclodextrin polymer; host–guest interaction; dye removal; wastewater treatment; electrospinning

1. Introduction

Contamination by dyes has led to many environmental problems [1–9]. In recent years, how to manage water pollution by an efficient, simple, and safe method has become a hot topic in the field of wastewater treatment research [10–12]. However, the treatment of dye wastewater is much more difficult than other kinds of wastewater due to the complexity and diversity of dye molecules [13–21]. Azo dyes are one of the most widely used dyes with chromogenic groups [22–24]. Azo dyes and their byproducts have been a focus of research attention due to their severe toxicological effects on human health: they are known to be genotoxic agents with carcinogenic properties [25–28] that may lead to birth defects [29] and food security issues [30–33]. Beyond this, dysfunction of the kidney, reproductive system, liver, brain, and central nervous system could also be exacerbated by azo dyes [34–36]. Thus, it is urgent that we learn how to properly deal with Azo dyes to achieve a safe and clean environment. Many studies have made great efforts to do this [37–40]. However, traditional technologies and methods cannot deal with the fact that different kinds of dyes need different treatments. Azo dyes are no exception. Therefore, selective chemical adsorption for different dyes has attracted wide interest. The selectivity of adsorbents is the key to the adsorption capacity and performance.

On the other hand, cyclodextrin, as a class of cyclic oligosaccharides, has a hollow circular cone with external hydrophilicity and internal hydrophobicity. This special structure has many special physical and chemical properties. It can selectively bind small organic molecules in an aqueous solution and the inclusion complexes formed have different degrees of stability [41–43]. Therefore, cyclodextrins and their derivatives are widely used in medicine, food, chemical engineering materials, and especially in wastewater treatment [44–48]. For example, Li et al. prepared a cyclodextrin-based material to remove malachite green [49]. The adsorption results fit the Langmuir model and the maximum adsorption capacity reached 91.9 mg/g. Yilmaz et al. synthesized two β-cyclodextrin-based polymers with the help of 4,4′-methylene-bis-phenyldiisocyanate (MDI) or hexamethylenediisocyanate (HMDI) [50]. These materials could remove azo dyes, as well as aromatic amines, while the dominant adsorption mechanism was host–guest interaction. Ozmen et al. synthesized three beta-cyclodextrins and a starch-based polymer using HMDI [51]. They have compared the adsorption capacity and the results showed high adsorption performance toward some azo dyes. Generally, researchers have made a lot of efforts towards dealing with azo dyes in the field of wastewater treatment. In addition, electrospinning technology demonstrated an effective method to prepare fiber materials due to its obvious advantages of simple operation and easy regulation [52–56]. At present, some cyclodextrin-based fiber systems via electrospun approach have been reported [57–60]. For examples, Cui et al. described the investigation of plasma-treated poly(ethylene oxide)-beta-cyclodextrin nanofibers to enhance the antibacterial activity [57]. Celebioglu et al. demonstrated the electrospinning of polymer-free nanofibrous structures from an inclusion complex between hydroxypropyl-beta-cydodextrin vitamin E [58]. The prepared vitamin E-contained web provided enhanced photostability for the sensitive vitamin E by the inclusion complexation even after exposure to UV light.

Based on previous reports, we have devoted our efforts to solving the increasingly serious azo dye contamination by novel selective composite fiber absorbents containing beta-cyclodextrin-based polymer (PCD) and ε-polycaprolactone (PCL). The electrospinning approach was an eco-friendly and simple preparation method. One of the indispensable advantages of the electrospinning membrane was the ultra-high specific surface area, which was extremely beneficial to adsorption. More importantly, the PCD was selected for its infinite long-chain and cavity structures. The obtained membrane contributed to the formation of more host–guest interaction due to a large number of free cyclodextrin cavities in the fiber surface. Thus, the excellent selective adsorption capability was foreseeable according to a previous report [61]. A few cyclodextrin cavities could be occupied by long-chain polymer molecules during the electrospinning progress, which is unfavorable for host–guest interactions and even results in a decrease in the undesirable adsorption effect. However, our PCL/(n%)PCD composite fibers have innumerable cavities, which could guarantee the selective adsorption capacity. Thus, it is obvious that the obtained PCL/(n%)PCD composite fibers can exhibit remarkable adsorption capacity towards azo dyes with the host–guest interaction. Moreover, the introduction of the β-cyclodextrin polymer could efficiently improve the mechanical strength and stability of the membrane. This indicated that the obtained composites have great potential to provide assistance with the problem of azo dye pollution in wastewater treatment.

2. Materials and Methods

2.1. Materials

Beta-cyclodextrin (98%, abbreviated as β-CD) and epichlorohydrin (C_3H_5ClO, 99.5%) were purchased from Aladdin Chemicals (Shanghai, China). Methylbenzene (99%), chloroform (99%), and N,N-dimethylformamide (99%, abbreviated as DMF) were obtained from Beijing Chemicals (analytical reagent grade, Beijing, China). Acetone (C_3H_6O, 99.5%) was purchased from Alfa Aesar

Chemicals (Shanghai, China). ε-Polycaprolactone (PCL average Mw~80000), methylene blue (MB) and 4-aminoazobenzene were purchased from Sinopharm Chemical Reagent Co., Ltd. (analytical reagent grade, Shanghai, China). Hydrochloric acid (HCl, 99%) and sodium hydroxide (NaOH, 99%) were obtained from Tianjin Kaitong Chemicals (Tianjin, China). Ultra-pure water was obtained using a Millipore Milli-Q water purification system with a resistivity of 18.2 MΩ·cm^{-1}. All chemicals were used as received without further purification.

2.2. Preparation of β-Cyclodextrin Polymer (PCD)

First, the used β-cyclodextrin polymer (PCD) was synthesized as in previous similar studies [62,63]. In brief, 10 g β-cyclodextrin was dissolved in 15 mL aqueous 15 wt% NaOH solution in a clean beaker, and the system was stirred by mechanical agitation for at least 24 h at 35 °C in a water bath. Subsequently, the 2 mL toluene solution was added to a beaker that was continuously stirred at 35 °C for two hours. Then we added 14.8 mL epichlorohydrin solution and the whole system was stirred for 3 h. After that the mixture system was added to 200 mL acetone solution and stirred at 50 °C overnight. The precursor was filtered and dissolved in water, before using hydrochloric acid to neutralize it. After seven days of dialysis with ultra-pure water, freeze-drying treatment at −48 °C was performed. The product, solid white PCD, was obtained and stored for further use.

2.3. Preparation of Electrospun Composite Fibers

The total mass of the electrospinning precursor was 10 g. The mixture solvent was made of chloroform and N,N-dimethylformamide with a volume ratio of 3:2. The 1.2 g ε-polycaprolactone in pellet form and 8.8 g mixture solvent were magnetically stirred for 4 h to obtain a uniform solution, in accordance with previous reports [64,65]. Through electrospinning, neat PCL fibers were obtained. In addition, a different mass of poly β-cyclodextrin powder was added to a PCL/(CCl$_4$/DMF) solution and formed a uniform spinning solution by magnetic stirring all night. During the following electrospinning, the flow rate was delivered at 1 mL·h^{-1}, while the potential difference was set to 15–30 kV and the distance was 15–30 cm from the point of the needle to the collector. By regulating the spinning conditions, we finally obtained the optimal conditions based on the analysis of SEM images. On the condition of constant content of PCL molecules in total mass, different masses of poly β-cyclodextrin (PCD) component (10, 20, 30, 40, 50 wt%) were mixed with PCL to obtain composite fibers abbreviated as PCL/(n%)PCD (n = 10, 20, 30, 40, and 50). The specific components and quantities of electrospinning solution in the different groups are shown in Table 1. All of the samples were rested in a vacuum drying oven for two days in order to volatilize the remaining solvent.

Table 1. The specific components and quantities used in electrospinning precursor solutions.

PCD Concentration (wt%)	0	10	20	30	40	50
PCL (g)	1.20	1.20	1.20	1.20	1.20	1.20
CCl$_4$ (mL)	4.17	4.11	4.02	3.92	3.78	3.60
DMF (mL)	2.78	2.74	2.68	2.61	2.52	2.40

2.4. Dye Removal Tests

The dyes methylene blue (MB) and 4-aminoazobenzene were used to estimate the adsorption properties of PCL/(n%)PCD (n = 10, 20, 30, 40, and 50) composite fibers with neat PCL fiber as the control group. UV–VIS absorption spectra were recorded for the process at wavelengths of 632 nm (MB) and 375 nm (4-aminoazobenzene) by a UV–VIS spectrometer. The freshly prepared definite samples (5 mg) were added to 50 mL dye solutions that contained MB (10 mg/L) and 4-aminoazobenzene (20 mg/L), respectively. The absorbance was measured, and corresponding concentrations and kinetic data were calculated by calibration curves. At the end of the adsorption process, all samples were washed with ethanol and DI water for several times and dried in a drying oven before further use.

In addition, the recycling capacity of PCL/(n%)PCD composites was investigated. The prepared samples were repeatedly used to remove the same fresh MB solution for eight consecutive cycles.

2.5. Characterization

The microstructures of all the obtained composite materials were characterized by a field-emission scanning electron microscopy (FE-SEM, Hitachi S-4800-II, Tokyo, Japan) with 5–15 kV accelerating voltage. Energy-dispersive X-ray spectrometry (EDXS) was utilized to distinguish the elements in membranes at an accelerating voltage of 200 kV by taking advantage of an Oxford Link-ISIS X-ray EDXS microanalysis system. In addition, themogravimetry-differential scanning calorimetry (TG-DSC) was carried out to estimate the thermal stability of samples in air with a NETZSCH STA 409 PC Luxxsi multaneous thermal analyzer (Netzsch Instruments Manufacturing Co., Ltd., Seligenstadt, Germany). FTIR spectra were measured to analyze the molecular absorption spectroscopy by a Fourier infrared spectroscopy (Thermo Nicolet Corporation, Madison, WI, USA) using the KBr tablet method. X-ray diffraction (XRD) analysis was performed on an X-ray diffractometer equipped with a Cu Kα X-ray radiation source and a Bragg diffraction setup (SMART LAB, Rigaku, Japan). Circular dichroism spectra were measured by a JASCO J-810 CD spectrometer (Jasco Inc., Easton, MD, USA). UV–VIS absorption was used to monitor the adsorption progress by a UV–VIS spectrometer (752-type, Sunny Hengping Scientific Instrument Co., Ltd., Shanghai, China) at room temperature.

3. Results and Discussion

3.1. Structural Characterization of the Composite Polymer Fibers

Firstly, we prepared a uniform PCL spinning solution by taking advantage of the CCl_4 and DMF mixed solvent. By attempting different parameters including voltage, type of stainless steel needle, distance between needle, aluminum foil, and injection rate, we finally confirmed the optimal conditions that need to be abided by in the following electrospinning. The illustration of the preparation and application in organic dyes of PCL/(n%)PCD composites is shown in Figure 1. Thus, we can get the PCL/(n%)PCD composite fibers by electrospinning under the optimal parameter conditions. In order to volatilize the excess solvent, all the samples have been put in a drying oven for two days. The adsorption capacities of a series of composite fibers were characterized by taking advantage of the methylene blue (MB) and 4-aminoazobenzene solution.

The optimal conditions of electrospinning were obtained under different mixed ratios of PCD, after attempts and characterization of different parameters. Based on this, the representative micromorphology images by SEM of neat PCL fibers and PCL/(n%)PCD composite fibers are depicted in Figure 2. The neat PCL fibers showed a homogeneous solid fiber structure, and multiple layers of fibers were stacked together in the form of an electrospun membrane. The physical properties of the spinning precursor solution changed through different ratios of PCD power. Therefore, the parameters of the electrospinning have also changed. After multiple trials and adjustment, the ideal conditions for fiber formation were obtained. The micromorphology pictures of fibers with different ratios of PCD, that is, PCL/(10%)PCD, PCL/(20%)PCD, PCL/(30%)PCD, PCL/(40%)PCD, and PCL/(50%)PCD, can be seen in Figure 2. The diameter of neat PCL fibers appeared at a centered position of 500–600 nm with the length in microns. In addition, with the addition of a PCD component in fibers, the fiber diameters decreased and reached a centered range of 200–400 nm for the obtained PCL/(40%)PCD composite fiber. A possible reason for diameter decrement was that the strong network between the neighboring chains could be temporarily destroyed during the electrospinning process, which enhanced the stretching of the jet [66]. As for the PCL/(50%)PCD composite fiber, more cross-linking fibers could be clearly observed, mainly due to the increasing viscosity of the precursor solution. Moreover, when the content of PCD in precursor solution exceeded 50%, the electrospun needle would clog and could not obtain continuous electrospun fibers.

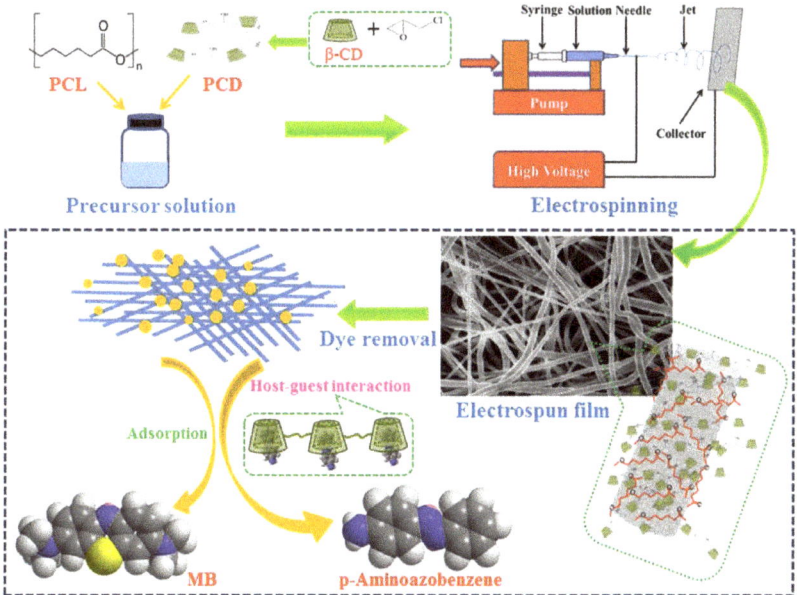

Figure 1. Schematic illustration of preparation and application in dye removal of PCL/(n%)PCD composite fibers by electrospinning.

It was well known that the thermal stability of composite materials is an important factor in their characterization and wider application. The thermal stabilities of the obtained neat PCL fiber and a series of PCL/(n%)PCD fiber composites were investigated by the thermogravimetry (TG) curves as shown in Figure 3. In the N_2 condition, all the samples were heated from room temperature to 800 °C through a temperature-programmed route. Before the temperature reached 300 °C, the thermogravimetric curves of PCL fiber remained stable and there was no significant weightlessness. The one-stage degradation was related to the decomposition of the carbon skeleton from 372 °C to 457 °C. The final weight loss was approximately 79.8 wt% at 800 °C. In the cases of present PCL/(n%)PCD composites, weight loss below 150 °C could be considered as removal of trace moisture vapor adsorbed by PCD fibers and/or a small amount of crystal water entrapped by PCD cavities. After that, the thermal degradation from 307 °C to about 370 °C corresponds to PCD molecules. This is followed by the degradation of neat PCL fibers from about 370 °C to 450 °C. With the increment of the PCD component, the initial degradation temperature and the final mass of residue of PCL/(10%)PCD obviously dropped to 17.4 wt%, while the other PCL/(n%)PCD composites was about 3 wt%. It was obvious that the weight loss originated from decomposition of PCL and PCD components.

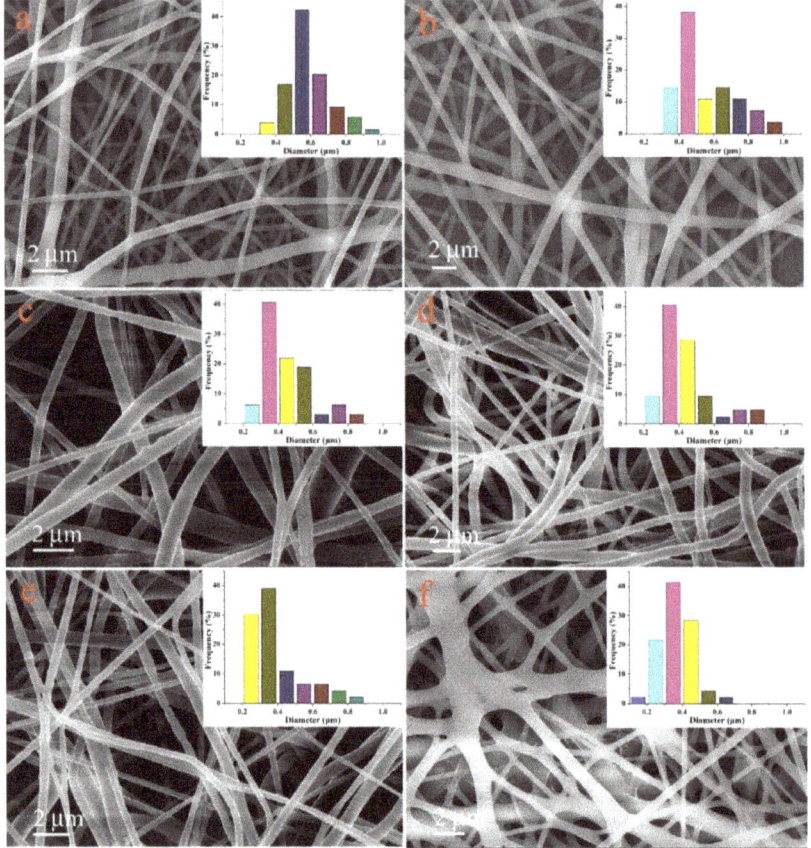

Figure 2. SEM images and diameter distribution histograms of neat PCL fibers (**a**), PCL/(10%)PCD (**b**), PCL/(20%)PCD (**c**), PCL/(30%)PCD (**d**), PCL/(40%)PCD (**e**), and PCL/(50%)PCD (**f**).

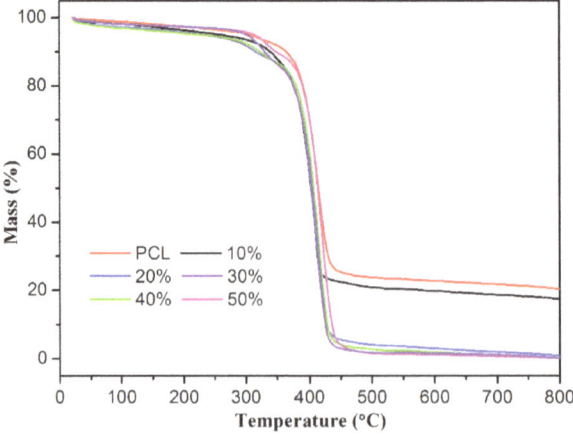

Figure 3. TG curves of neat PCL fibers and composite fiber composites.

XRD data are frequently used to characterize and confirm the presence of both components. As seen in Figure 4, the two diffraction peaks of neat PCL fibers appeared at 2θ values of 22° and 24° can be indexed to (110) and (200) reflections, which confirms its orthorhombic crystal structure. In addition, PCD powder elicits many small and cluttered characteristic peaks, which can be attributed to the cage structure of native beta-CD. It was apparent that all samples of obtained PCL/(n%)PCD composite fibers show the same characteristic peak and only have two components without any other impurities. However, slightly broadened characteristic peaks of PCL molecules in PCL/(n%)PCD fibers can be clearly observed. We also noted that the peak shifted slightly to the right. Such results can imply that there are some interactions between PCL and PCD molecules. The content of PCD incorporated into PCL fibers was little in our work. Therefore, it is not sufficient to cause the obvious XRD spectral characteristic peak change of PCL/(n%)PCD composite fibers. XRD results complement the TG findings and indicate the presence of a physical mixture in the obtained composites as well.

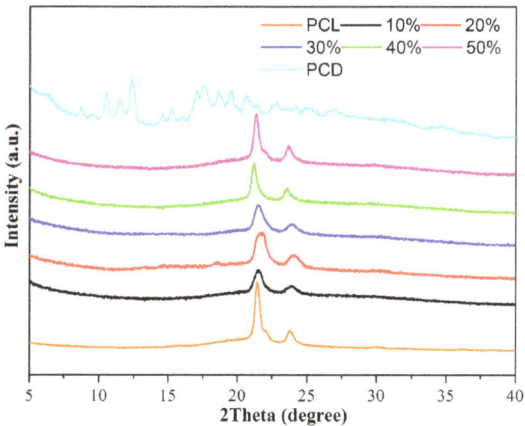

Figure 4. XRD patterns of the prepared neat PCL fibers and composite fiber composites.

FT-IR analyses were performed for conclusive evidence, and the results are shown in Figure 5. The conformational changes of the PCL fibers and PCL/(n%)PCD fibers were characterized by Fourier transform infrared spectroscopy. It can be seen that the characteristic peaks of pure PCL spectra are mainly typical ester bonds and hydrocarbon bonds. The additional peaks at 1728 cm^{-1}, 1243 cm^{-1}, and 1046 cm^{-1} represent the vibration peaks of C=O, C–O–C, and C–C groups. In addition, the strong wide peak at 3355 cm^{-1} can be ascribed to the association of hydrogen bonds formed by the –OH group. In addition, the absorption peak at 1033 cm^{-1} represents C–O–C and C–O stretching vibration of the beta-CD cross-linked polymer cavity. In addition, composite fibers of PCL with increasing content of PCD addition (0, 10, 20, 30, 40, and 50 wt%) can show similar changes of type and position of characteristic peak. Based on the above, the designed PCL/(n%)PCD composite fiber samples were successfully synthesized. In addition, the microstructures of the obtained PCL fiber and PCL/(n%)PCD composite fibers were investigated using N$_2$ adsorption–desorption isotherms. The obtained properties of the samples were generalized in Table 2. It could be clearly observed that the as-obtained PCL fibers showed a specific surface area of 7.50 m^2·g^{-1}. In addition, with the increment of the PCD component in the composite fibers, the values of specific surface area obviously increased and reached 11.52 m^2·g^{-1} for PCL/(50%)PCD composite fibers, demonstrating the formation of more anchoring sites facilitating the next adsorption of dye molecules. Meanwhile, the pore size and pore volume of all samples were calculated via BJH methods. The obtained PCL/(50%)PCD composite fibers also exhibited enhanced pore size and pore volume, meaning that larger pore diameters and pore volumes in composite fibers could demonstrate lots of micro/nanoscale channels, thereby making them effective for the next adsorption experiment.

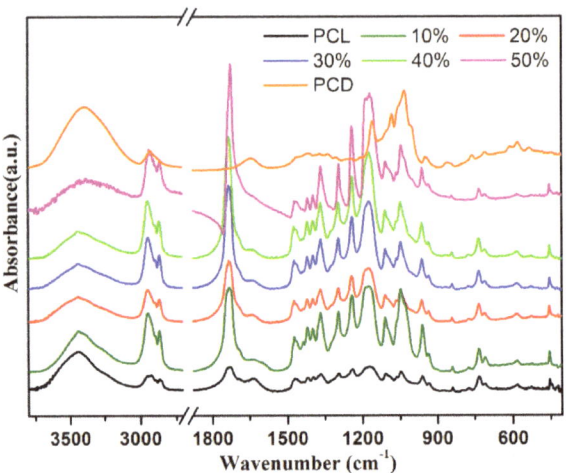

Figure 5. FT-IR spectra of neat PCL fibers and PCL/(n%)PCD (n = 10, 20, 30, 40, 50) composites.

Table 2. Physical data of as-prepared composite fibers.

Samples	Specific Surface Area (m^2·g^{-1})	Average Pore Diameter (nm)	Pore Volume (cm^3·g^{-1})
PCL fiber	7.50	20.1	0.008231
PCL/(10%)PCD	7.82	20.5	0.008464
PCL/(20%)PCD	7.96	21.0	0.008576
PCL/(30%)PCD	8.45	22.5	0.009125
PCL/(40%)PCD	10.85	23.6	0.009277
PCL/(50%)PCD	11.52	23.6	0.009446

The tensile properties and stress–strain plots of several PCL/(n%)PCD composites were conducted at room temperature with neat PCL fibers as a control for comparison, as shown in Figure 6. It is evident that the neat PCL fiber has a high elongation at break (above 470%). Correspondingly, the elongation at break of our functionalized PCL/(n%)PCD samples has decreased markedly with the increase in PCD content. The elongation of PCL/(10%)PCD fibers at break was 230%, while the value of PCL/(50%)PCD fibers was only 74%. The elongation at break of PCL/(20%)PCD, PCL/(30%)PCD and PCL/(40%)PCD was 171%, 150%, and 127%, respectively. Compare with neat PCL fibers, the elongation at break eventually declined sharply by six times. In addition, the ultimate tensile strength of neat PCL was 2.25 MPa. In hybridization cases of PCL/(n%)PCD samples, the fracture stress presented a trend of first increasing, then decreasing, and next increasing along with the change of content of PCD. It could be seen that the fracture stress generally increased and the final PCL/(50%)PCD composites were destroyed when the fracture stress reached 3.41 MPa. Clearly, the introduction of PCD significantly weakened the elongation at break but prominently increased the ultimate tensile strengths of present composite fibers. This change phenomena seemed similar to previous report about electrospun composite poly(ethylene glycol)/poly(caprolactone) nanofibrous membrane [67]. One explanation could be that the content of PCD increased with the decrease of solvents, resulting in the presence of hard segments or clusters. It could be clearly seen that the obtained PCL/(n%)PCD composite materials showed good tensile strength, with PCD component playing a key role in improving it [68,69].

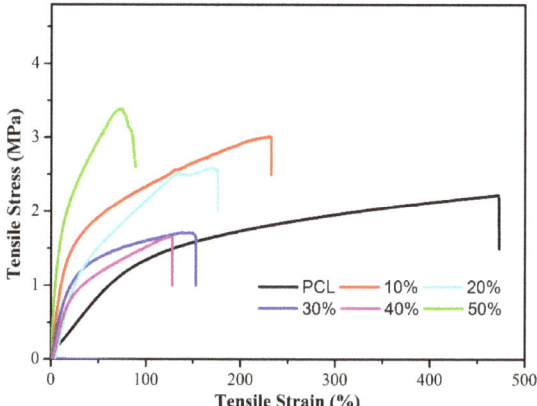

Figure 6. Stress–strain plots of neat PCL fibers and PCL/(n%)PCD (n = 10, 20, 30, 40, 50) composites.

3.2. Dye Removal Performance of the Composite Fibers

In order to characterize the selective dye removal performances of the present obtained composite fiber absorbents, the obtained PCL/(n%)PCD composite fibers were investigated with respect to their uptake of removing MB and 4-aminoazobenzene (AA) as typical models [70,71]. The mechanism of removing organic dye molecules was adsorption and host–guest interaction for MB and 4-aminoazobenzene, respectively. In addition, neat PCL was used as a control group and the whole adsorption progress was monitored by taking advantage of the UV–VIS spectra. The adsorption properties of PCL/(n%)PCD composites toward MB and 4-aminoazobenzene are shown in Figure 7. Clearly, the PCL/(n%)PCD samples showed better dye uptake than neat PCL toward the two organic dyes. However, the adsorption uptake of PCL/(n%)PCD composite fibers became better with the increment of the content of PCD compared with PCL. The adsorption kinetics data distinctly demonstrated the above view. The pseudo-first-order model and pseudo-second-order model adsorption equations were used to further evaluate adsorption kinetics by fitting the experimental data. All the fitted results are summarized in Table 3.

The pseudo-first-order model can be demonstrated by Equation (1) [53]:

$$\log(q_e - q_t) = \log q_e - \frac{k_1}{2.303} t \tag{1}$$

where t is the adsorption time, q_e is the adsorption capacity at equilibrium, k_1 is the pseudo-first-order model rate constant, and q_t is the adsorption capacity at time t.

The pseudo-second-order model can be demonstrated by Equation (2) [72]:

$$\frac{t}{q_t} = \frac{1}{k_2 q_e^2} + \frac{t}{q_e} \tag{2}$$

where q_e is the adsorption uptake at equilibrium, k_2 is the pseudo-second-model rate constant, and q_t is the adsorption uptake at time t.

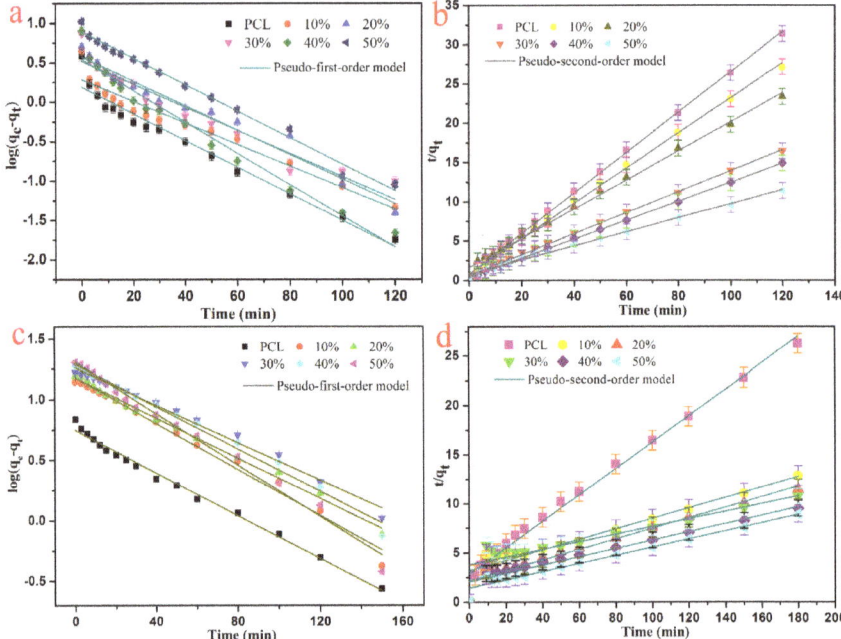

Figure 7. Kinetic adsorption of MB (**a,b**) and 4-aminoazobenzene (**c,d**).

Table 3. The fitting results achieved for kinetic adsorption data using the pseudo-first-order model and pseudo-second-order model equation.

MB	Pseudo-First-Order Model			Pseudo-Second-Order Model		
	q_e (mg/g)	R^2	K_1 (min^{-1})	q_e (mg/g)	R^2	K_2 (g/mg min)
PCL	3.8246	0.9337	0.0393	3.8879	0.9989	0.0811
PCL/(10%)PCD	4.4238	0.9129	0.0317	4.4767	0.9979	0.0568
PCL/(20%)PCD	5.1414	0.9498	0.0337	5.4077	0.9936	0.0206
PCL/(30%)PCD	7.3002	0.9404	0.0376	7.4862	0.9980	0.0303
PCL/(40%)PCD	8.0465	0.9398	0.0550	8.2740	0.9991	0.0373
PCL/(50%)PCD	10.5238	0.9837	0.0399	11.1632	0.9931	0.0102
4-aminoazobenzene	Pseudo-first-order model			Pseudo-second-order model		
	q_e (mg/g)	R^2	K_1 (min^{-1})	q_e (mg/g)	R^2	K_2 (g/mg min)
PCL	6.8517	0.9916	0.0203	7.4427	0.9868	0.0064
PCL/(10%)PCD	13.9828	0.9872	0.0218	18.7512	0.9981	0.0090
PCL/(20%)PCD	15.6643	0.9941	0.0191	18.6324	0.9931	0.0014
PCL/(30%)PCD	16.7213	0.9986	0.0178	25.7070	0.9986	0.0004
PCL/(40%)PCD	18.9147	0.9895	0.0201	23.6183	0.9978	0.0009
PCL/(50%)PCD	20.1740	0.9870	0.0244	24.0674	0.9943	0.0012

In the case of the MB solution, the pseudo-second-order model had a higher correlation coefficient ($R^2 > 0.99$) than the pseudo-first-order model ($R^2 > 0.93$). The obtained values of adsorption uptake were almost equal to those fitted from the pseudo-second-order model. In addition, neat PCL fibers showed low adsorption uptake, while the PCL/(n%)PCD samples all greatly improved, as shown in Figure 7 and Table 2. The dye removal efficiency of neat PCL fibers only reached 3.8246 mg/g. With the increment of PCD content, the adsorption uptake of composite fibers enhanced significantly, from 4.4238 mg/g to 10.5238 mg/g. The composite fibers showed a good adsorption performance for MB mainly due to the following two reasons. Firstly, PCL, as the most basic component in electrospun fibers, has almost no efficient groups for adsorption. Secondly, the unique cavity structure of PCD has great potential to identify and select organic molecules. Thus, MB molecules could not enter into the cavity of PCD. Only a small number of hydroxyl groups could form hydrogen bonds to facilitate the adsorption of the MB dye solution. However, we could still come to the conclusion that the introduction of PCD was conducive to adsorption.

By contrast, PCL/(n%)PCD samples exhibited outstanding adsorption uptake to 4-aminoazobenzene. It was clearly observed that the pseudo-second-order ($R^2 > 0.99$) was more accurate than the pseudo-first-order model. Thus, the adsorption progress was more consistent with the pseudo-second-order dynamic model. In Table 3, the dye removal efficiency of PCL/(10%)PCD could reach 18.7512 mg/g, which was twice as high as the PCL fibers' adsorption uptake. While more and more PCD molecules were added and the proportion in spinning membranes was increasing, the dye removal efficiency of the obtained composite fibers increased continuously and rapidly. The PCL/(50%)PCD composites finally reached 24.0674 mg/g. Apparently, the adsorption uptake of PCL/(n%)PCD toward 4-aminoazobenzene had appreciable performance due to the addition of host–guest inclusion complexation. The 4-aminoazobenzene molecules could be included in the cavity structure of PCD through host–guest interaction and/or being bound by the fiber surface through electrostatic interaction and hydrogen bonds. The formed self-assembled structures were stable and highly efficient.

In order to further prove that the driving force of removal mechanism relative to 4-aminoazobenzene consisted of host–guest interaction, we collected the data of UV–VIS and circular dichroism spectra to characterize the PCL/(50%)PCD membrane before and after the adsorption process, as shown in Figure 8. Obviously, as-prepared PCL/(50%)PCD membrane have no significant characteristic peak in Figure 8a. After adsorption of 4-aminoazobenzene, the maximum absorption peak of the composites appears at 400 nm, which was attributed to π–π* electron transition of the 4-aminoazobenzene group [73–77]. However, the additional characteristic peak of 4-aminoazobenzene was at 370 nm. Therefore, the peak position of PCL/(50%)PCD membrane had a red shift after adsorption, which could result from an interaction between the hydroxyl of PCD and the chromogenic group of 4-aminoazobenzene. The circular dichroism spectra of PCL/(50%)PCD membrane have shown similar results. The intensity of the signal was notable at 400 nm, with one positive Cotton effect in Figure 8b. In addition, the images of SEM with C/O/N elemental mapping of PCL/(50%)PCD composite fibers after adsorption of 4-aminoazobenzene have also been measured and are shown in Figure 9. Obviously, a large quantity of N element was well distributed onto the obtained fibers (Figure 9d), which further confirmed the presence and the good distribution of 4-aminoazobenzene in the obtained composite fiber. It could be speculated that hydrophilic 4-aminoazobenzene molecules was successfully anchored on the surface of PCL/(50%)PCD fibers by intermolecular host–guest interaction and/or electrostatic interaction/hydrogen bonding, which could be expected to exert adsorption activity and good stability in the next recovery and reuse process. Thus, combined with UV–VIS spectra, the presence of a 4-aminoazobenzene group in the obtained composite materials was further confirmed. So, it could be considered that the host–guest reaction occurred and the 4-aminoazobenzene moiety was located inside the cavity of PCD molecules via host–guest interaction and/or anchored the surface of fiber via electrostatic interaction/hydrogen bonding. In addition, it should be noted that the signal intensity of the circular dichroism spectra was lower, which could be mainly due to two reasons. Firstly, partial PCD did not have enough contact with 4-aminoazobenzene to form an inclusion complex because some cavities of PCD were inside the fibers. Secondly, partial 4-aminoazobenzene molecules only stayed on the surface of the membrane due to intermolecular hydrogen bonding.

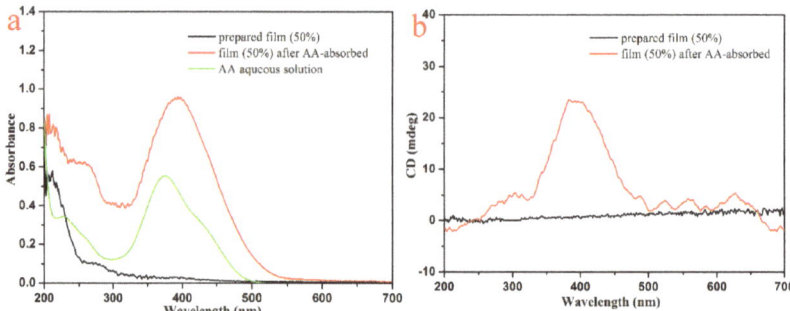

Figure 8. UV–VIS (**a**) and circular dichroism spectra (**b**) of the PCL/(50%)PCD composite fibers.

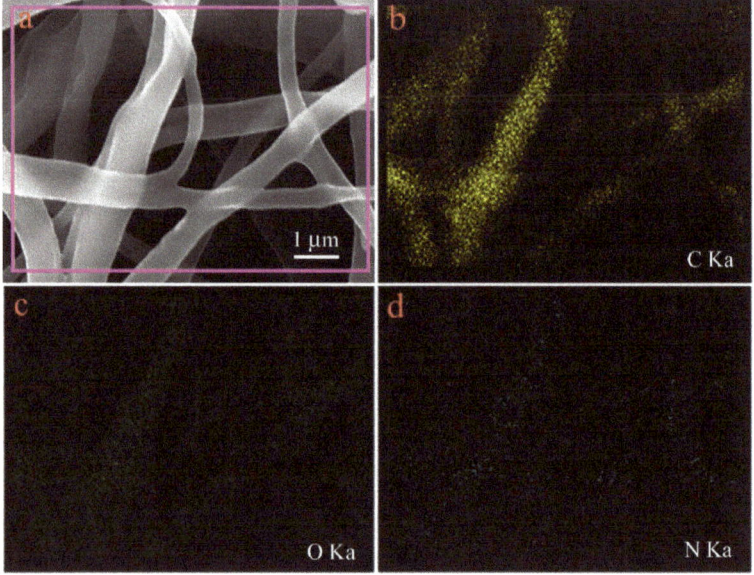

Figure 9. SEM image (**a**) with C/O/N elemental mapping (**b**–**d**) of AA-absorbed PCL/(50%)PCD composite fibers.

Adsorption reutilization research into PCL/(50%)PCD's removal of MB was conducted and the results are shown in Figure 10, with graphic illustration of used nanocomposite after eight reutilization cycles. The obtained PCL/(n%)PCD composite fibers could be recovered and reused by a simple washing and drying process. The membrane became blue after the adsorption of MB molecules, followed by desorption in ethanol. Rapid and simple desorption and regeneration created favorable conditions for reutilization. In the case of PCL/(50%)PCD, the removal of MB dye could still reach 78% after repeated adsorption over eight cycles. The loss of adsorption may result from a slightly deformed fiber structure and a few cavities of PCD molecules being occupied. All in all, the composites consisting of PCL and PCD molecules have good stability in the field of adsorption of organic dyes. It should be noted that the obtained PCL/(n%)PCD composites can hardly exhibit outstanding adsorption capacity in a three-dimensional matrix such as a hydrogel structure. However, the present PCL/(n%)PCD composites have demonstrated a capacity for selectivity adsorption and excellent stability towards present two kinds of dyes. In addition, the electrospinning nanocomposites formed by poly β-cyclodextrin have relatively more cavities than β-cyclodextrin monomer molecules.

Thus, generous cavities were highly beneficial to the field of selectivity adsorption. Moreover, the introduction of PCD molecules made the composites have better mechanical strength, which also contributed to the improvement of stability. All in all, the PCL/(n%)PCD electrospinning membranes not only have remarkable selectivity adsorption, but also admirable stability, which is important for the prospect of industrialization in the field of wastewater treatment and self-assembled nanomaterials [78–80].

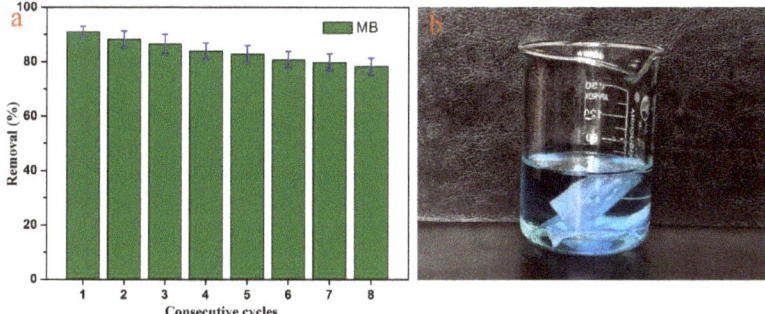

Figure 10. Relative adsorption reutilization studies towards removal of MB for different consecutive cycles at room temperature of PCL/(50%)PCD composites (**a**) and the photograph of PCL/(50%)PCD composite after eight reutilization cycles (**b**).

4. Conclusions

In summary, we have successfully prepared electrospun PCL/(n%)PCD (n = 10, 20, 30, 40, 50) composite fiber materials via a simple and low-cost method. It could be seen that the prepared PCL/(n%)PCD composites showed uniform fiber nanostructures and had been well characterized. According to the strain–stress plots, PCD molecules were advantageous to improve the mechanical strength of the obtained fiber films. In addition, the introduction of PCD led to excellent uptake of selectivity adsorption in the obtained electrospun composite films with a high specific surface area. In addition, the improvement of mechanical strength also enhances the stability of electrospinning membranes. This research work has proposed a new design of electrospun composites with a cyclodextrin component and suggested new possibilities in selective adsorption for dye removal.

Author Contributions: T.J. and Q.L. conceived and designed the experiments; R.G., R.W., and J.Y. performed the experiments; L.Z., H.H., X.Z., and J.Z. analyzed the data; L.Z., H.H., and Q.P. contributed reagents/materials/analysis tools; R.G., R.W., and T.J. wrote and revised the paper.

Funding: This research was funded by the National Natural Science Foundation of China (nos. 21872119, 21473153), the Support Program for the Top Young Talents of Hebei Province, the China Postdoctoral Science Foundation (no. 2015M580214), the Research Program of the College Science and Technology of Hebei Province (no. ZD2018091), and the Scientific and Technological Research and Development Program of Qinhuangdao City (no. 201701B004).

Conflicts of Interest: The authors declare no conflict of interest.

References

1. De Oliveira, G.A.R.; de Lapuente, J.; Teixidó, E.; Porredón, C.; Borràs, M.; de Oliveira, D.P. Textile dyes induce toxicity on zebrafish early life stages. *Environ. Toxicol. Chem.* **2016**, *35*, 429–434. [CrossRef] [PubMed]
2. Chong, M.N.; Jin, B.; Chow, C.W.; Saint, C. Recent developments in photocatalytic water treatment technology: A review. *Water Res.* **2010**, *44*, 2997–3027. [CrossRef] [PubMed]
3. Oliveira, G.A.R.; Leme, D.M.; de Lapuente, J.; Brito, L.B.; Porredón, C.; Rodrigues, L.B.; Brull, N.; Serret, J.T.; Borras, M.; Disner, G.R.; et al. A test battery for assessing the ecotoxic effects of textile dyes. *Chem. Biol. Interact.* **2018**, *291*, 171–179. [CrossRef] [PubMed]

4. Li, D.; Li, Q.; Mao, D.; Bai, N.; Dong, H. A versatile bio-based material for efficiently removing toxic dyes, heavy metal ions and emulsified oil droplets from water simultaneously. *Bioresour. Technol.* **2017**, *245*, 649–655. [CrossRef] [PubMed]
5. Zhu, W.; Li, Y.; Dai, L.; Li, J.W.; Li, X.; Li, W. Bioassembly of fungal hyphae/carbon nanotubes composite as a versatile adsorbent for water pollution control. *Chem. Eng. J.* **2018**, *339*, 214–222. [CrossRef]
6. Zhao, H.; Jiao, T.F.; Zhang, L.X.; Zhou, J.X.; Zhang, Q.R.; Peng, Q.M.; Yan, X.H. Preparation and adsorption capacity evaluation of graphene oxide-chitosan composite hydrogels. *Sci. China Mater.* **2015**, *58*, 811–818. [CrossRef]
7. Li, K.K.; Jiao, T.F.; Xing, R.R.; Zou, G.Y.; Zhou, J.X.; Zhang, L.X.; Peng, Q.M. Fabrication of tunable hierarchical MXene@AuNPs nanocomposites constructed by self-reduction reactions with enhanced catalytic performances. *Sci. China Mater.* **2018**, *61*, 728–736. [CrossRef]
8. Guo, R.; Jiao, T.F.; Li, R.F.; Chen, Y.; Guo, W.C.; Zhang, L.X.; Zhou, J.X.; Zhang, Q.R.; Peng, Q.M. Sandwiched Fe_3O_4/carboxylate graphene oxide nanostructures constructed by layer-by-layer assembly for highly efficient and magnetically recyclable dye removal. *ACS Sustain. Chem. Eng.* **2018**, *6*, 1279–1288. [CrossRef]
9. Zhao, X.; Ma, K.; Jiao, T.; Xing, R.; Ma, X.; Hu, J.; Huang, H.; Zhang, L.; Yan, X. Fabrication of hierarchical layer-by-layer assembled diamond based core-shell nanocomposites as highly efficient dye absorbents for wastewater treatment. *Sci. Rep.* **2017**, *7*, 44076. [CrossRef]
10. Guo, H.; Jiao, T.; Zhang, Q.; Guo, W.; Peng, Q.; Yan, X. Preparation of graphene oxide-based hydrogels as efficient dye adsorbents for wastewater treatment. *Nanoscale Res. Lett.* **2015**, *10*, 272. [CrossRef]
11. Sun, S.; Wang, C.; Han, S.; Jiao, T.; Wang, R.; Yin, J.; Li, Q.; Wang, Y.; Geng, L.; Yu, X.; et al. Interfacial nanostructures and acidichromism behaviors in self-assembled terpyridine derivatives Langmuir-Blodgett films. *Colloid Surf. A Physicochem. Eng. Asp.* **2019**, *564*, 1–9. [CrossRef]
12. Xu, Y.; Ren, B.; Wang, R.; Zhang, L.; Jiao, T.; Liu, Z. Facile preparation of rod-like MnO nanomixtures via hydrothermal approach and highly efficient removal of methylene blue for wastewater Treatment. *Nanomaterials* **2019**, *9*, 10. [CrossRef] [PubMed]
13. Chen, K.; Li, J.; Zhang, L.; Xing, R.; Jiao, T.; Gao, F.; Peng, Q. Facile synthesis of self-assembled carbon nanotubes/dye composite films for sensitive electrochemical determination of Cd(II) ions. *Nanotechnology* **2018**, *29*, 445603. [CrossRef] [PubMed]
14. Paul, D. Research on heavy metal pollution of river Ganga: A review. *Ann. Agrar. Sci.* **2017**, *15*, 278–286. [CrossRef]
15. Dargo, H.; Gabbiye, N.; Ayalew, A. Removal of methylene blue dye from textile wastewater using activated carbon prepared from rice husk. *Can. Entomol.* **2014**, *9*, 317–325.
16. Dotto, G.L.; Santos, J.M.N.; Tanabe, E.H.; Bertuol, D.A.; Foletto, E.L.; Lima, E.C.; Pavan, F.A. Chitosan/polyamide nanofibers prepared by Forcespinning® technology: A new adsorbent to remove anionic dyes from aqueous solutions. *J. Clean Prod.* **2017**, *144*, 120–129. [CrossRef]
17. Zhang, D.; Zeng, F. Synthesis of an Ag-ZnO nanocomposite catalyst for visible light-assisted degradation of a textile dye in aqueous solution. *Res. Chem. Intermed.* **2010**, *36*, 1055–1063. [CrossRef]
18. Fei, B.L.; Zhong, J.K.; Deng, N.P.; Wang, J.H.; Liu, Q.B.; Li, Y.G.; Mei, X. A novel 3D heteropoly blue type photo-Fenton-like catalyst and its ability to remove dye pollution. *Chemosphere* **2018**, *197*, 241–250. [CrossRef]
19. Fei, B.L.; Deng, N.P.; Wang, J.H.; Liu, Q.B.; Lang, G.Y.; Li, Y.G.; Mei, X. A heteropoly blue as environmental friendly material: An excellent heterogeneous Fenton-like catalyst and flocculent. *J. Hazard. Mater.* **2017**, *340*, 326–335. [CrossRef]
20. Jin, X.; Li, N.; Weng, X.; Li, C.; Chen, Z. Green reduction of graphene oxide using eucalyptus leaf extract and its application to remove dye. *Chemosphere* **2018**, *208*, 417–424. [CrossRef]
21. Zhang, J.; Dang, L.; Zhang, M.; Zhao, S.; Lu, Q. Micro/nanoscale magnesium silicate with hierarchical structure for removing dyes in water. *Mater. Lett.* **2017**, *196*, 194–197. [CrossRef]
22. Farrokhi, M.; Hosseini, S.C.; Yang, J.K.; Shirzadsiboni, M. Application of $ZnO-Fe_3O_4$ nanocomposite on the removal of azo dye from aqueous solutions: kinetics and equilibrium studies. *Water Air Soil Pollut.* **2014**, *225*, 2113. [CrossRef]
23. Ince, N.H.; Stefan, M.I.; Bolton, J.R. UV/H_2O_2 degradation and toxicity reduction of textile azo dyes: remazol black-B, a case study. *J. Adv. Oxid. Technol.* **2017**, *2*, 442–448. [CrossRef]
24. Harisha, S.; Keshavayya, J.; Swamy, B.E.K.; Viswanath, C.C. Synthesis, characterization and electrochemical studies of azo dyes derived from barbituric acid. *Dyes Pigment.* **2017**, *136*, 742–753. [CrossRef]

25. Levine, W.G. Metabolism of azo dyes: Implication for detoxication and activation. *Drug Metab. Rev.* **1991**, *23*, 253–309. [CrossRef] [PubMed]
26. Gupta, V.K.; Jain, R.; Mittal, A.; Saleh, T.A.; Nayak, A.; Agarwal, S.; Sikarwas, S. Photo-catalytic degradation of toxic dye amaranth on TiO_2/UV in aqueous suspensions. *Mater. Sci. Eng. C Mater.* **2012**, *32*, 12–17. [CrossRef] [PubMed]
27. Koutsogeorgopoulou, L.; Maravelias, C.; Methenitou, G.; Loutselinis, A. Immunological aspects of the common food colorants, amaranth and tartrazine. *Vet. Hum. Toxicol.* **1998**, *40*, 1–4.
28. Kadirvelu, K.; Palanival, M.; Kalpana, R.; Rajeswari, S. Activated carbon from an agricultural by-product, for the treatment of dyeing industry wastewater. *Bioresour. Technol.* **2000**, *74*, 263–265. [CrossRef]
29. Bantle, J.A.; Fort, D.J.; Rayburn, J.R.; Deyoung, D.J.; Bush, S.J. Further validation of FETAX: Evaluation of the developmental toxicity of five known mammalian teratogens and non-teratogens. *Drug Chem. Toxicol.* **1990**, *13*, 267–282. [CrossRef]
30. Mittal, A. Adsorption kinetics of removal of a toxic dye, Malachite Green, from wastewater by using hen feathers. *J. Hazard. Mater.* **2006**, *133*, 196–202. [CrossRef]
31. Sweeney, E.A.; Chipman, J.K.; Forsythe, S.J. Evidence for Direct-Acting Oxidative Genotoxicity by Reduction Products of Azo Dyes. *Environ. Health Perspect.* **1994**, *102*, 119–122. [PubMed]
32. Gottlieb, A.; Shaw, C.; Smith, A.; Wheatley, A.; Forsythe, S. The toxicity of textile reactive azo dyes after hydrolysis and decolourisation. *J. Biotechnol.* **2003**, *101*, 49–56. [CrossRef]
33. Papic, S.; Koprivanac, N.; Metes, A. Optimizing Polymer-Induced Flocculation Process to Remove Reactive Dyes from Wastewater. *Environ. Technol. Lett.* **2000**, *21*, 97–105. [CrossRef]
34. Kadirvelu, K.; Kavipriya, M.; Karthika, C.; Radhika, M.; Vennilamani, N.; Pattabhi, S. Utilization of various agricultural wastes for activated carbon preparation and application for the removal of dyes and metal ions from aqueous solutions. *Bioresour. Technol.* **2003**, *87*, 129–132. [CrossRef]
35. Shen, D.Z.; Fan, J.; Zhou, W.; Gao, B.Y.; Yue, Q.Y.; Qi, K. Adsorption kinetics and isotherm of anionic dyes onto organo-bentonite from single and multisolute systems. *J. Hazard. Mater.* **2009**, *172*, 99–107. [CrossRef] [PubMed]
36. Wang, C.; Yin, J.; Wang, R.; Jiao, T.; Huang, H.; Zhou, J.; Zhang, L.; Peng, Q. Facile preparation of self-assembled polydopamine-modified electrospun fibers for highly effective removal of organic dyes. *Nanomaterials* **2019**, *9*, 116. [CrossRef]
37. Zheng, Y.; Yu, S.; Shuai, S.; Zhou, Q.; Chen, Q.; Liu, M.; Gao, C.J. Color removal and COD reduction of biologically treated textile effluent through submerged filtration using hollow fiber nanofiltration membrane. *Desalination* **2013**, *314*, 89–95. [CrossRef]
38. Balapure, K.; Bhatt, N.; Madamwar, D. Mineralization of reactive azo dyes present in simulated textile waste water using down flow microaerophilic fixed film bioreactor. *Bioresour. Technol.* **2015**, *175*, 1–7. [CrossRef]
39. Konicki, W.; Aleksandrzak, M.; Moszyński, D.; Mijowsika, E. Adsorption of anionic azo-dyes from aqueous solutions onto graphene oxide: Equilibrium, kinetic and thermodynamic studies. *J. Colloid Interface Sci.* **2017**, *496*, 188–200. [CrossRef] [PubMed]
40. Ferreira, G.M.D.; Hespanhol, M.C.; Rezende, J.D.P.; Pires, A.C.D.S.; Gurgel, L.V.A.; Silva, L.H.M. Adsorption of red azo dyes on multi-walled carbon nanotubes and activated carbon: A thermodynamic study. *Colloid Surf. A* **2017**, *529*, 531–540. [CrossRef]
41. Li, L.; Guo, X.; Fu, L.; Prudhomme, R.K.; Lincoln, S.F. Complexation Behavior of α-, β-, and γ-Cyclodextrin in Modulating and Constructing Polymer Networks. *Langmuir* **2008**, *24*, 8290–8296. [CrossRef] [PubMed]
42. Guo, X.; Wang, J.; Li, L.; Chen, Q.; Li, Z.; Pham, D.T.; May, B.L.; Prud'homme, R.K.; Easton, C.J. Tunable polymeric hydrogels assembled by competitive complexation between cyclodextrin dimers and adamantyl substituted poly(acrylate)s. *AIChE J.* **2010**, *56*, 3021–3024. [CrossRef]
43. Wang, J.; Pham, D.T.; Guo, X.; Li, L.; Pham, D.T.; Luo, Z.F.; Ke, H.L.; Zheng, L.; Prud'homme, R.K. Polymeric Networks Assembled by Adamantyl and β-Cyclodextrin Substituted Poly(acrylate)s: Host-Guest Interactions, and the Effects of Ionic Strength and Extent of Substitution. *Ind. Eng. Chem. Res.* **2009**, *49*, 609–612. [CrossRef]
44. Theron, J.; Walker, J.A.; Cloete, T.E. Nanotechnology and Water Treatment: Applications and Emerging Opportunities. *Criti. Rev. Microbiol.* **2008**, *34*, 43–69. [CrossRef] [PubMed]

45. Leudjo, T.A.; Pillay, K.; Yangkou, M.X. Nanosponge cyclodextrin polyurethanes and their modification with nanomaterials for the removal of pollutants from waste water: A review. *Carbohyd. Polym.* **2017**, *159*, 94–107. [CrossRef] [PubMed]
46. Khaoulani, S.; Chaker, H.; Cadet, C.; Bychkov, E.; Cherifaouali, L.; Bengueddach, A.; Fourmentin, S. Wastewater treatment by cyclodextrin polymers and noble metal/mesoporous TiO_2 photocatalysts. *C. R. Chim.* **2015**, *18*, 23–31. [CrossRef]
47. Yamasaki, H.; Makihata, Y.; Fukunaga, K. Efficient phenol removal of wastewater from phenolic resin plants using crosslinked cyclodextrin particles. *J. Chem. Technol. Biotechnol.* **2010**, *81*, 1271–1276. [CrossRef]
48. Chai, K.; Ji, H. Dual functional adsorption of benzoic acid from wastewater by biological-based chitosan grafted β-cyclodextrin. *Chem. Eng. J.* **2012**, *203*, 309–318. [CrossRef]
49. Li, J.; Loh, X.J. Cyclodextrin-based supramolecular architectures: Syntheses, structures, and applications for drug and gene delivery. *Adv. Drug Deliv. Rev.* **2008**, *60*, 1000–1017. [CrossRef]
50. Yilmaz, E.; Memon, S.; Yilmaz, M. Removal of direct azo dyes and aromatic amines from aqueous solutions using two β-cyclodextrin-based polymers. *J. Hazard. Mater.* **2010**, *174*, 592–597. [CrossRef]
51. Ozmen, E.Y.; Sezgin, M.; Yilmaz, A.; Yilmaz, M. Synthesis of beta-cyclodextrin and starch based polymers for sorption of azo dyes from aqueous solutions. *Bioresour. Technol.* **2008**, *99*, 526–531. [CrossRef] [PubMed]
52. Liu, Y.; Hou, C.; Jiao, T.; Song, J.; Zhang, X.; Xing, R.; Zhou, J.; Zhang, L.; Peng, Q. Self-assembled AgNP-containing nanocomposites constructed by electrospinning as efficient dye photocatalyst materials for wastewater treatment. *Nanomaterials* **2018**, *8*, 35. [CrossRef] [PubMed]
53. Xing, R.; Wang, W.; Jiao, T.; Ma, K.; Zhang, Q.; Hong, W.; Qiu, H.; Zhou, J.; Zhang, L.; Peng, Q. Bioinspired polydopamine sheathed nanofibers containing carboxylate graphene oxide nanosheet for high-efficient dyes scavenger. *ACS Sustain. Chem. Eng.* **2017**, *5*, 4948–4956. [CrossRef]
54. Wang, C.; Sun, S.; Zhang, L.; Yin, J.; Jiao, T.; Zhang, L.; Xu, Y.; Zhou, J.; Peng, Q. Facile preparation and catalytic performance characterization of AuNPs-loaded hierarchical electrospun composite fibers by solvent vapor annealing treatment. *Colloid Surf. A Physicochem. Eng. Asp.* **2019**, *561*, 283–291. [CrossRef]
55. Zhou, J.; Liu, Y.; Jiao, T.; Xing, R.; Yang, Z.; Fan, J.; Liu, J.; Li, B.; Peng, Q. Preparation and enhanced structural integrity of electrospun poly(ε-caprolactone)-based fibers by freezing amorphous chains through thiol-ene click reaction. *Colloid Surf. A Physicochem. Eng. Asp.* **2018**, *538*, 7–13. [CrossRef]
56. Huang, X.; Jiao, T.; Liu, Q.; Zhang, L.; Zhou, J.; Li, B.; Peng, Q. Hierarchical electrospun nanofibers treated by solvent vapor annealing as air filtration mat for high-efficiency PM2.5 capture. *Sci. China Mater.* **2019**, *62*, 423–436. [CrossRef]
57. Cui, H.Y.; Bai, M.; Lin, L. Plasma-treated poly(ethylene oxide) nanofibers containing tea tree oil/beta-cyclodextrin inclusion complex for antibacterial packaging. *Carbohyd. Polym.* **2018**, *179*, 360–369. [CrossRef] [PubMed]
58. Celebioglu, A.; Uyar, T. Antioxidant Vitamin E/cyclodextrin inclusion complex electrospun nanofibers: Enhanced water solubility, prolonged shelf life, and photostability of Vitamin E. *J. Agric. Food Chem.* **2017**, *65*, 5404–5412. [CrossRef] [PubMed]
59. Aytac, Z.; Ipek, S.; Durgun, E.; Tekinay, T.; Uyar, T. Antibacterial electrospun zein nanofibrous web encapsulating thymol/cyclodextrin-inclusion complex for food packaging. *Food Chem.* **2017**, *233*, 117–124. [CrossRef]
60. Lin, L.; Dai, Y.J.; Cui, H.Y. Antibacterial poly(ethylene oxide) electrospun nanofibers containing cinnamon essential oil/beta-cyclodextrin proteoliposomes. *Carbohyd. Polym.* **2017**, *178*, 131–140. [CrossRef]
61. Zhang, H.R.; Chen, G.; Wang, L.; Ding, L.; Tian, Y.; Jin, W.Q.; Zhang, H.Q. Study on the inclusion complexes of cyclodextrin and sulphonated azo dyes by electrospray ionization mass spectrometry. *Int. J. Mass Spectrom.* **2006**, *252*, 1–10. [CrossRef]
62. Koopmans, C.; Ritter, H. Formation of physical hydrogels via host–guest interactions of β-cyclodextrin polymers and copolymers bearing adamantyl groups. *Macromolecules* **2008**, *41*, 7418–7422. [CrossRef]
63. Jiang, Q.; Zhang, Y.; Zhuo, R.; Jiang, X. Supramolecular host–guest polycationic gene delivery system based on poly(cyclodextrin) and azobenzene-terminated polycations. *Colloid Surf. B* **2016**, *147*, 25–35. [CrossRef] [PubMed]
64. Bauer, A.J.P.; Grim, Z.B.; Li, B.B. Hierarchical polymer blend fibers of high structural regularity prepared by facile solvent vapor annealing treatment. *Macromol. Mater. Eng.* **2018**, *303*, 1700489. [CrossRef]

65. Liu, J.Z.; Bauer, A.J.P.; Li, B.B. Solvent vapor annealing: An efficient approach for inscribing secondary nanostructures onto electrospun fibers. *Macromol. Rapid Comm.* **2014**, *35*, 1503–1508. [CrossRef]
66. Narayanan, G.; Aguda, R.; Hartman, M.; Chung, C.C.; Boy, R.; Gupta, B.S.; Tonelli, A.E. Fabrication and characterization of poly(ε-caprolactone)/α-cyclodextrin pseudorotaxane nanofibers. *Biomacromolecules* **2015**, *17*, 271–279. [CrossRef] [PubMed]
67. Chen, C.H.; Chen, S.H.; Shalumon, K.T.; Chen, J.P. Prevention of peritendinous adhesions with electrospun polyethylene glycol/polycaprolactone nanofibrous membranes. *Colloids Surf. B* **2015**, *133*, 221–230. [CrossRef] [PubMed]
68. Narayanan, G.; Ormond, B.R.; Gupta, B.S.; Tonelli, A.E. Efficient wound odor removal by β-cyclodextrin functionalized poly (ε-caprolactone) nanofibers. *J. Appl. Polym. Sci.* **2015**, *132*, 42782. [CrossRef]
69. Narayanan, G.; Gupta, B.S.; Tonelli, A.E. Enhanced mechanical properties of poly (ε-caprolactone) nanofibers produced by the addition of non-stoichiometric inclusion complexes of poly (ε-caprolactone) and α-cyclodextrin. *Polymer* **2015**, *76*, 321–330. [CrossRef]
70. Huang, X.N.; Huang, Y.G.; Chai, X.S.; Wei, W. Study of adsorption kinetics for fluorescent whitening agent on fiber surfaces. *Sci. China Chem.* **2008**, *51*, 473–478. [CrossRef]
71. Chen, B.L.; Zhou, D.D.; Zhu, L.Z.; Shen, X.Y. Sorption characteristics and mechanisms of organic contaminant to carbonaceous biosorbents in aqueous solution. *Sci. China Chem.* **2008**, *51*, 464–472. [CrossRef]
72. Zhan, F.; Wang, R.; Yin, J.; Han, Z.; Zhang, L.; Jiao, T.; Zhou, J.; Zhang, L.; Peng, Q. Facile solvothermal preparation of Fe_3O_4–Ag nanocomposite with excellent catalytic performance. *RSC Adv.* **2019**, *9*, 878–883. [CrossRef]
73. Huo, S.; Duan, P.; Jiao, T.; Peng, Q.; Liu, M. Self-assembled luminescent quantum dots to generate full-color and white circularly polarized light. *Angew. Chem. Int. Ed.* **2017**, *56*, 12174–12178. [CrossRef] [PubMed]
74. Xing, R.; Liu, K.; Jiao, T.; Zhang, N.; Ma, K.; Zhang, R.; Zou, Q.; Ma, G.; Yan, X. An injectable self-assembling collagen-gold hybrid hydrogel for combinatorial antitumor photothermal/photodynamic therapy. *Adv. Mater.* **2016**, *28*, 3669–3676. [CrossRef] [PubMed]
75. Luo, X.; Ma, K.; Jiao, T.; Xing, R.; Zhang, L.; Zhou, J.; Li, B. Graphene oxide-polymer composite Langmuir films constructed by interfacial thiol-ene photopolymerization. *Nanoscale Res. Lett.* **2017**, *12*, 99. [CrossRef] [PubMed]
76. Zhang, R.Y.; Xing, R.R.; Jiao, T.F.; Ma, K.; Chen, C.J.; Ma, G.H.; Yan, X.H. Synergistic in vivo photodynamic and photothermal antitumor therapy based on collagen-gold hybrid hydrogels with inclusion of photosensitive drugs Colloids and Surfaces A: Physicochemical and Engineering Aspects. *ACS Appl. Mater. Interfaces* **2016**, *8*, 13262–13269. [CrossRef] [PubMed]
77. Xing, R.; Jiao, T.; Liu, Y.; Ma, K.; Zou, Q.; Ma, G.; Yan, X. Co-assembly of graphene oxide and albumin/photosensitizer nanohybrids towards enhanced photodynamic therapy. *Polymers* **2016**, *8*, 181. [CrossRef]
78. Liu, K.; Yuan, C.Q.; Zou, Q.L.; Xie, Z.C.; Yan, X.H. Self-assembled Zinc/cystine-based chloroplast mimics capable of photoenzymatic reactions for sustainable fuel synthesis. *Angew. Chem. Int. Ed.* **2017**, *56*, 7876–7880. [CrossRef]
79. Liu, K.; Xing, R.R.; Li, Y.X.; Zou, Q.L.; Möhwald, H.; Yan, X.H. Mimicking primitive photobacteria: Sustainable hydrogen evolution based on peptide-porphyrin co-assemblies with self-mineralized reaction center. *Angew. Chem. Int. Ed.* **2016**, *55*, 12503–12507. [CrossRef]
80. Liu, K.; Xing, R.R.; Chen, C.J.; Shen, G.Z.; Yan, L.Y.; Zou, Q.L.; Ma, G.H.; Möhwald, H.; Yan, X.H. Peptide-induced hierarchical long-range order and photocatalytic activity of porphyrin assemblies. *Angew. Chem. Int. Ed.* **2015**, *54*, 500–505. [CrossRef]

© 2019 by the authors. Licensee MDPI, Basel, Switzerland. This article is an open access article distributed under the terms and conditions of the Creative Commons Attribution (CC BY) license (http://creativecommons.org/licenses/by/4.0/).

Article

Facile Preparation of Self-Assembled Polydopamine-Modified Electrospun Fibers for Highly Effective Removal of Organic Dyes

Cuiru Wang [1,2,3], Juanjuan Yin [3], Ran Wang [3], Tifeng Jiao [1,2,3,*], Haiming Huang [2], Jingxin Zhou [3], Lexin Zhang [3] and Qiuming Peng [1,*]

1. State Key Laboratory of Metastable Materials Science and Technology, Yanshan University, Qinhuangdao 066004, China; wcr2016ysdx@163.com
2. School of Environment and Civil Engineering, Dongguan University of Technology, Dongguan 523808, China; huanghaiming52hu@163.com
3. Hebei Key Laboratory of Applied Chemistry, School of Environmental and Chemical Engineering, Yanshan University, Qinhuangdao 066004, China; jjy1729@163.com (J.Y.); wr1422520780@163.com (R.W.); zhoujingxin@ysu.edu.cn (J.Z.); zhanglexin@ysu.edu.cn (L.Z.)
* Correspondence: tfjiao@ysu.edu.cn (T.J.); pengqiuming@ysu.edu.cn (Q.P.); Tel.: +86-335-8056854 (T.J.)

Received: 4 December 2018; Accepted: 16 January 2019; Published: 18 January 2019

Abstract: Polydopamine (PDA) nanoparticles can be used as an adsorbent with excellent adsorption capacity. However, nanosized adsorbents are prone to aggregation and thus are severely limited in the field of adsorption. In order to solve this problem, we utilized polydopamine in-situ oxidation self-polymerization on the surface of polycaprolactone (PCL)/polyethylene oxide (PEO) electrospun fiber after solvent vapor annealing (SVA) treatment, and successfully designed and prepared a PCL/PEO@PDA composite membrane. The SVA treatment regulated the microscopic morphology of smooth PCL/PEO electrospun fibers that exhibited a pleated microstructure, increasing the specific surface area, and providing abundant active sites for the anchoring of PDA nanoparticles. The PCL/PEO@PDA composite obtained by chemical modification of PDA demonstrated numerous active sites for the adsorption of methylene (MB) and methyl orange (MO). In addition, the PCL/PEO@PDA composites were reusable several times with good reutilization as adsorbents. Therefore, we have developed a highly efficient and non-agglomerated dye adsorbent that exhibits potential large-scale application in dye removal and wastewater purification.

Keywords: electrospinning; solvent vapor annealing; structural regularity; polydopamine; dye removal

1. Introduction

In recent years, the rapid development of industrialization has caused serious water pollution problems, which have received widespread attention [1,2]. Among the various pollutants, organic dyes are one of the main sources of water pollution. Therefore, finding a convenient method for effectively removing organic dyes from wastewater is a serious challenge [3–5]. A variety of physical, chemical, and biological techniques have been reported for the removal of dyes from wastewater including adsorption, coacervation, membrane separation, etc. [6]. Adsorption is an effective and economical method for purifying dye wastewater due to its high efficiency, simplicity of operation, and insensitivity to contaminants [7,8]. Various adsorbents such as natural materials, activated carbon, hydrotalcite, nano-oxide particles, and biological adsorbents are used for wastewater purification [9,10]. Due to its simple preparation method, high porosity, large specific surface area, and high adsorption efficiency, nano-adsorbents exhibit significant adsorption capacity [11].

For instance, Luo et al. [12] utilized a submerged circulation impinging stream reactor (SCISR) to prepare a millimeter-sized cellulose bead adsorbent named MCB-AC, which was applied to wastewater treatment. However, some nano-adsorbents were prone to agglomeration and were difficult to regenerate, thus severely limiting their use in the field of adsorption [13]. Consequently, it is of great significance to find organic dye adsorbents with high adsorption capacity and recyclability.

Electrospinning technology has become one of the most effective methods for preparing fibers due to its obvious advantages of simple operation and easy regulation [14,15]. At present, the construction of nanostructures of various materials such as polymers, inorganic substances, and multi-component composites has been realized by electrospinning technology, which simultaneously regulates the size and morphology of the nanostructures [16–18]. Electrospun fibers have the advantages of large specific surface area, porosity, and excellent flexibility, showing great potential in the fields of catalysis, drug carriers, and ultrafiltration [19,20]. In addition, the fiber material exhibited good adsorption capacity due to its microporous structure. For example, Miao et al. [21] combined electrospinning technology and a hydrothermal reaction to prepare SiO_2@ALOOH core/shell fiber and applied it to wastewater purification. Obviously, the electrospun fiber membrane met the stringent requirements in the field of decontaminating wastewater. It should be noted that polydopamine (PDA) has been reported as an effective adsorbent [22,23]. The polymer's abundant reactive functional groups such as amino, imino, and phenolic hydroxyl groups seemed beneficial for secondary reactions with other ions. In addition, PDA-coated nanomaterials increased the numerous active sites and improved the adsorption performance. Dong et al. [23] prepared a PDA-coated graphene oxide (GO/PDA) composite adsorbent using the synergistic effect of PDA and graphene oxide that showed a higher adsorption capacity than pure GO and PDA. Polycaprolactone (PCL), which has good biocompatibility and degradability, could be used as an electrospinning material. Furthermore, the addition of a polyethylene oxide (PEO) component with abundant oxygen-containing groups was proposed to regulate the phase structures and interfacial active sites at suitable conditions. It has been reported that the PCL/PEO fibers prepared by electrospinning carried considerable free amorphous PCL chains [24–26]. During solvent vapor annealing (SVA) treatment, the free amorphous PCL chains absorbed acetone vapor faster than the chains in the crystallization zone. However, PEO, playing the role of mini dividers, limited the growth of semi-crystalline PCL. Thus, the swollen amorphous PCL chains were deposited on the crystalline lamellae of preexisting PCL or PEO while the PEO phase remained largely unchanged. Once the desiccator of the SVA treatment was turned on, the acetone vapor quickly left the fiber system, creating an obvious morphological change due to the amorphous chains that preferentially crystallized on the edge of the pre-existing crystallites [25].

In order to solve the abovementioned difficulties, we utilized the in-situ oxidation self-polymerization of polydopamine by the good adsorption capacity of electrospun fibers and polydopamine by using electrospun polycaprolactone (PCL)/polyethylene oxide (PEO) as the substrate. This was combined with the SVA method to regulate the surface morphology of the substrate, and we successfully designed and prepared a PCL/PEO@PDA composite adsorbent. The SVA treatment controlled the layered structures, which increased the specific surface area to provide more active sites for the PDA coating. It was found that the prepared PCL/PEO@PDA composite showed excellent dye adsorption capacity due to its wrinkled morphology and improved hydrophilicity via the PDA layer. In addition, the obtained composite adsorbents could be utilized several times with good stability and recyclability, demonstrating wide applications in wastewater treatment and composite materials.

2. Experimental Method

2.1. Materials

Both polycaprolactone (PCL, Average M_n = 80,000) and polyethylene oxide (PEO, average M_v = 600,000) were purchased from Sigma–Aldrich. Chloroform (analytical grade, 99.0%) and acetone (analytical grade, 99.5%) were obtained from Tianzheng Chemical Reagent (Tianjin, China).

Tris (hydroxymethyl) aminomethane hydrochloride (Tris HCL, super purity, 99%, Aladdin), sodium hydroxide (NaOH, AR, Kermel Reagent, Tianjin, China), dopamine hydrochloride (98%, Aladdin), methylene blue (MB), methyl orange (MO), Safranine T (ST), and Rhodamine B (RhB) were from Aladdin without further purification. All aqueous solutions were prepared using ultrapure water throughout the experiment. All chemicals were used as received.

2.2. Preparation of PCL/PEO@PDA Nanocomposites

In our previous work, PCL and PEO were simultaneously dissolved in chloroform, and magnetically stirred overnight at room temperature to obtain a uniform electrospinning precursor solution for the preparation of electrospun fibers [24]. The electrospinning parameters were 15 kV voltage, 1 mL/h flow rate, and 25 cm between the needle and the receiving plate. The prepared PCL/PEO composite fiber was directly separated from the receiving plate and placed in a hermetical desiccator room containing 200 mL acetone at room temperature for five days to optimize the surface morphology. Subsequently, polydopamine modification was performed. Specifically, a 10 mM Tris-HCl buffer solution was prepared to which NaOH was added to adjust the pH to 8.5. Then, dopamine (DA) was added to the above pH-adjusted buffer solution to obtain 2.0 mg/mL of dopamine aqueous solution. Once DA was added, the color of the solution changed from colorless to dark brown in an instant, which meant the formation of polydopamine (PDA) and was similar to previous reports [27,28]. The SVA-treated PCL/PEO was immersed into the aqueous dopamine solution and stirred for different times (5 h, 30 h, and 45 h) to gain PCL/PEO@PDA composite films with different PDA modification time intervals. The PCL/PEO@PDA composite membranes were removed from the aqueous dopamine solution and washed several times with ultrapure water to remove free polydopamine molecules. Finally, it was dried in a vacuum oven at room temperature for 24 h and stored for subsequent use.

2.3. Adsorption Capacity Test

The adsorption activity of PCL/PEO@PDA composite membranes to methylene blue (MB) and methyl orange (MO) dyes was measured by a 721 visible spectrophotometer (Shanghai Yidian Analytical Instrument Co. Ltd., Shanghai, China). The entire adsorption experiment was carried out by magnetic stirring at room temperature (298 K). First, four concentration-absorbance calibration curves were established by measuring the absorbance of MB (662 nm), MO (463 nm), RhB (554 nm), and ST (518 nm) solutions at different concentrations according to previous reports [29]. Subsequently, 15 mg of the prepared PCL/PEO@PDA composites as adsorbent materials were added to 50 mL of dye solutions (MB, 10 mg/L; MO, 50 mg/L; RhB, 5 mg/L; and ST, 30 mg/L), respectively. The absorbance of the dye solution was measured at different intervals until the absorbance stabilized. The concentration of the dye solution was calculated from the calibration curve established above. The SVA-treated PCL/PEO composite fiber was used as a reference for the adsorption experiments. Finally, the PCL/PEO@PDA composite adsorbent under optimal PDA modification time was subjected to recycling experiments. A total of 15 mg of the prepared PCL/PEO@PDA was added to a freshly prepared MO (100 mg/L) solution for the adsorption experiments; after reaching adsorption saturation, the PCL/PEO@PDA adsorbent was removed directly and washed several times with ultrapure water and ethanol to further adsorb another fresh MO (50 mg/L) solution. The above experimental procedure was repeated eight times to complete the cyclic adsorption test.

2.4. Characterization

The morphologies of the samples were examined via a scanning electron microscope (SEM, FEI Corporate, Hillsboro, OR, USA) with gold plasma deposition. Fourier transform infrared spectroscopy (Thermo Nicolet Corporation) was performed by the KBr pellet method. The thermal stability of the as-prepared samples was investigated by thermogravimetry-differential scanning calorimetry (TG-DSC) under an argon atmosphere using a 409 PC Luxxsi thermal analysis instrument (Netzsch Instruments Manufacturing Co., Ltd., Seligenstadt, Germany). We obtained

X-ray photoelectron spectroscopy (XPS) data by monitoring a Thermo Scientific ESCALab 250Xi (Netzsch Instruments Manufacturing Co., Ltd., Seligenstadt, Germany) equipped with 200 W of monochromatic AlKα radiation. The 500 μm X-ray spot was used for XPS analysis. The base pressure in the analysis chamber was about 3×10^{-10} mbar. Typically, the hydrocarbon C1s line at 284.8 eV from adventitious carbon is used for energy referencing. Both the survey scan and individual high-resolution scan were recorded. The N_2 adsorption–desorption properties were measured at 77.3 K by a Quadrasorb analyzer (Quantachrome Instruments, Boynton Beach, FL, USA). Before the measurement, the samples were outgassed at 40 °C under vacuum for 4 h. The surface areas were calculated by the Brunauer–Emmett–Teller (BET) method.

3. Results and Discussion

3.1. Characterization of PCL/PEO@PDA Composite Adsorbents

Figure 1 illustrates the preparation and dye adsorption process of PCL/PEO@PDA as a dye adsorbent. The experimental process was mainly divided into four parts: electrospinning, SVA treatment, chemical modification, and dye adsorption. The electrospinning PCL/PEO fiber membrane was prepared according to the previous work and subjected to subsequent SVA treatment. The SVA-treated PCL/PEO fiber membrane was then chemically modified with polydopamine to obtain the PCL/PEO@PDA dye adsorbent. Finally, the prepared PCL/PEO@PDA composite materials were used as dye adsorbents to adsorb several dyes such as methylene blue and methyl orange. The SEM images of the SVA-treated PCL/PEO and PCL/PEO@PDA with different chemical modification times (5 h, 30 h, and 45 h) are shown in Figure 2. Our previous work reported significant changes in the morphology of electrospun PCL/PEO fiber membranes before and after SVA treatment [24]. The as-prepared electrospun PCL/PEO fiber demonstrated a straight and smooth structure, as shown in our previous report [24]. It was interesting to note that the SVA-treated fiber exhibited a curved and wrinkled structure, as shown in Figure 2a,a'. Figure 2b–d clearly illustrated the differences in the surface morphologies of PCL/PEO@PDA for different PDA modification intervals. For a modification time of 5 h, as shown in Figure 2b,b', it was obvious that the surface of the wrinkled structure was partially covered by a PDA layer. As the PDA modification time increased, numerous PDA nanoparticles anchored and accumulated on the surface of the wrinkled PCL/PEO composite fiber, which can be clearly seen in Figure 2d,d'. At the same time, as shown in Figure 2e, the UV-vis spectra of the PDA solution and PCL/PEO@PDA-45 fiber composite showed the same peaks at 218 and 280 nm, indicating the formation of PDA particles due to the DA oxidation and the polymerization reaction process [30]. Previous studies have shown that the adhesion behavior of PDA nanoparticles on the surface of solid membranes is mainly due to the noncovalent interactions (such as hydrogen bonding and electrostatic forces) between the hydrophilic amino-groups of PDA and the functional active groups on the surface of the composites [31–34]. The results of the above SEM topography strongly demonstrate the successful preparation of PCL/PEO@PDA composites. The modification of PDA could improve the hydrophilicity, stability, and adsorption capacity of the prepared composite.

We surveyed the thermogravimetric (TG) curves of the initial samples and subsequent PDA modified samples with different reaction time intervals to investigate the thermal stability of the PCL/PEO@PDA composites. As shown in Figure 3, all samples exhibited a slow weight loss below 250 °C, which corresponded to the removal of the moisture remaining in the samples [35]. In addition, the weight of all samples tended to be stable above 550 °C. The PCL/PEO composite fibers before and after SVA treatment exhibited similar thermal stability, and the qualities' retention ratios were 1.0% and 2.5%, respectively. The qualities' retention ratios of the above two samples were close to zero because the PCL and PEO polymers were thermally decomposed at high temperatures into volatiles such as H_2O, CO_2, etc. [36]. A sharp weight loss between 380 °C and 440 °C originated from the thermal decomposition of the alkyl chains and various functional groups in these samples [37].

It was reported that pure PDA demonstrated an approximate 62% weight retention ratio at 550 °C [30]. For the present synthesized composite fiber materials, as the chemical modification time increased, the weight retention ratios reached 22% for the PCL/PEO@PDA-45 composite. The decreased weight loss of the PCL/PEO@PDA composites after SVA treatment indicated that substantial PDA nanoparticles had successfully anchored on the surface of the SVA-treated PCL/PEO fibers and significantly improved stability.

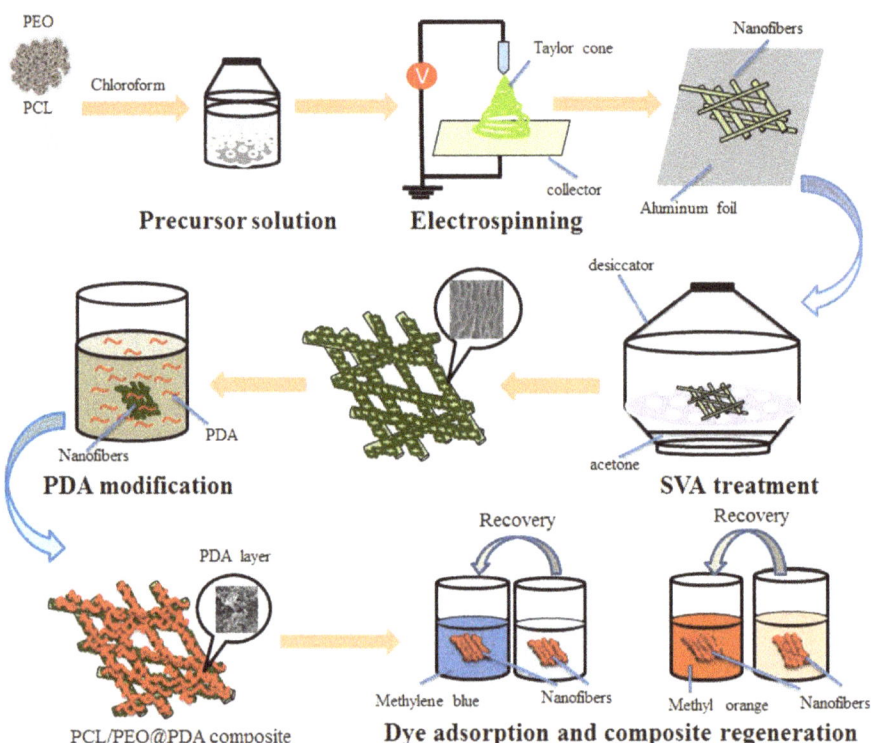

Figure 1. Schematic illustration of the preparation and adsorption dye process of the polycaprolactone/polyethylene oxide@polydopamine (PCL/PEO@PDA) composites as dye absorbents.

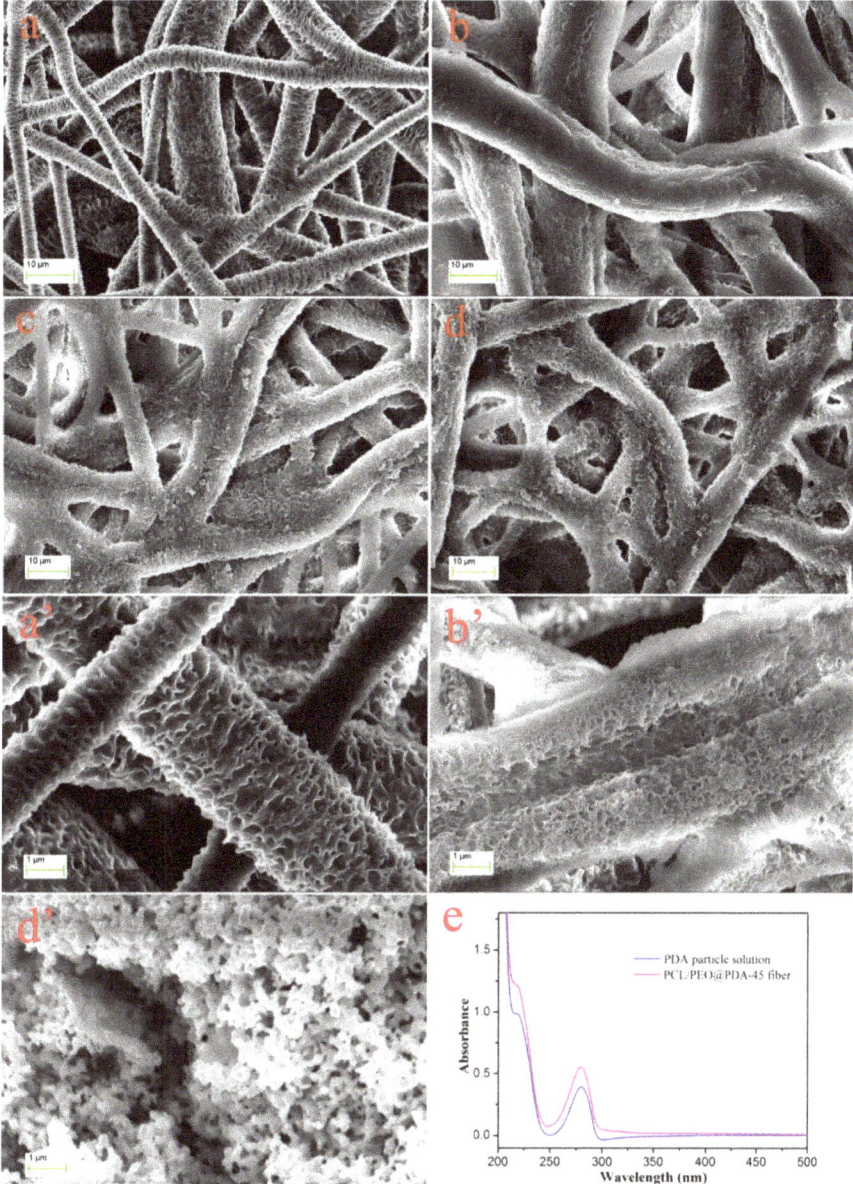

Figure 2. SEM images of the solvent vapor annealing (SVA)-treated PCL/PEO fibers (**a,a′**) and PCL/PEO@PDA of different modification time intervals (**b,b′**), 5 h; (**c**), 30 h; (**d,d′**), 45 h; (**e**) UV-vis spectra of the PDA solution and PCL/PEO@PDA-45 fiber composite.

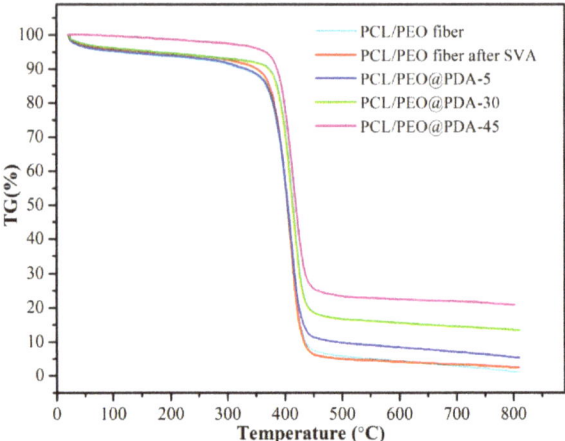

Figure 3. Thermogravimetry (TG) curves of the prepared PCL/PEO electrospun fibers before and after SVA treatment and the PCL/PEO@PDA composites of different PDA modification intervals (5 h, 30 h, and 45 h).

Next, the X-ray photoelectron spectroscopy (XPS) was measured to investigate the interfacial elemental compositions of the obtained composite fiber materials. First, the survey XPS spectra of different fiber composites are shown in Figure 4a. Some characteristic peaks appeared such as general C1s and O1s as well as additional N1s peaks in the PCL/PEO@PDA-45 composite. It should be noted that the as-obtained PCL/PEO fiber showed the relative atomic ratios of C and O elements with values of 72.6% and 26.5%, respectively. After SVA treatment, the relative atomic ratios of C and O elements showed the values of 77.3% and 21.7%, respectively. In addition, the atomic ratios of C, O, and N elements for the PCL/PEO@PDA-45 composite changed with values of 73.2%, 19.6%, and 4.9%, respectively. Due to the presence of the N element and decrement of the O element, it could be reasonably speculated that the synthesized PDA particles were successfully anchored on the surface of the composite fibers. Figure 4b shows that the C1s signals of the SVA-treated PCL/PEO fiber were mainly located at 284.1, 284.8, 285.5, 286.5, and 288.7 eV, corresponding to the C–C & C=C & C–H, C–OH, C–O, C=O, and O=C–O bonds, respectively [33,38,39]. The O1s deconvolution data are shown in Figure 4c, which clarified three peaks at 531.7, 532.7, and 533.5 eV, representing the bonds of C=O, C–O, and –O–H, respectively [33,40]. In addition, Figure 4d–f illustrate the deconvolutions of the C1s, O1s, and N1s peaks in the PCL/PEO@PDA-45 composite. The C1s peak was deconvoluted to five peaks at 283.0, 283.9, 284.9, 286.0, and 288.0 eV, corresponding to the C–Si, C–C & C=C & C–H, C–N & C–OH, C=O, and O=C–O groups [26,33,38,39]. The presence of C–Si was due to the silicon plate during the testing procedure. The O1s deconvolution exhibited peaks of C=O, C–O, H_2O, and –O–H species at positions of 530.0, 531.6, 532.6, and 533.6 eV, respectively [26,33,40]. The possible H_2O at 532.6 eV originated from the residual moisture in the sample. The N1s deconvolution showed amine and C–N species at 398.6 eV and 400.4 eV, respectively, indicating that the PDA layer was apparently present on the fiber surface [26,37,38,41,42]. In addition, the microstructures of the SVA-treated PCL/PEO and PCL/PEO@PDA-45 samples were investigated using N_2 adsorption–desorption isotherms. The obtained properties of the samples are generalized in Table 1. As shown in Table 1, the as-obtained PCL/PEO fiber showed the specific surface area of 8.5467 m^2g^{-1}. After SVA treatment for 45 h, the value of the specific surface area increased obviously and reached 15.3133 m^2g^{-1}, demonstrating the formation of more anchoring sites for the next modification of the PDA coating. In addition, the obtained PCL/PEO@PDA-45 after chemical modification of PDA showed a decreased BET specific surface area (9.0741 m^2g^{-1}) than the SVA-treated PCL/PEO fibers (15.3133 m^2g^{-1}), indicating that the PCL/PEO@PDA-45 composite fiber anchored numerous PDA particles in interfacial adsorption sites, facilitating the next adsorption of dye molecules. Meanwhile, the pore size and pore volume of the two samples were calculated via BJH methods. The obtained PCL/PEO@PDA-45 composite also exhibited decreased pore size and pore volume, meaning that larger pore diameters and pore volumes in the SVA-treated PCL/PEO fibers could demonstrate lots of micro/nanoscale channels for PDA nanoparticles to transfer into the composites, thereby making it effective for the next adsorption capacities experiment [43,44].

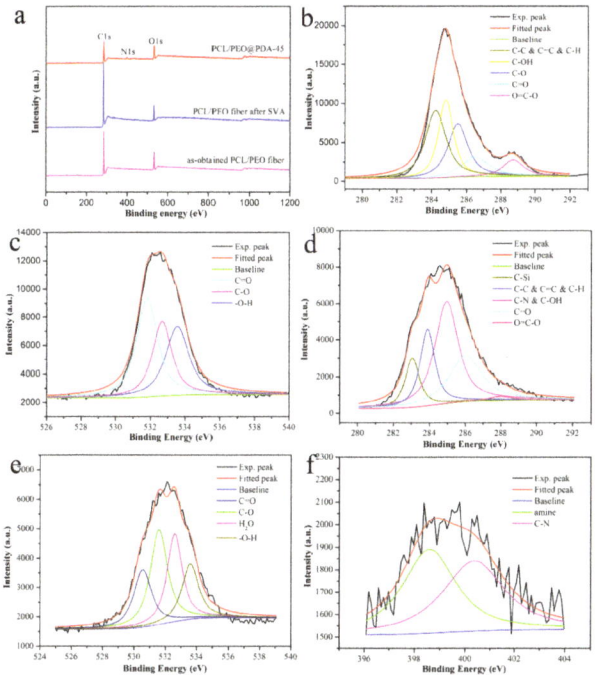

Figure 4. Survey XPS spectra of the fiber composites (**a**) and the peak deconvolutions: (**b**,**c**), C1s and O1s in the PCL/PEO fiber after SVA treatment, respectively; (**d**–**f**), C1s, O1s, and N1s in the PCL/PEO@PDA-45 composite, respectively.

Table 1. Physical data of as-prepared different fiber composites.

Samples	Specific Surface Area (m^{2}g^{-1})	Average Pore Diameter (nm)	Pore Volume (cm^{3}g^{-1})
As-obtained PCL/PEO fiber	8.5467	24.1263	0.008736
PCL/PEO fiber after SVA	15.3133	47.0420	0.012631
PCL/PEO@PDA-45	9.0741	27.1665	0.008920

In order to further investigate the obtained composite fiber, the infrared curves of the initial materials and modified composites with different chemical modification time are shown in Figure 5. The infrared spectrum of the initial PEO exhibited triplet peaks of C–O–C stretching vibration at 1149, 1101, and 1060 cm^{-1} with the maximum peak at 1101 cm^{-1} [45]. Meanwhile, the characteristic peaks at 964, 1470, 2884, and 3438 cm^{-1} were derived from the initial PEO, corresponding to the rock vibration of CH$_2$, the CH$_2$ stretching vibration, and the terminal hydroxyl group, respectively [45–48]. As for the PDA curve, the peak at 1610 cm^{-1} was attributed to the aromatic rings stretching vibrations and the N–H bending vibrations, and the peak at 3400 cm^{-1} was attributed to the catechol –OH groups and N–H groups [30]. As shown in the spectral curves of the electrospun PCL/PEO and SVA-treated PCL/PEO, characteristic peaks at 2945, 2871 cm^{-1}, and 1723 cm^{-1} were attributed to asymmetric and symmetric CH$_2$ stretching vibration and ester carbonyl stretching vibration, respectively [48].

After the SVA-treated PCL/PEO was chemically modified by PDA, the obtained PCL/PEO@PDA showed new characteristic peaks at 1558, 1550, 1500 cm^{-1}, corresponding to the amide bands, amines, and aromatic benzene rings in the PDA component, respectively, and accompanying the absorption peak at 1723 cm^{-1}, which slightly shifted to 1730 cm^{-1} [33,45,49]. All characteristic peaks were correspondingly displayed in the infrared curve of the PCL/PEO@PDA composite. The above infrared data further demonstrates that the PDA component was successfully anchored on the surface of the substrate SVA treated PCL/PEO material.

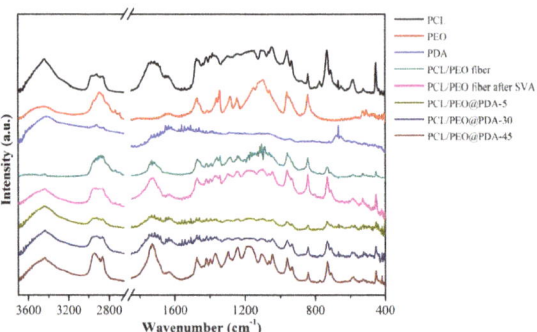

Figure 5. FT-IR spectra of the initial materials and the obtained composite fibers.

3.2. Adsorption Dye Performance of PCL/PEO@PDA Composite

Adsorption kinetics is of great significance in the evaluation of adsorption efficiency. A variety of adsorbents evaluated by pseudo-first-order and pseudo-second-order kinetic models have been reported, as listed in Table 2 [12,13,22,23,33,50]. The adsorption behaviors of several adsorbents towards methylene blue (MB) and methyl orange (MO) in Table 2 indicated that different adsorbents show selective adsorption properties for different dyes. For examples, magnetic cellulose beads (MCB-AC) synthesized by Luo et al. [12] showed certain adsorption capacity for MB and MO (1.8 mg/g and 1.3 mg/g, respectively), which could be well regenerated and reused. Fu et al. [13] reported the synthesis of independent PDA microspheres and the selective adsorption/separation capacities of organic dyes, where the obtained PDA microsphere adsorbent showed a better selective adsorption of MB (147.0 mg/g) than MO (almost zero). Zhou et al. [22] reported the magnetic core-shell structure of Fe$_3$O$_4$@PDA nanoparticles which selectively adsorbed MB (10.0 mg/g) rather than MO (2.1 mg/g) due to electrostatic attraction between dye molecules and adsorbents. Thus, both MB and MO molecules were regarded as typical model dyes for the adsorption test in present study system. The adsorption properties of our prepared PCL/PEO@PDA-45 composite and the SVA-treated PCL/PEO fiber as adsorbents for MB and MO aqueous solutions were estimated by pseudo-first-order and pseudo-second-order kinetic models. From the adsorption curves seen in Figure 6, it was found that the composite films chemically modified by PDA exhibited improved adsorption capacities than the substrate composite film. This result can be attributed to the affluent amino and hydroxyl groups on the surface of the PDA to provide more adsorption activity points for the dye molecules [33]. The following classical kinetic model equations illustrate the above adsorption mechanism:

The pseudo-first-order kinetic model Equation (1) is expressed:

$$\log(q_e - q_t) = \log q_e - \frac{k_1}{2.303} t \tag{1}$$

The pseudo-second-order kinetic model Equation (2) is expressed as:

$$\frac{t}{q_t} = \frac{1}{k_2 q_e^2} + \frac{t}{q_e} \tag{2}$$

where q_e (mg/g) is the adsorption amount when the adsorption process reaches equilibrium; q_t (mg/g) is the adsorption amount when the adsorption time t (min); k_1 (min^{-1}) is the pseudo-first-order rate constants; and k_2 (g/mg·min) is the pseudo-second-order rate constants. The obtained adsorption kinetic related data (Figure 6) are shown in Table 3. The adsorption amount (q_e) of the dye at the equilibrium of adsorption indicated that a good fit of the adsorption experimental curves had been obtained. It can be seen from Figure 6a and Table 3 that the adsorption efficiency of the PCL/PEO@PDA adsorbent for the MB dye enhanced with the increment of the chemical modification time of PDA. In addition, the correlation coefficient R^2 corresponding to the pseudo-first-order and pseudo-second-order kinetic model indicated that the adsorption process of the PCL/PEO@PDA-45 composite adsorbent for MO dye was more consistent with the pseudo-second-order kinetic model, while the adsorption kinetics of other adsorbents for the MB, MO, and ST dyes were more consistent with the pseudo-first-order kinetic model. It is worth mentioning that the obtained PCL/PEO@PDA-45 composite adsorbent exhibited better fitted adsorption efficiency for MO (60.22 mg/g) than that of MB (14.85 mg/g). That is to say, the chemically modified PCL/PEO@PDA-45 composite adsorbent exhibited excellent selective adsorption capacity for the anionic dye (MO) versus the cationic dye (MB) [13]. As shown in Table 2, for examples, Dong et al. [23] reported the adsorption performance of PDA/GO composite adsorbents toward various organic dyes, such as MB (1800.0 mg/g), MO (30.0 mg/g), RhB (100.0 mg/g), and neutral red (1400.0 mg/g), showing an extremely high adsorption capacity. Xing et al. [33] investigated the self-assembly of hierarchical poly(vinyl alcohol)/poly(acrylic acid)/carboxylate graphene oxide nanosheets@polydopamine (PVA/PAA/GO-COOH@PDA) composite and adsorption performances for MB (34.1 mg/g), RhB (8.4 mg/g) and Congo red (12.9 mg/g), which showed selectively MB-adsorbed process due to strong π-π stacking and electrostatic interaction. Liu et al. [50] reported that poly(catechol-tetraethylenepentamine-cyanuric chloride)@hydrocellulose(PCEC-C) composite absorbent effectively removed MO (37.2 mg/g) at 298 K due to electrostatic interaction, hydrogen bonding, and π-π stacking interaction. In our present system, to further explore the adsorption performance of the prepared composite fibers, we also studied the adsorption process of the PCL/PEO@PDA-45 composite and SVA-treated PCL/PEO fiber on typical cationic dyes (RhB and ST). The obtained results in Figure 6 and Table 3 showed that the PCL/PEO@PDA-45 composite adsorbent exhibited common adsorption properties for RhB and ST (7.7316 and 9.6628 mg/g, respectively). It is well known that most forces between adsorbents and dyes are ion interactions, electrostatic attraction, π–π stacking interactions, and host–guest interactions [13,23]. However, space steric hindrance (such as the aromatic ring of ST, and the long-chain alkyl chain of RhB) could offset the electrostatic attraction and π–π stacking interactions, and so on [50]. In addition, the Eschenmoser salt between the ortho position of the catechol phenolic hydroxyl group in PDA and the Eschenmoser structure of the dye (such as MB) assisted the 1,4-Michael addition reaction, increasing the adsorption capacity of the composite adsorbent containing a PDA component [23]. Thus, in combination with the above several adsorption systems, the PCL/PEO@PDA-45 composite adsorbent prepared in this study exhibited a selective adsorption performance on the used ion dyes. In all, the obtained PCL/PEO@PDA composite adsorbent selectively and effectively removed MO due to stronger π-π stacking and electrostatic interactions [13,23,33,50], friendly adsorbed MB because of the Eschenmoser salt formed by the 1,4-Michael addition reaction. At the same time, the PCL/PEO@PDA composite adsorbent exhibited a limited adsorption capacity on RhB and ST. The above obtained experimental results showed that present composite adsorbents could potentially be used in the wide fields of wastewater treatment.

Table 2. Comparison of the adsorption capacity of other adsorbents in the previous studies at 298 K.

NO.	Materials	q_e (mg/g) MB	q_e (mg/g) MO	Characteristics	Ref.
1	Magnetic cellulose beads (MCB-AC)	1.8	1.3	Selective magnetic response, environmentally friendly process, reusable spherical beads.	[12]
2	PDA microspheres	147.0	almost zero	Selective adsorption of cationic dyes, economical adsorption and separation.	[13]
3	Fe_3O_4@PDA NPs	10.0	2.1	Selective adsorption capacity for cationic dyes, magnetic core-shell structure, easily magnetic separation.	[22]
4	PDA/GO	1800.0	30.0	Controllable PDA layer thickness, high surface area structure, excellent adsorption performance.	[23]
5	poly(vinyl alcohol)/poly(acrylic acid)/carboxylate graphene oxide@polydopamine (PVA/PAA/GO-COOH@PDA)	31.3	-	Environmentally friendly and controllable preparation method, larger specific surface area, excellent reusability.	[33]
6	poly(catechol-tetraethylenepentamine-cyanuric chloride)@hydrocellulose (PCEC-C)	-	37.2	Selective adsorption of anionic dyes, simple method and better adsorption stability.	[50]
7	PCL/PEO@PDA-45	14.8	60.2	Selective adsorption of MO, convenient and controllable method, excellent stability and reuse.	Present work

Table 3. Adsorption kinetic parameters of SVA treated PCL/PEO and PCL/PEO@PDA adsorbents toward dyes at 298 K.

MB	Pseudo-First-Order Model			Pseudo-Second-Order Model		
	q_e (mg/g)	R^2	K_1 (min^{-1})	q_e (mg/g)	R^2	K_2 (g/mg·min)
PCL/PEO after SVA	2.5533	0.9949	0.1270	2.6183	0.9987	0.3809
PCL/PEO@PDA-5	6.0916	0.8968	0.0459	6.7700	0.9931	0.1457
PCL/PEO@PDA-30	7.1766	0.9738	0.0184	7.5614	0.9904	0.1297
PCL/PEO@PDA-45	13.5342	0.9576	0.0335	14.8522	0.9966	0.0669
RhB						
PCL/PEO after SVA	1.5026	0.9344	0.1217	1.5018	0.9923	1.1646
PCL/PEO@PDA-45	7.2790	0.9668	0.0878	7.7316	0.9997	0.0286
ST						
PCL/PEO after SVA	2.1838	0.9566	0.0059	1.8543	0.9358	0.0115
PCL/PEO@PDA-45	8.9755	0.9788	0.0162	9.6628	0.9826	0.0041
MO						
PCL/PEO after SVA	29.6755	0.9963	0.0432	30.8547	0.9980	0.0323
PCL/PEO@PDA-45	60.2260	0.9983	0.0046	59.2417	0.8960	0.0136

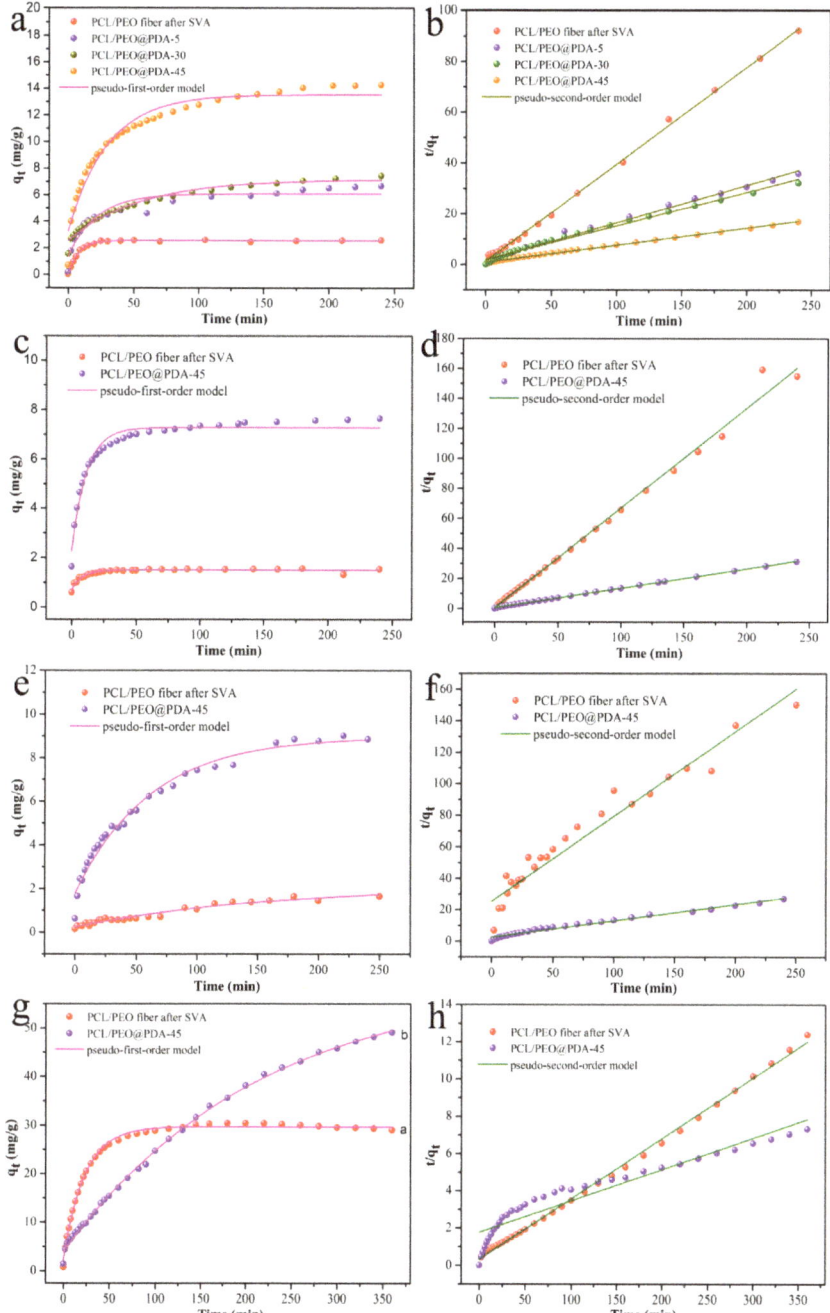

Figure 6. Adsorption kinetics curves of the prepared composite fibers on methylene blue (MB) (**a**,**b**), Rhodamine B (RhB) (**c**,**d**), Safranine T (ST) (**e**,**f**), and methyl orange (MO) (**g**,**h**) at 298 K.

Good stability and reusability of the adsorbent are important considerations for practical production. Adsorbents with excellent reusability have been reported in previous studies [51–58].

For example, Lu et al. [41] investigated the reuse efficiency of the PDA/PEI@PVA/PEI composite adsorbent on Ponceau S (PS) of up to 91% after ten times cycling, suggesting excellent stability and recycling ability. It should be noted that the reported PDA/PEI@PVA/PEI composite adsorbent showed an inconspicuous loss of the soluble PEI component in the adsorbent material during the adsorption measurements. Next, the adsorption stability and reusability of the PCL/PEO@PDA-45 composite was investigated. The cyclic adsorption experiment was carried out eight times in succession, and each adsorption time was the same as that in Figure 6c for four hours. The dye-adsorbed PCL/PEO@PDA composite was thoroughly washed with ultrapure water and ethanol to remove the adsorbed dye molecules and by-products generated during the adsorption process as much as possible, which was utilized for further cyclic adsorption. The cyclic adsorption efficiency of the PCL/PEO@PDA composite adsorbent towards MO dye is shown in Figure 7. The dye removal efficiency was calculated using the following equation:

$$\text{Removal\%} = \frac{C_0 - C_t}{C_0} \times 100 \tag{3}$$

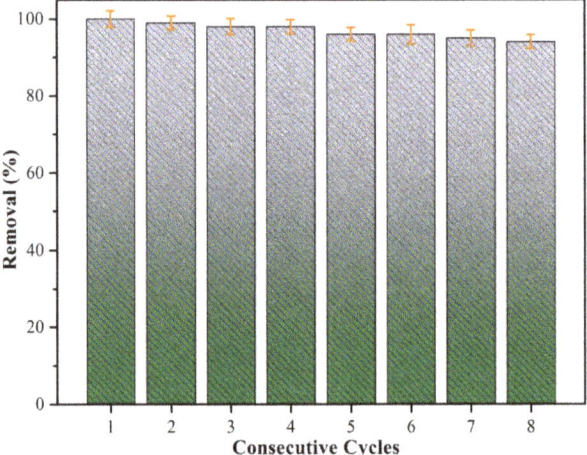

Figure 7. Regeneration studies of the PCL/PEO@PDA-45 composite toward MO for different consecutive cycles at 298 K.

After eight adsorption-desorption cycles, the removal efficiency of the PCL/PEO@PDA composite adsorbent on MO dye decreased from 99% to 93%, indicating excellent stability and recyclability of the adsorbent. The slight decrease in the removal efficiency of the PCL/PEO@PDA composite adsorbent was ascribed to the loss of trace amounts of PDA nanoparticles anchored on the adsorbent surface during the washing process or fewer residual dye molecules on the adsorbent surface [33]. It could be reasonably speculated that the possible loss of the soluble PEO component in our PCL/PEO@PDA-45 composite adsorbent showed an inconspicuous state during the adsorption measurements. The above cyclic adsorption results indicated that the PCL/PEO@PDA composite showed significant adsorption efficiency, stability, and recyclability [59–65].

4. Conclusions

In summary, we successfully designed and prepared PCL/PEO@PDA composites with significantly excellent adsorption capacity and stability. The SVA-treated PCL/PEO fibers exhibited a pleated microstructure, which increased the specific surface area of the smooth electrospun fibers and provided more active attachment sites and spaces for the subsequent chemical modification of polydopamine. The synthesized PCL/PEO@PDA composite after the chemical modification of PDA possessed numerous active sites to anchor the next dye molecules during the adsorption process. Furthermore, the PCL/PEO@PDA composite adsorbent showed a better adsorption capacity than the SVA-treated PCL/PEO composite. Simultaneously, the PCL/PEO@PDA composite adsorbent selectively adsorbed the anionic dye MO, exhibiting better adsorption efficiency than the adsorption toward the used cationic dyes (MB, RHB, and ST). In addition, the PCL/PEO@PDA composite could be separated from the dye solutions to avoid possible agglomeration, exhibiting significant regenerative and reproducible capability for dye removal. The current research work provided a new approach to designing and preparing PDA-based composite adsorbents, showing potential practical applications in the field of wastewater treatment.

Author Contributions: T.J. and Q.P. conceived and designed the experiments; C.W., R.W., and J.Y. performed the experiments; H.H., J.Z., and L.Z. analyzed the data; L.Z. and Q.P. contributed reagents/materials/analysis tools; C.W. and T.J. wrote the paper.

Funding: We greatly appreciate the financial support of the National Natural Science Foundation of China (Nos. 21872119 and 21473153), the Support Program for the Top Young Talents of Hebei Province, China Postdoctoral Science Foundation (No. 2015M580214), the Research Program of the College Science & Technology of Hebei Province (No. ZD2018091), and the Scientific and Technological Research and Development Program of Qinhuangdao City (No. 201701B004).

Conflicts of Interest: The authors declare no conflict of interest.

References

1. Robinson, T.; McMullan, G.; Marchant, R.; Nigam, P. Remediation of dyes in textile effluent: A critical review on current treatment technologies with a proposed alternative. *Bioresour. Technol.* **2001**, *77*, 247–255. [CrossRef]
2. Szlinder-Richert, J.; Usydus, Z.; Malesa-Ciećwierz, M.; Polak-Juszczak, L.; Ruczynska, W. Marine and farmed fish on the Polish market: Comparison of the nutritive value and human exposure to PCDD/Fs and other contaminants. *Chemosphere* **2011**, *85*, 1725–1733. [CrossRef] [PubMed]
3. Dadfarnia, S.; Shabani, A.H.; Moradi, S.E.; Emami, S. Methyl red removal from water by iron based metal-organic frameworks loaded onto iron oxide nanoparticle adsorbent. *Appl. Surf. Sci.* **2015**, *330*, 85–93. [CrossRef]
4. Zhang, Y.R.; Shen, S.L.; Wang, S.Q.; Huang, J.; Su, P.; Wang, Q.R.; Zhao, B.X. A dual function magnetic nanomaterial modified with lysine for removal of organic dyes from water solution. *Chem. Eng. J.* **2014**, *239*, 250–256. [CrossRef]
5. Wang, Y.; Li, Z.; He, Y.; Li, F.; Liu, X.Q.; Yang, J.B. Low-temperature solvothermal synthesis of graphene–TiO_2 nanocomposite and its photocatalytic activity for dye degradation. *Mater. Lett.* **2014**, *134*, 115–118. [CrossRef]
6. Qu, X.; Alvarez, P.J.; Li, Q. Applications of nanotechnology in water and wastewater treatment. *Water Res.* **2013**, *47*, 3931–3946. [CrossRef] [PubMed]
7. Rafatullah, M.; Sulaiman, O.; Hashim, R.; Ahmad, A. Adsorption of methylene blue on low-cost adsorbents: A review. *J. Hazard. Mater.* **2010**, *177*, 70–80. [CrossRef]
8. Ren, T.; Si, Y.; Yang, J.M.; Ding, B.; Yang, X.X.; Hong, F.; Yu, J.Y. Polyacrylonitrile/polybenzoxazine-based Fe_3O_4@carbon nanofibers: Hierarchical porous structure and magnetic adsorption property. *J. Mater. Chem.* **2012**, *22*, 15919–15927. [CrossRef]
9. Rostamian, R.; Najafi, M.; Rafati, A.A. Synthesis and characterization of thiol-functionalized silica nano hollow sphere as a novel adsorbent for removal of poisonous heavy metal ions from water: Kinetics, isotherms and error analysis. *Chem. Eng. J.* **2011**, *171*, 1004–1011. [CrossRef]
10. Mahdavi, S.; Jalali, M.; Afkhami, A. Heavy metals removal from aqueous solutions using TiO_2, MgO, and Al_2O_3 nanoparticles. *Chem. Eng. Commun.* **2013**, *200*, 448–470. [CrossRef]
11. Aditya, D.; Rohan, P.; Suresh, G. Nano-adsorbents for wastewater treatment: A review. *Res. J. Chem. Environ.* **2011**, *15*, 1033–1040.
12. Luo, X.G.; Zhang, L. High effective adsorption of organic dyes on magnetic cellulose beads entrapping activated carbon. *J. Hazard. Mater.* **2009**, *171*, 340–347. [CrossRef]
13. Fu, J.W.; Xin, Q.Q.; Wu, X.C.; Chen, Z.H.; Yan, Y.; Liu, S.J.; Wang, M.H.; Xu, Q. Selective adsorption and separation of organic dyes from aqueous solution on polydopamine microspheres. *J. Colloid Interface Sci.* **2016**, *461*, 292–304. [CrossRef]

14. Wang, X.F.; Yu, J.Y.; Sun, G.; Ding, B. Electrospun nanofibrous materials: A versatile medium for effective oil/water separation. *Mater. Today* **2016**, *19*, 403–414. [CrossRef]
15. Greiner, A.; Wendorff, J.H. Electrospinning: A fascinating method for the preparation of ultrathin fibers. *Angew. Chem. Int. Ed.* **2007**, *46*, 5670–5703. [CrossRef]
16. Hu, X.L.; Liu, S.; Zhou, G.Y.; Huang, Y.B.; Xie, Z.G.; Jing, X.B. Electrospinning of polymeric nanofibers for drug delivery applications. *J. Control. Release* **2014**, *185*, 12–21. [CrossRef]
17. Liu, J.Z.; Bauer, A.J.P.; Li, B.B. Solvent vapor annealing: An efficient approach for inscribing secondary nanostructures onto electrospun fibers. *Macromol. Rapid Commun.* **2014**, *35*, 1503–1508. [CrossRef]
18. Lu, Q.P.; Yu, Y.F.; Ma, Q.L.; Chen, B.; Zhang, H. 2D Transition-metal-dichalcogenide-nanosheet-based composites for photocatalytic and electrocatalytic hydrogen evolution reactions. *Adv. Mater.* **2016**, *28*, 1917–1933. [CrossRef]
19. Lin, J.Y.; Ding, B.; Yang, J.M.; Yu, J.Y.; Sun, G. Subtle regulation of the micro-and nanostructures of electrospun polystyrene fibers and their application in oil absorption. *Nanoscale* **2012**, *4*, 176–182. [CrossRef]
20. Si, Y.; Wang, X.Q.; Li, Y.; Chen, K.; Wang, J.Q.; Yu, J.Y.; Wang, H.J.; Ding, B. Optimized colorimetric sensor strip for mercury (II) assay using hierarchical nanostructured conjugated polymers. *J. Mater. Chem. A* **2014**, *2*, 645–652. [CrossRef]
21. Miao, Y.E.; Wang, R.Y.; Chen, D.; Liu, Z.Y.; Liu, T.X. Electrospun self-standing membrane of hierarchical SiO$_2$@γ-AlOOH (Boehmite) core/sheath fibers for water remediation. *ACS Appl. Mater. Interfaces* **2012**, *4*, 5353–5359. [CrossRef]
22. Zhou, Z.W.; Liu, R. Fe$_3$O$_4$@ polydopamine and derived Fe$_3$O$_4$@ carbon core–shell nanoparticles: Comparison in adsorption for cationic and anionic dyes. *Colloid Surf. A-Physicochem. Eng. Asp.* **2017**, *522*, 260–265. [CrossRef]
23. Dong, Z.H.; Wang, D.; Liu, X.; Pei, X.F.; Chen, L.W.; Jin, J. Bio-inspired surface-functionalization of graphene oxide for the adsorption of organic dyes and heavy metal ions with a superhigh capacity. *J. Mater. Chem. A* **2014**, *2*, 5034–5040. [CrossRef]
24. Wang, C.R.; Sun, S.X.; Zhang, L.X.; Yin, J.J.; Jiao, T.F.; Zhang, L.; Xu, Y.L.; Zhou, J.X.; Peng, Q.M. Facile preparation and catalytic performance characterization of AuNPs-loaded hierarchical electrospun composite fibers by solvent vapor annealing treatment. *Colloid Surf. A-Physicochem. Eng. Asp.* **2019**, *561*, 283–291. [CrossRef]
25. Bauer, A.J.P.; Grim, Z.B.; Li, B. Hierarchical polymer blend fibers of high structural regularity prepared by facile solvent vapor annealing treatment. *Macromol. Mater. Eng.* **2018**, *303*, 1700489. [CrossRef]
26. Huang, X.X.; Jiao, T.F.; Liu, Q.Q.; Zhang, L.X.; Zhou, J.X.; Li, B.B.; Peng, Q.M. Hierarchical electrospun nanofibers treated by solvent vapor annealing as air filtration mat for high-efficiency PM2.5 capture. *Sci. China Mater.* **2019**, *62*, 423–436. [CrossRef]
27. Farnad, N.; Farhadi, K.; Voelcker, N.H. Polydopamine nanoparticles as a new and highly selective biosorbent for the removal of copper(II) ions from aqueous solutions. *Water Air Soil Pollut.* **2012**, *223*, 3535–3544. [CrossRef]
28. Yu, X.; Fan, H.L.; Liu, Y.; Shi, Z.J.; Jin, Z.X. Characterization of carbonized polydopamine nanoparticles suggests ordered supramolecular structure of polydopamine. *Langmuir* **2014**, *30*, 5497–5505. [CrossRef]
29. Gan, L.; Shang, S.M.; Hu, E.L.; Yuen, C.W.M.; Jiang, S.X. Konjac glucomannan/graphene oxide hydrogel with enhanced dyes adsorption capability for methyl blue and methyl orange. *Appl. Surf. Sci.* **2015**, *357*, 866–872. [CrossRef]
30. Cheng, C.; Li, S.; Zhao, W.H.; Wei, Q.; Nie, S.Q.; Sun, S.D.; Zhao, C.S. The hydrodynamic permeability and surface property of polyethersulfone ultrafiltration membranes with mussel-inspired polydopamine coatings. *J. Membr. Sci.* **2012**, *417*, 228–236. [CrossRef]
31. Lee, H.; Dellatore, S.M.; Miller, W.M.; Messersmith, P.B. Mussel-inspired surface chemistry for multifunctional coatings. *Science* **2007**, *318*, 426–430. [CrossRef]
32. Lee, H.; Scherer, N.F.; Messersmith, P.B. Single-molecule mechanics of mussel adhesion. *Proc. Natl. Acad. Sci. USA* **2006**, *103*, 12999–13003. [CrossRef]
33. Xing, R.R.; Wang, W.; Jiao, T.F.; Ma, K.; Zhang, Q.R.; Hong, W.; Qiu, H.; Zhou, J.X.; Zhang, L.X.; Peng, Q.M. Bioinspired polydopamine sheathed nanofibers containing carboxylate graphene oxide nanosheet for high-efficient dyes scavenger. *ACS Sustain. Chem. Eng.* **2017**, *5*, 4948–4956. [CrossRef]
34. Tang, L.; Livi, K.J.T.; Chen, K.L. Polysulfone membranes modified with bioinspired polydopamine and silver nanoparticles formed in situ to mitigate biofouling. *Environ. Sci. Technol. Lett.* **2015**, *2*, 59–65. [CrossRef]

35. Ji, Y.; Ghosh, K.; Shu, X.Z.; Li, B.Q.; Sokolov, J.C.; Prestwich, G.D.; Clark, R.A.F.; Rafailovich, M.H. Electrospun three-dimensional hyaluronic acid nanofibrous scaffolds. *Biomaterials* **2006**, *27*, 3782–3792. [CrossRef]
36. Yang, X.H.; Shao, C.L.; Guan, H.Y.; Li, X.L.; Gong, J. Preparation and characterization of ZnO nanofibers by using electrospun PVA/zinc acetate composite fiber as precursor. *Inorg. Chem. Commun.* **2004**, *7*, 176–178. [CrossRef]
37. Konwer, S.; Boruah, R.; Dolui, S.K. Studies on conducting polypyrrole/graphene oxide composites as supercapacitor electrode. *J. Electron. Mater.* **2011**, *40*, 2248. [CrossRef]
38. Contarini, S.; Howlett, S.P.; Rizzo, C.; De Angelis, B.A. XPS study on the dispersion of carbon additives in silicon carbide powders. *Appl. Surf. Sci.* **1991**, *51*, 177–183. [CrossRef]
39. Jiang, H.; Wang, X.B.; Li, C.Y.; Li, J.S.; Xu, F.J.; Mao, C.; Yang, W.T.; Shen, J. Improvement of hemocompatibility of polycaprolactone film surfaces with zwitterionic polymer brushes. *Langmuir* **2011**, *27*, 11575–11581. [CrossRef]
40. Manakhov, A.; Kedroňová, E.; Medalová, J.; Černochová, P.; Obrusník, A.; Michlíček, M.; Shtansky, D.V.; Zajíčková, L. Carboxyl-anhydride and amine plasma coating of PCL nanofibers to improve their bioactivity. *Mater. Des.* **2017**, *132*, 257–265. [CrossRef]
41. Zhu, Z.G.; Wu, P.; Liu, G.J.; He, X.F.; Qi, B.Y.; Zeng, G.F.; Wang, W.; Sun, Y.H.; Cui, F.Y. Ultrahigh adsorption capacity of anionic dyes with sharp selectivity through the cationic charged hybrid nanofibrous membranes. *Chem. Eng. J.* **2017**, *313*, 957–966. [CrossRef]
42. Ye, W.C.; Chen, Y.; Zhou, Y.X.; Fu, J.J.; Wu, W.C.; Gao, D.Q.; Zhou, F.; Wang, C.M.; Xue, D.S. Enhancing the catalytic activity of flowerike Pt nanocrystals using polydopamine functionalized graphene supports for methanol electrooxidation. *Electrochim. Acta* **2014**, *142*, 18–24. [CrossRef]
43. Zhao, X.N.; Jiao, T.F.; Xing, R.R.; Huang, H.; Hu, J.; Qu, Y.; Zhou, J.X.; Zhang, L.X.; Peng, Q.M. Preparation of diamond-based AuNP-modified nanocomposites with elevated catalytic performances. *RSC Adv.* **2017**, *7*, 49923–49930. [CrossRef]
44. Zhao, X.N.; Jiao, T.F.; Ma, X.L.; Huang, H.; Hu, J.; Qu, Y.; Zhou, J.X.; Zhang, L.X.; Peng, Q.M. Facile fabrication of hierarchical diamond-based AuNPs-modified nanocomposites via layer-by-layer assembly with enhanced catalytic capacities. *J. Taiwan Inst. Chem. Eng.* **2017**, *80*, 614–623. [CrossRef]
45. Duan, B.; Dong, C.H.; Yuan, X.Y.; Yao, K.D. Electrospinning of chitosan solutions in acetic acid with poly (ethylene oxide). *J. Biomater. Sci. Polym. Ed.* **2004**, *15*, 797–811. [CrossRef] [PubMed]
46. Lu, C.H.; Chiang, S.W.; Du, H.D.; Li, J.; Gan, L.; Zhang, X.; Chu, X.D.; Yao, Y.W.; Li, B.H.; Kang, F.Y. Thermal conductivity of electrospinning chain-aligned polyethylene oxide (PEO). *Polymer* **2017**, *115*, 52–59. [CrossRef]
47. Canbolat, M.F.; Celebioglu, A.; Uyar, T. Drug delivery system based on cyclodextrin-naproxen inclusion complex incorporated in electrospun polycaprolactone nanofibers. *Colloids Surf. B Biointerfaces* **2014**, *115*, 15–21. [CrossRef] [PubMed]
48. Tavares, M.R.; de Menezes, L.R.; Dutra Filho, J.C.D.; Cabral, L.M.; Tavares, M.I.B. Surface-coated polycaprolactone nanoparticles with pharmaceutical application: Structural and molecular mobility evaluation by TD-NMR. *Polym. Test.* **2017**, *60*, 39–48. [CrossRef]
49. Zhu, B.; Edmondson, S. Polydopamine-melanin initiators for surface-initiated ATRP. *Polymer* **2011**, *52*, 2141–2149. [CrossRef]
50. Liu, Q.; Liu, Q.Z.; Wu, Z.T.; Wu, Y.; Gao, T.T.; Yao, J.S. Efficient removal of methyl orange and alizarin red S from pH-unregulated aqueous solution by the catechol–amine resin composite using hydrocellulose as precursor. *ACS Sustain. Chem. Eng.* **2017**, *5*, 1871–1880. [CrossRef]
51. Guo, R.; Jiao, T.F.; Li, R.F.; Chen, Y.; Guo, W.C.; Zhang, L.X.; Zhou, J.X.; Zhang, Q.R.; Peng, Q.M. Sandwiched Fe_3O_4/carboxylate graphene oxide nanostructures constructed by layer-by-layer assembly for highly efficient and magnetically recyclable dye removal. *ACS Sustain. Chem. Eng.* **2018**, *6*, 1279–1288. [CrossRef]
52. Liu, Y.; Hou, C.; Jiao, T.; Song, J.; Zhang, X.; Xing, R.; Zhou, J.; Zhang, L.; Peng, Q. Self-assembled AgNP-containing nanocomposites constructed by electrospinning as efficient dye photocatalyst materials for wastewater treatment. *Nanomaterials* **2018**, *8*, 35. [CrossRef] [PubMed]
53. Li, K.; Jiao, T.; Xing, R.; Zou, G.; Zhou, J.; Zhang, L.; Peng, Q. Fabrication of tunable hierarchical MXene@AuNPs nanocomposites constructed by self-reduction reactions with enhanced catalytic performances. *Sci. China Mater.* **2018**, *61*, 728–736. [CrossRef]

54. Zhou, J.; Gao, F.; Jiao, T.; Xing, R.; Zhang, L.; Zhang, Q.; Peng, Q. Selective Cu(II) ion removal from wastewater via surface charged self-assembled polystyrene-Schiff base nanocomposites. *Colloid Surf. A-Physicochem. Eng. Asp.* **2018**, *545*, 60–67. [CrossRef]
55. Chen, K.; Li, J.; Zhang, L.; Xing, R.; Jiao, T.; Gao, F.; Peng, Q. Facile synthesis of self-assembled carbon nanotubes/dye composite films for sensitive electrochemical determination of Cd(II) ions. *Nanotechnology* **2018**, *29*, 445603. [CrossRef] [PubMed]
56. Luo, X.; Ma, K.; Jiao, T.; Xing, R.; Zhang, L.; Zhou, J.; Li, B. Graphene oxide-polymer composite Langmuir films constructed by interfacial thiol-ene photopolymerization. *Nanoscale Res. Lett.* **2017**, *12*, 99. [CrossRef] [PubMed]
57. Huo, S.; Duan, P.; Jiao, T.; Peng, Q.; Liu, M. Self-assembled luminescent quantum dots to generate full-color and white circularly polarized light. *Angew. Chem. Int. Ed.* **2017**, *56*, 12174–12178. [CrossRef]
58. Xing, R.; Liu, K.; Jiao, T.; Zhang, N.; Ma, K.; Zhang, R.; Zou, Q.; Ma, G.; Yan, X. An injectable self-assembling collagen-gold hybrid hydrogel for combinatorial antitumor photothermal/photodynamic therapy. *Adv. Mater.* **2016**, *28*, 3669–3676. [CrossRef]
59. Guo, H.; Jiao, T.; Zhang, Q.; Guo, W.; Peng, Q.; Yan, X. Preparation of graphene oxide-based hydrogels as efficient dye adsorbents for wastewater treatment. *Nanoscale Res. Lett.* **2015**, *10*, 272. [CrossRef]
60. Zhan, F.; Wang, R.; Yin, J.; Han, Z.; Zhang, L.; Jiao, T.; Zhou, J.; Zhang, L.; Peng, Q. Facile solvothermal preparation of Fe$_3$O$_4$–Ag nanocomposite with excellent catalytic performance. *RSC Adv.* **2019**, *9*, 878–883. [CrossRef]
61. Sun, S.; Wang, C.; Han, S.; Jiao, T.; Wang, R.; Yin, J.; Li, Q.; Wang, Y.; Geng, L.; Yu, X.; et al. Interfacial nanostructures and acidichromism behaviors in self-assembled terpyridine derivatives Langmuir-Blodgett films. *Colloid Surf. A-Physicochem. Eng. Asp.* **2019**, *564*, 1–9. [CrossRef]
62. Xu, Y.; Ren, B.; Wang, R.; Zhang, L.; Jiao, T.; Liu, Z. Facile preparation of rod-like MnO nanomixtures via hydrothermal approach and highly efficient removal of methylene blue for wastewater Treatment. *Nanomaterials* **2019**, *9*, 10. [CrossRef]
63. Liu, K.; Yuan, C.Q.; Zou, Q.L.; Xie, Z.C.; Yan, X.H. Self-assembled Zinc/cystine-based chloroplast mimics capable of photoenzymatic reactions for sustainable fuel synthesis. *Angew. Chem. Int. Ed.* **2017**, *56*, 7876–7880. [CrossRef] [PubMed]
64. Liu, K.; Xing, R.R.; Li, Y.X.; Zou, Q.L.; Möhwald, H.; Yan, X.H. Mimicking primitive photobacteria: Sustainable hydrogen evolution based on peptide-porphyrin co-assemblies with self-mineralized reaction center. *Angew. Chem. Int. Ed.* **2016**, *55*, 12503–12507. [CrossRef]
65. Liu, K.; Xing, R.R.; Chen, C.J.; Shen, G.Z.; Yan, L.Y.; Zou, Q.L.; Ma, G.H.; Möhwald, H.; Yan, X.H. Peptide-induced hierarchical long-range order and photocatalytic activity of porphyrin assemblies. *Angew. Chem. Int. Ed.* **2015**, *54*, 500–505. [CrossRef]

© 2019 by the authors. Licensee MDPI, Basel, Switzerland. This article is an open access article distributed under the terms and conditions of the Creative Commons Attribution (CC BY) license (http://creativecommons.org/licenses/by/4.0/).

Article

Dy(III) Doped BiOCl Powder with Superior Highly Visible-Light-Driven Photocatalytic Activity for Rhodamine B Photodegradation

Jun Yang [1,†], **Taiping Xie** [2,3,†], **Chenglun Liu** [3,4,*] **and Longjun Xu** [3,*]

1. College of Materials and Chemical Engineering, Chongqing University of Arts and Sciences, Yongchuan 402160, China; bbyangjun@foxmail.com
2. Chongqing Key Laboratory of Extraordinary Bond Engineering and Advanced Materials Technology (EBEAM), Yangtze Normal University, Chongqing 408100, China; deartaiping@163.com
3. State Key Laboratory of Coal Mine Disaster Dynamics and Control, Chongqing University, Chongqing 400044, China
4. College of Chemistry and Chemical Engineering, Chongqing University, Chongqing 401331, China
* Correspondence: xlclj@cqu.edu.cn (C.L.); xulj@xqu.edu.cn (L.X.)
† These authors contributed equally to this work.

Received: 15 August 2018; Accepted: 4 September 2018; Published: 6 September 2018

Abstract: Dy-doped BiOCl powder photocatalyst was synthesized A one–step coprecipitation method. The incorporation of Dy^{3+} replaced partial Bi^{3+} in BiOCl crystal lattice system. For Rhodamine B (RhB) under visible light irradiation, 2% Dy doped BiOCl possessed highly efficient photocatalytic activity and photodegradation efficiency. The photodegradation ratio of RhB could reach 97.3% after only 30 min of photocatalytic reaction; this was more than relative investigations have reported in the last two years. The main reason was that the 4f electron shell of Dy in the BiOCl crystal lattice system can generate a special electronic shell structure that facilitated the transfer of electron from valance band to conduction band and separation of the photoinduced charge carrier. Apart from material preparation, this research is expected to provide important references for RhB photodegradation in practical applications.

Keywords: Dy^{3+}; BiOCl; photocatalyst; RhB photodegradation; doping modification

---

1. Introduction

Photocatalytic technology using visible light irradiation is an environmentally-friendly approach towards environmental pollutant treatment. It has attracted considerable attention due to inexhaustible visible light from solar light energy. Over the past few decades, some visible-light-driven photocatalysts were engineered and fabricated for the photodegradation of organic wastewater. For example, $Fe^{(0)}$ doped $g-C_3N_4/MoS_2$, fluorinated Bi_2WO_6, and TiO_2 with interface defects [1–3] were synthesized, specifically aiming at Rhodamine B (RhB) photodegradation. However, the corresponding photocatalytic reactions were very slow and took several hours to degrade less than 98% RhB, thus impeding their practical application due to the high time costs. If the used light source was simulated visible light, and not solar light, the long times of the photocatalytic reaction would increase energy consumption and costs. Therefore, the challenges we face are how to enhance photocatalytic reaction kinetics, to shorten reaction time, and to boost photocatalytic efficiency under identical visible light irradiation.

Bi-based photocatalysts are important visible-light-responsive photocatalysts and have recently attracted increasing attention. Considering the stability of Bi^{3+}, Bi^{3+}-containing compounds, such as Bi_2O_3, $BiVO_4$, Bi_2WO_6, $BiPO_4$, $BiFeO_3$, and BiOX (X = Cl, Br, I) [4], were synthesized for photocatalytic reactions. Most of these compounds, especially BiOX (X = Cl, Br, I), possessed a layered structure and a plate-like appearance that could produce an internal electric field [5,6], which facilitated the migration of photoinduced carriers to some extent. Among these Bi^{3+}-based photocatalysts, studies regarding the structure and properties of BiOCl were found. In fact, the photogenerated electrons and holes of BiOCl have not been easily exploited and utilized under visible light irradiation [4,7].

Foreign ion doping has been widely adopted to increase the visible light absorption for single phase photocatalysts, because this process can generate a doping level between the conduction band (CB) and valence band (VB) [8]. Consequently, the energy required to excite electrons is decreased, and the light response of semiconductors is enhanced. Doping modification also steered the charges migrating in a special manner for semiconductors, thus leading to an augmented transfer efficiency of carriers. Hence, doping modifications were commonly employed to boost the photocatalytic activity of Bi-based semiconductors.

Dy-doped ZnO nanoparticles using a photocatalyst were also investigated [9–11]. This rare earth metal (Dy) was used as an efficient dopant into the interstitial sites of the ZnO crystal structure. It has been found that rare-earth-metal-doping modification could reduce the electron–hole pair recombination, which is the precondition for efficient photocatalytic applications. Dy could also be incorporated in the crystal lattice of ZnO and thus form ZnO nanoparticles, which could be tuned for optical, morphological and photocatalytic properties. Dy^{3+} ions are well known as an activated dopant for different inorganic crystal lattices, producing visible light by appropriately adjusting yellow and blue emissions.

To the best of our knowledge, no studies have investigated the application of in situ synthesized Dy-doped BiOCl powder photocatalyst for RhB photodegradation in aqueous environments. The as-synthesized Dy-doped BiOCl possessed good photocatalytic activity for the removal of RhB. A total of 97.3% RhB could be degraded in only 30 min of photocatalytic reaction under visible light irradiation.

2. Experimental Section

All reagents were of analytical grade purity and were used directly without further purification. Deionized water was used in all experimental processes.

2.1. Materials Syntheses

One gram of polyvinylpyrrolidone K30 (PVP K30, Aladdin, Shanghai, China) was completely dissolved in 100 mL distilled water via agitation to form a homogeneous solution that was divided into two parts. A total of 5 mmol Bi $(NO_3)_3 \cdot 5H_2O$ (Sigma-Aldrich, Hongkong, China) and a proper mol% amount of Dy $(NO_3)_3 \cdot 6H_2O$ (0.5%, 1%, 1.5%, 2%, 2.5%) (Sigma-Aldrich) were then completely dissolved in the above homogeneous solution by stirring for 30 min. A total of 5 mmol KCl (JZ Chemical, Taoyuan, Taiwan) was also dissolved in the above homogeneous solution by stirring. Then, KCl solution was slowly added into Bi $(NO_3)_3 \cdot 5H_2O$ and Dy $(NO_3)_3 \cdot 6H_2O$ mixed suspension under magnetic stirring conditions, and then continuously stirred for 4 h. The mixed solution was filtered. The obtained filter cake was washed several times using 300 mL deionized water and 150 mL absolute ethyl alcohol. The washed filter cake was dried at 80 °C for 10 h to obtain the resultant Dy-doped BiOCl. The pure BiOCl was synthesized using the similar process without adding Dy $(NO_3)_3 \cdot 6H_2O$.

2.2. Materials Characterization

X-ray diffraction (XRD) measurements were conducted using standard powder diffraction procedures. The samples were smear-mounted on a glass slide and analyzed at a scan rate of 4° (2θ) min^{-1} using monochromatic Cu Kα radiation (MAC Science, MXP18, Tokyo, Japan) at 30 kV and 20 mA. The recorded specific peak intensities and 2θ values were further identified by a computer database system (JCPDS). The chemical compositions of the samples were determined with an X-ray photoelectron spectra (XPS, Physical Electronic ESCA PHI 1600, Chanhassan, MN, USA) at an excitation energy of 1486.6 eV of Al Kα. The C 1s (284.5 eV) signal served as a calibration standard for the Bi and Dy species and their spectra over a wide region. XPS signals of the above species were recorded with a cylindrical mirror analyzer (CMA). The Raman scattering measurements were performed on BiDyxOCl powder samples using an INVIA Raman microprobe (Renishaw Instruments, Wotton-under-Edge, England). The microprobe had an excitation source (488 nm) that was well equipped with a Peltier cooled charge coupled device detector. The morphology, microstructure, and particle size of the as-prepared samples were characterized by field–emission scanning electron microscopy (FE–SEM) (Hitachi, S–4700 Type II, Tokyo, Japan) with a resolution of 0.1 nm and using a high-resolution transmission electron microscopy (HR–TEM) (Hitachi H–7500, Honshū, Japan) at 100 kV, after dispersing the samples on a carbon film supported on a copper grid. The pore volume and surface area of samples were calculated from the nitrogen adsorption–desorption isotherms measured at −196 °C using an ASAP 2010 instrument (micromeritics with surface area deviation of 1%) (ASAP-2010, Micromeritics, Norcross, GA, USA). The optical properties of samples were examined using an ultraviolet–visible diffuse reflectance spectrophotometer (UV-vis DRS, TU1901, Beijing, China).

2.3. Photocatalytic Test

One hundred milligrams of powdered photocatalyst were added to 10.0 mg·L^{-1} RhB solution (100.0 mL). The solution was placed in the dark for 1 h, while stirring (500 r·min^{-1}) to reach to adsorption–desorption equilibrium (See Figure S1). Single wavelength light-emitting diode (LED) visible light (λ = 470 nm) was used as the visible light source (power = 140 W). At given irradiation time intervals, a series of the reaction solution was sampled and the absorption spectrum was measured.

3. Results and Discussion

3.1. Phase Analyses

Figure 1 displays the XRD crystal diffraction patterns of pure BiOCl and Dy-doped BiOCl samples. It can be seen that the diffraction peaks are obviously broadened, indicating that the grain size was smaller, which was due to the size effect of PVP. The smaller grain size contributed to the growth of the final products of smaller particles.

The diffraction peaks of all samples were fully indexed into BiOCl (JCPDS card number: PDF#06-0248) for the tetragonal system [12]. After the introduction of Dy, the diffraction peak of BiOCl had no obvious displacement, which was similar to the investigation of Eu-doped BiOCl. Similar phenomena were observed through investigations of Cu-, Co-, and Fe-doped BiOCl [13–16]. The Dy$_2$O$_3$ phase or other impurity peaks were not observed. This was indicated that Dy^{3+} substituted for Bi^{3+} in the BiOCl crystal lattice. The grain sizes of pure BiOCl and (0.5%, 1%, 1.5%, 2% and 2.5%) Dy-doped BiOCl were 9.9, 8.5, 8.3, 7.8, 7.8, and 7.7 nm, determined via calculation using the Scherrer Equation. It can be seen that, with an increase in the Dy doping amount, the grain size of BiOCl gradually decreased, which was due to the distortion of the crystal cell structure caused by the larger ionic radius of the Bi^{3+} ion (r_{Bi}^{3+} = 1.17 Å) in the crystal lattice, substituted by Dy^{3+} with a smaller ion radius (r_{Dy}^{3+} = 0.91 Å).

The Raman spectra of pure BiOCl and Dy doped BiOCl samples are depicted in Figure 2. The intensity of the symmetric vibration peaks in the Raman spectrum was stronger than that of the asymmetric vibration peaks. The peak at 143.3 cm^{-1} in the spectrum was assignable to symmetrical stretching vibration of the Bi-Cl bond. The peak at 199.6 cm^{-1} was ascribable to the symmetric expansion vibration of the Bi-Cl bond [17]. In addition, the weak and wide peak at 398.0 cm^{-1} was attributable to the oxygen atom vibration peak in the BiOCl system. The asymmetric stretching vibration peak of the Bi-Cl bond should appear at 60.0 cm^{-1}, which was not detected here, because the peak intensity of the asymmetric vibration was too weak. Meanwhile, the Raman peaks of BiOCl at 84.0 cm^{-1} and the peaks of BiDy$_{2.0}$OCl at 88.3 cm^{-1} were observed, which was due to the crystal lattice distortion caused by Dy doping into the BiOCl crystal lattice. With the increase in Dy doping amount, the Raman peak at 199.6 cm^{-1} gradually moved to a low wave number, which resulted from grain size reduction with the increase in the Dy doping amount.

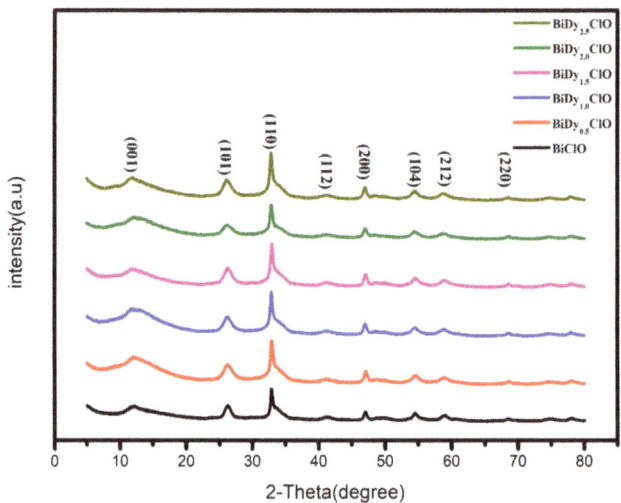

Figure 1. XRD patterns of pure BiOCl and Dy-doped BiOCl.

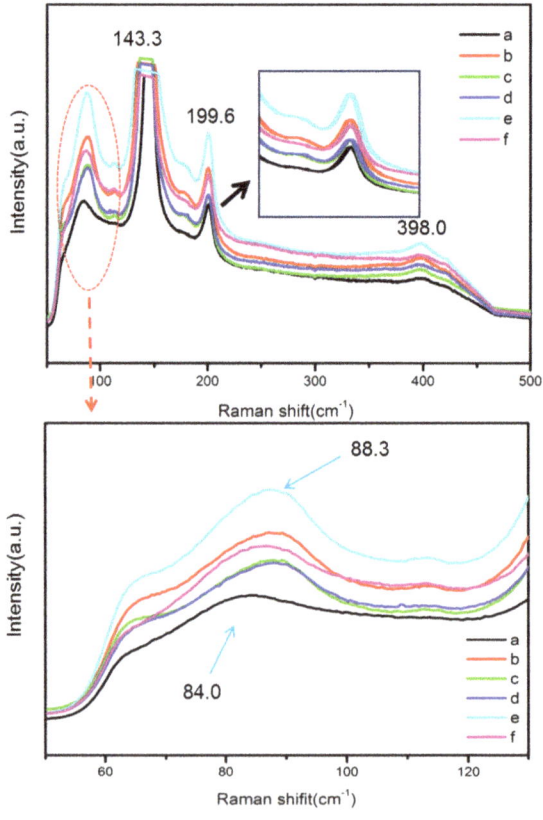

Figure 2. The Raman spectra of (**a**) BiClO; (**b**) BiDy$_{0.5}$ClO; (**c**) Bi0Dy$_{1.0}$ClO; (**d**) BiDy$_{1.5}$ClO; (**e**) BiDy$_{2.0}$ClO; and (**f**) BiDy$_{2.5}$ClO.

The surface composition and chemical state of a 2% Dy-doped BiOCl sample was analyzed by X-ray photoelectron spectroscopy (XPS), as shown in Figure 3. The Dy element was not detected via common or etch methods, which was attributed to the incorporation of Dy entering into the crystal lattice of BiOCl. Figure 3a shows the XPS full spectra of the original sample and the sample with 30 s of denudation. Only Bi, O, Cl and C were found in the full XPS spectrum of the original sample. The Dy element still could not be detected after etching sample for 30 s. It can be further confirmed that the introduction of Dy^{3+} replaced Bi^{3+} in the BiOCl crystal lattice system.

Figure 3b–d shows the high-resolution spectra of the etching sample. The peaks at 159.09 eV and 164.35 eV could be ascribable to Bi $4f_{7/2}$ and Bi $4f_{5/2}$ spin-orbital Bi^{3+} in BiOCl [18]. The O1s peak at 532.9 eV was from the oxygen atom of the Bi-O bond. The peak of the Cl 2p photoelectron peak appeared at 198.95 eV, corresponding to Cl^- in BiOCl.

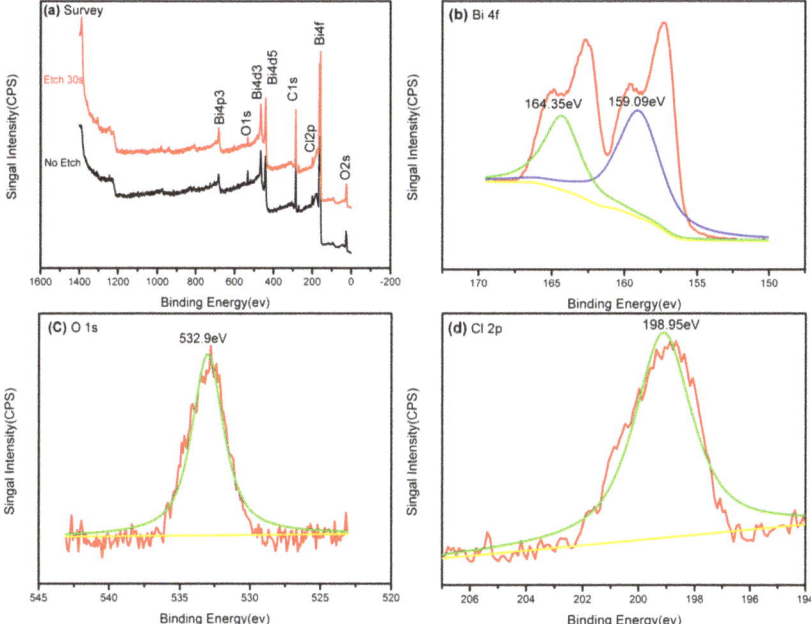

Figure 3. XPS spectra of 2% Dy doped BiOCl (a) and the corresponding high-resolution XPS spectra of Bi (b); O (c); Cl (d).

The adsorption–desorption isotherms and the pore size distribution curves (See inset) for pure BiOCl and $BiDy_{2.0}OCl$ are shown in Figure 4. The most probable pore-size distributions were 2.76 nm and 2.30 nm for pure BiOCl and $BiDy_{2.0}OCl$, respectively. The introduction of Dy increased the number of micropore and mesopores, which can be confirmed via the adsorption–desorption isotherms. The BET surface areas for BiOCl and $BiDy_{2.0}OCl$ were 4.15 $m^2 \cdot g^{-1}$ and 9.45 $m^2 \cdot g^{-1}$, respectively. Incorporation of Dy could increase the surface area. The smaller ionic radii Dy substituted for Bi could reduce the grain size of BiOCl and further increase its special surface area. The larger the surface area, the more surface-active sites for a catalyst.

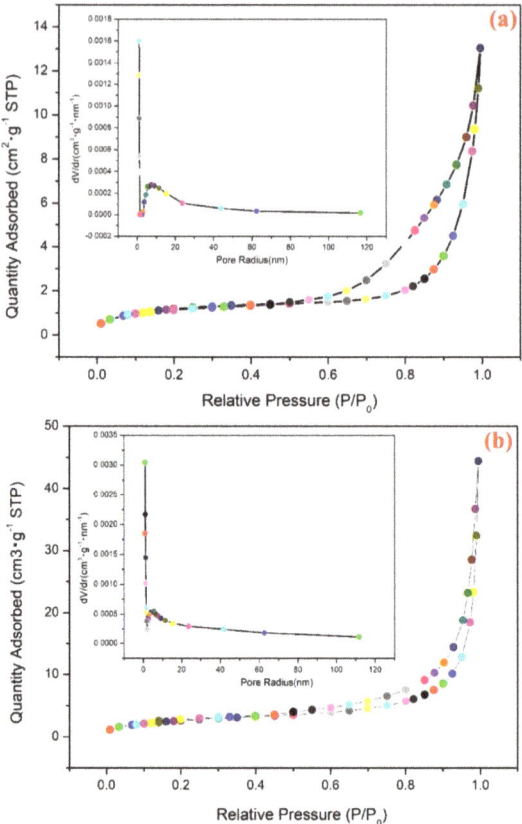

Figure 4. The adsorption–desorption isotherms and the pore size distribution curves (inset). (**a**) BiOCl and (**b**) BiDy$_{2.0}$OCl.

3.2. Micromorphology Analyses

An SEM image of pure BiOCl is shown in Figure 5a. The image presents a spherical-flower structure with a size of 1–2 μm. With the increase in Dy doping amount from 0.5 to 2.5%, the surface structure remained similar, which revealed that the incorporation of Dy did not change the surface morphology and crystal structure of BiOCl. This further confirmed that Dy^{3+} entered into the crystal lattice structure of BiOCl.

Figure 6 shows pure BiOCl and 2% Dy-doped BiOCl transmission electron microscope (TEM) diagrams and the corresponding high transmission electron microscope (HR-TEM, Hitachi H–7500, Honshū, Japan) diagrams, which are illustrations of the selected area electron diffraction (SAED). From Figure 5a,c, it can be seen that the morphologies of pure BiOCl and 2% Dy-doped BiOCl are spherical in structure, which was consistent with the results of the scanning electron microscope test. The results of HRTEM revealed that the crystal lattice stripe was clearly visible and highly consistent, indicating that the crystal structure of the sample was complete and that the crystallinity was good. Further tests indicated that the spacing of the crystal lattice stripe was 0.275 nm [19] (Figure 6b,d), which corresponded to the space between the (110) surface of the BiOCl of the tetragonal system, which was also in accordance with the XRD results. Meanwhile, selective electron diffraction of BiOCl and 2% Dy-doped BiOCl was carried out. As shown in Figure 6b,d, clear diffraction points could be observed. The sample belonged to single crystal, corresponding to BiOCl and Dy-doped BiOCl (110) and (200) surfaces, respectively, thus indicating that the samples had good crystallinity.

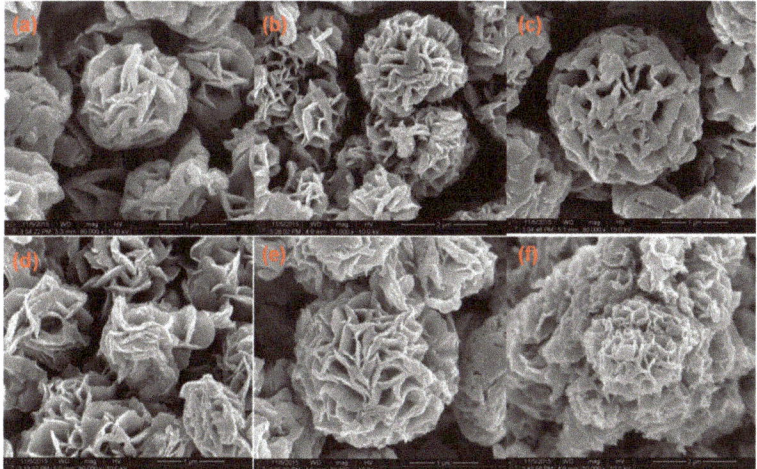

Figure 5. SEM images of BiOCl (**a**) and (0.5%, 1%, 1.5%, 2% and 2.5%) Dy-doped BiOCl (**b**–**f**).

3.3. Optical Properties

Figure 7 displays UV–vis diffuse reflectance spectroscopy results and the corresponding band gap energy diagrams of pure BiOCl and Dy-doped BiOCl.

As can be seen from Figure 7a, with the increase in Dy doping amount, the absorption band of the samples presented a red shift, and the absorption capacity of light increased gradually. Through calculation of the band gap width (Figure 7b), it was found that, when the doping amount was 2.5%, the band gap energy decreased to 3.31 eV from 3.42 eV for pure BiOCl. There was no change in the color of the sample with the introduction of Dy, which implied that Dy showed a change in the electronic structure of the BiOCl, was altered when Dy^{3+} entered the BiOCl crystal lattice and replaced Bi^{3+}.

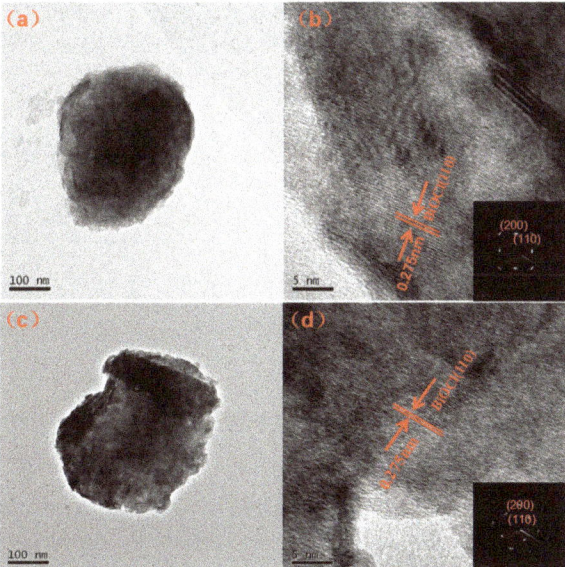

Figure 6. TEM images of BiOCl (**a**), 2% Dy doped BiOCl (**c**) and corresponding HRTEM (**b**,**d**); Inset: SAED pattern of the BiOCl and 2% Dy-doped BiOCl.

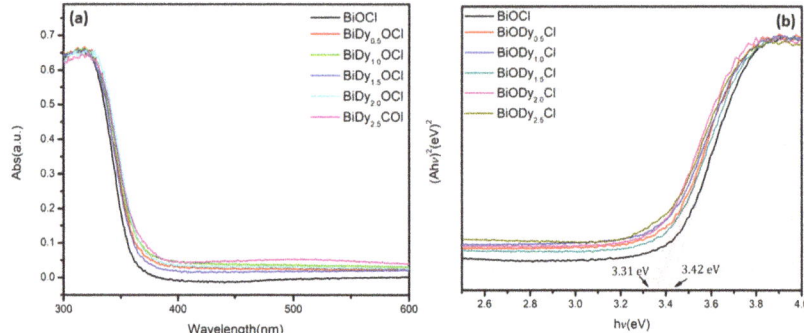

Figure 7. UV–vis diffuses reflectance spectra BiOCl with different Dy contents (**a**) and the corresponding band gap energy (**b**).

The effect of Mn doping on the electronic structure of BiOCl was thoroughly studied using the first principle [20]. It was found that Mn doping could make the whole energy level of BiOCl move to a low energy level. The incorporation of Mn was in the middle of the band gap and the bottom of conduction band of the BiOCl system, which produced a new impurity level, further reducing the width of the band gap.

Rare earth metal elements possess unique optical properties because of their unique electronic structures. The main reason for this is the existence of the 4f electron shell [21].

The electronic shell structure of rare earth metal elements could be expressed as: $4f^N 5s^2 5p^6$. The electron at 4f shell was not the outermost electron, but indeed the photoactive electrons were at the 4f layer [21]. Therefore, the doping of rare earth metal ions would affect their electronic structure and change their optical properties. Here, the reduction in band gap energy was attributed to the introduction of Dy^{3+} ions which could generate a new electron energy level in the middle of the band gap of BiOCl.

The conduction band of BiOCl was mainly composed of Bi 6p, and the valence band was mainly composed of O 2p, Cl 3p and a small amount of Bi 6s [22]. Electrons were passed from O 2p and Cl 3p to the 4f electronic shell of Dy instead of being directly transmitted to Bi 6p, which can be seen in Figure 8. This was a "springboard" between the valence band and the conduction band, which facilitated the easy jump to the conduction band for photoexcited electrons.

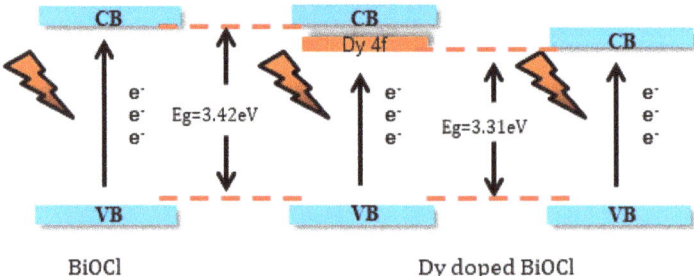

Figure 8. The schematic of reduction in the band gap of BiOCl.

3.4. Photocatalytic Activity and Corresponding Mechanism

In order to find the optimal Dy doping amount, the photocatalytic performances of pure BiOCl and Dy-doped BiOCl (0.5%, 1%, 1.5%, 2% and 2.5%) samples were investigated via RhB photodegradation under visible light irradiation. This was also the original intention of this modification study for BiOCl. Here, a single wavelength LED light source was used as simulation visible light due to its low light intensity and securing wavelength. The experimental results are shown in Figure 9.

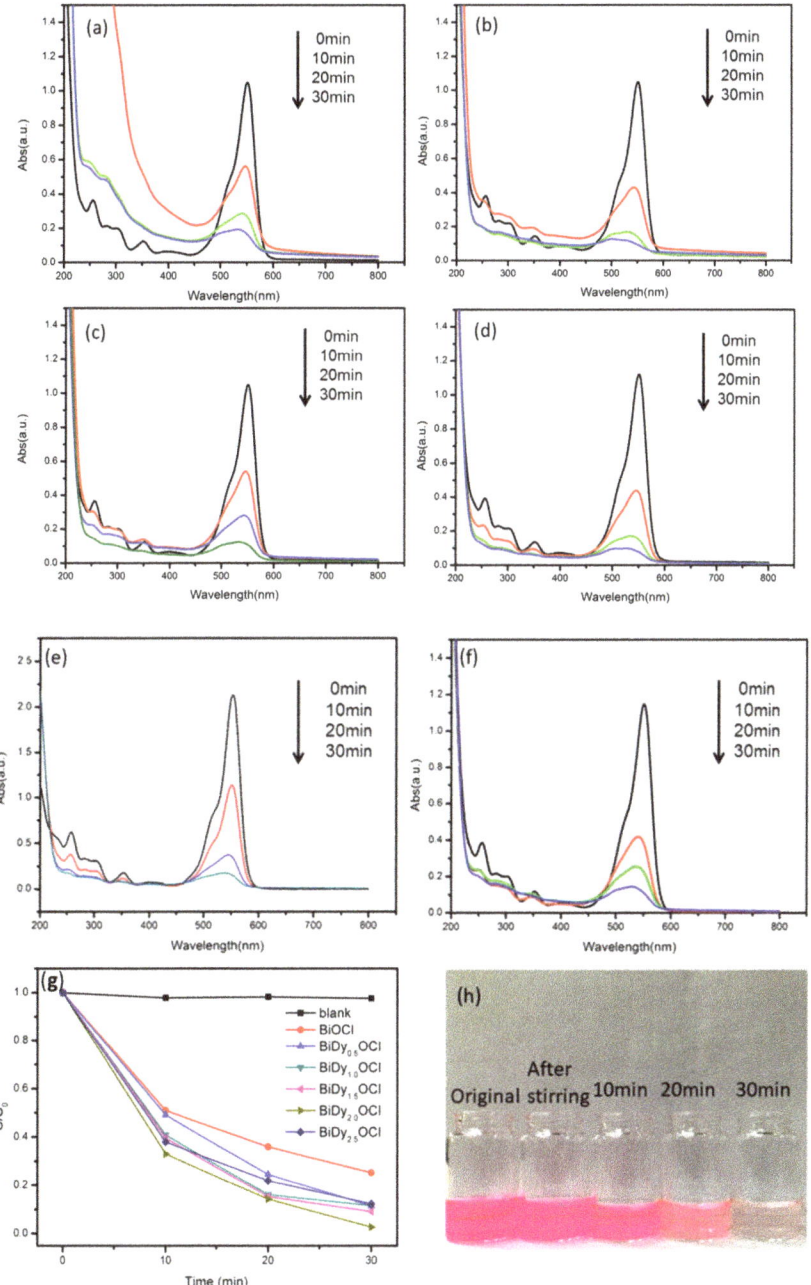

Figure 9. Time-dependent UV-vis absorption spectra of the RhB in the presence of various Dy-doped BiOCl photocatalysts (**a**–**f**) and the corresponding degradation ratio of RhB (**g**). The color change of RhB using 2% Dy doped BiOCl in photodegradation process (**h**).

Figure 9a–f shows the time-dependent UV-vis absorption spectra of RhB in the presence of the as-prepared samples. Figure 9g,h shows the photocatalytic degradation ratio of RhB versus visible light irradiation time and the color change of RhB using 2% Dy-doped BiOCl as a photocatalyst. It was seen that the 2% Dy-doped BiOCl possessed the best photocatalytic activity under identical visible light irradiation. Using 2% Dy-doped BiOCl, the RhB photodegradation ratio reached 97.3%, which was 1.3 times more than pure BiOCl under identical light irradiation (see Table 1). With the increase in the Dy doping amount from 0.5 to 1.5%, photocatalytic efficiency increased tremendously. However, when the Dy doping amount was 2.5%, the RhB photodegradation ratio decreased, which indicated that excessive Dy doping was detrimental to the enhancement of the photocatalytic activity of BiOCl. In addition, the reaction rate constants were determined from the RhB degradation kinetic curves (Figure 10). The reaction rate constant for RhB degradation using 2% Dy-doped BiOCl as photocatalyst was greatest, ca. 0.084 min^{-1}, which was more than 1.3 times that of pure BiOCl.

Table 1. The photodegradation ratio for RhB using BiOCl and Dy doped BiOCl samples. A total of 100.0 mg of powder photocatalyst was put into 10.0 mg·L^{-1} of RhB solution (100.0 mL).

Time (min)	BiOCl	BiDy$_{0.05}$OCl	BiDy$_{0.1}$OCl	BiDy$_{0.15}$OCl	BiDy$_{0.2}$OCl	BiDy$_{0.25}$OCl
10	49.0%	51.0%	59.0%	60.3%	67.0%	62.0%
20	64.0%	74.5%	83.8%	84.7%	85.7%	78.1%
30	74.8%	88.6%	88.5%	90.9%	97.3%	87.7%

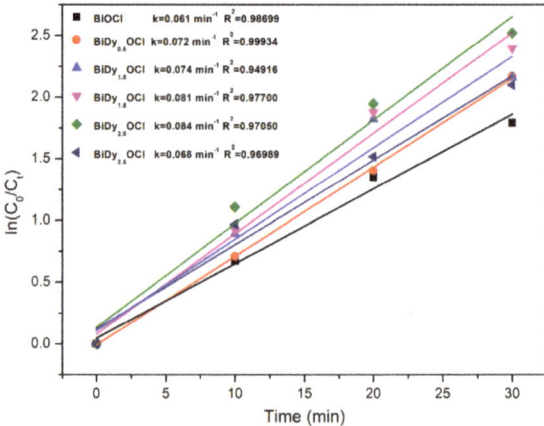

Figure 10. Kinetic linear fitting curve for RhB degradation using different photocatalyst samples.

It is worth mentioning that the photocatalytic activity of 2% Dy-doped BiOCl for RhB photodegradation was outstanding. To the best of our knowledge, the photocatalytic ratio reached 97.3% after only 30 min of photocatalytic reaction, and the efficiency was superior to that of existing literature reports (see Table S1).

Overall the incorporation of Dy can significantly enhance the photocatalytic activity of BiOCl.

First, the band gap energy of BiOCl decreased with Dy doping. In addition, the introduction of the 4f electron shell of Dy was equivalent to adding a "springboard" between the valance band and the covalent band of BiOCl, so that the valence band electrons could be more easily transferred to the conduction band through the 4f electric sublayer of Dy under visible light irradiation.

Secondly, the electron shell was relatively stable in the semi-full state, and when the Dy^{3+} trapped an electron, the 4f electron shell in the half-full electron state was destroyed, and the stability was reduced, and the electrons were released to the stable state. The released electrons reacted with the oxygen adsorbed on the surface of the sample to produce a photocatalytic active substance. In fact, a dye sensitized electron was possibly captured by Dy^{3+}, so the 4f electron shell could be used either as an electronic conductor or as a collection of electrons. The two processes could facilitate the separation of photoinduced electron–hole pairs and further improve photocatalytic activity.

Finally, when the Dy doping amount was 2.5%, the photocatalytic activity decreased. The incorporation of Dy provided a "springboard" for the low energy light activation electron transition, but excessive doping amounts would give rise to scattered distributions for the produced impurity energy, which would generate a recombination center for the electrons and holes.

In fact, environmental pollution resulting from dye wastewater is becoming more and more serious, which prompted humanity to realize the importance of green chemistry (environmentally-friendly chemistry) [23] that advocates for existing chemistry technologies and methods to be used to reduced or stop hazards to human health, community safety and the ecological environment. At present, some reports [24–26] have opened new frontiers in the field of catalysis. A super membrane technology (membrane-grafted catalyst) could reduce the emissions of reaction by-products and the recovery of residual solvents, which would be a good research direction.

4. Conclusions

Dy-doped BiOCl powder photocatalyst was synthesized using a one-step coprecipitation method. The incorporation of Dy^{3+} was successfully substituted for a part of Bi^{3+} in the BiOCl crystal lattice system. Two-percent Dy-doped BiOCl possessed the best photocatalytic activity and photodegradation efficiency for RhB under visible light irradiation. The photodegradation ratio of RhB reached 97.3% within 30 min of photocatalytic reaction under visible light irradiation. The main reason for this is that the 4f electron shell of Dy in the BiOCl crystal lattice system could generate a special electronic shell structure that facilitated the transfer of electrons from the valance band to the conduction band and also the separation of the photoinduced charge carrier. This work hopes to provide an important reference for RhB photodegradation in practical applications using this photocatalyst.

Supplementary Materials: The following are available online at http://www.mdpi.com/2079-4991/8/9/697/s1, Figure S1: The degradation rate of RhB with $BiDy_{2.0}OCl$ in dark, Table S1: Comparison of photodegradation ratios using different photocatalysts under visible light irradiation (reported in the last two years).

Author Contributions: Supervision, C.L. and L.X.; Funding Acquisition, J.Y. and C.L. In addition, J.Y. and T.X. contributed equally to this work.

Funding: This research was funded by the Chongqing Municipal Education Commission (KJ1711286/KJ1711266) and the Chongqing Basic Science and Advanced Technology Research Program (CSTC2015jcyjBX0015).

Acknowledgments: The authors want to thank the technological guidance of Kuen-Song Lin employed by the Chemical Engineering and Materials Science/Environmental Technology Research Center of Yuan Ze University in Taiwan. We want to thank the financial support from the Scientific and Technological Research Program of the Chongqing Municipal Education Commission (KJ1711286/KJ1711266) and the Chongqing Basic Science and Advanced Technology Research Program (CSTC2015jcyjBX0015). We gratefully acknowledge the many important contributions from the researchers of all the reports cited in this paper.

Conflicts of Interest: The authors declare no conflict of interest.

References

1. Wang, X.; Hong, M.Z.; Zhang, F.W.; Zhuang, Z.Y.; Yu, Y. Recyclable nanoscale zero valent iron doped g-C_3N_4/MoS_2 for efficient photocatalytic of RhB and Cr(VI) driven by visible light. *ACS Sustain. Chem. Eng.* **2016**, *4*, 4055–4063. [CrossRef]
2. Fu, H.B.; Zhang, S.C.; Xu, T.G.; Zhu, Y.F.; Chen, J.M. Photocatalytic degradation of RhB by fluorinated Bi_2WO_6 and distributions of the intermediate products. *Environ. Sci. Technol.* **2008**, *42*, 2085–2091. [CrossRef] [PubMed]
3. Zhuang, J.D.; Dai, W.X.; Tian, Q.F.; Li, Z.H.; Xie, L.Y.; Wang, J.X.; Liu, P. Photocatalytic degradation of RhB over TiO_2 bilayer films: Effect of defects and their location. *Langmuir* **2010**, *26*, 9686–9694. [CrossRef] [PubMed]
4. He, R.G.; Xu, D.F.; Cheng, B.; Yu, J.G.; Ho, W.K. Review on nanoscale Bi-based photocatalysts. *Nanoscale Horizons* **2018**. [CrossRef]
5. Jiang, J.; Zhao, K.; Xiao, X.Y.; Zhang, L.Z. Synthesis and facet-dependent photoreactivity of BiOCl single crystalline nanosheets. *J. Am. Chem. Soc.* **2012**, *134*, 4473–4476. [CrossRef] [PubMed]

6. Xie, T.P.; Xu, L.J.; Liu, C.L.; Yang, J.; Wang, M. Magnetic composite BiOCl–SrFe$_{12}$O$_{19}$: A novel p-n type heterojunction with enhanced photocatalytic activity. *Dalton Trans.* **2014**, *43*, 2211–2220. [PubMed]
7. Li, H.; Li, J.; Ai, Z.H.; Jia, F.L.; Zhang, L.Z. Oxygen vacancy-mediated photocatalysis of BiOCl: Reactivity, selectivity, and perspectives. *Angew. Chem. Int. Ed.* **2018**, *57*, 122–138.
8. Xie, T.P.; Wang, Y.; Liu, C.L.; Xu, L.J. New insights into sensitization mechanism of the Doped Ce (IV) into strontium Titanate. *Materials* **2018**, *11*, 646. [CrossRef] [PubMed]
9. Bhatia, S.; Verma, N.; Kumar, R. Morphologically-dependent photocatalytic and gas sensing application of Dy-doped ZnO nanoparticles. *J. Alloys Compd.* **2017**, *726*, 1274–1285.
10. Yayapao, O.; Thongtem, T.; Phuruangrat, A.; Thongtem, S. Sonochemical synthesis of Dy-doped ZnO nanostructures and their photocatalytic properties. *J. Alloys Compd.* **2013**, *576*, 72–79. [CrossRef]
11. Khataee, A.; Soltani, R.D.C.; Hanifehpour, Y.; Safarpour, M.; Ranjbar, H.G.; Joo, S.W. Synthesis and characterization of dysprosium-doped ZnO nanoparticles for photocatalysis of a textile dye under visible light irradiation. *Ind. Eng. Chem. Res.* **2014**, *53*, 1924–1932. [CrossRef]
12. Gao, X.Y.; Zhang, X.C.; Wang, Y.W.; Peng, S.Q.; Yue, B.; Fan, C.M. Photocatalytic degradation of carbamazepine using hierarchical BiOCl microspheres: Some key operating parameters, degradation intermediates and reaction pathway. *Chem. Eng. J.* **2015**, *273*, 156–165. [CrossRef]
13. Gao, M.C.; Zhang, D.F.; Pu, X.P.; Shao, X.; Li, H.; Lv, D.D. Combustion synthesis and enhancement of BiOCl by doping Eu^{3+} for photodegradation of organic dye. *J. Am. Ceram. Soc.* **2016**, *99*, 881–887. [CrossRef]
14. Zhang, L.; Wang, W.; Sun, S.; Sun, Y.; Gao, E.; Xu, J. Water splitting from dye wastewater: A case study of BiOCl/copper (II) phthalocyanine composite photocatalyst. *Appl. Catal. B* **2013**, *132*, 315–320. [CrossRef]
15. Wang, C.Y.; Zhang, Y.J.; Wang, W.K.; Pei, D.N.; Huang, G.X.; Chen, J.J.; Zhang, X.; Yu, H.Q. Enhanced photocatalytic degradation of bisphenol A by Co doped BiOCl nanosheets under visible light irradiation. *Appl. Catal. B* **2018**, *221*, 320–328. [CrossRef]
16. Huang, C.J.; Hu, J.L.; Cong, S.; Zhao, Z.G.; Qiu, X.Q. Hierarchical BiOCl microflowers with improved visible-light-driven photocatalytic activity by Fe(III) modification. *Appl. Catal. B* **2015**, *174*, 105–112. [CrossRef]
17. Li, Y.W.; Zhao, Y.; Wu, G.J.; Ma, H.M.; Zhao, J.Z. Bi superlattice nanopolygons at BiOCl (001) nanosheet assembled architectures for visible-light photocatalysis. *Mater. Res. Bull.* **2018**, *101*, 39–47. [CrossRef]
18. Hu, J.L.; Fan, W.J.; Ye, W.Q.; Huang, C.J.; Qiu, X.Q. Insights into the photosensitivity activity of BiOCl under visible light irradiation. *Appl. Catal. B* **2014**, *158*, 182–189. [CrossRef]
19. Yang, Y.; Teng, F.; Kan, Y.D.; Yang, L.M.; Liu, Z.L.; Gu, W.H.; Zhang, A.; Hao, W.Y.; Teng, Y.R. Investigation of the charges separation and transfer behavior of BiOCl/BiF$_3$ heterojunction. *Appl. Catal. B* **2017**, *205*, 412–420. [CrossRef]
20. Zhang, X.C.; Fan, C.M.; Wang, Y.W.; Wang, Y.F.; Liang, Z.H.; Han, P.D. DFT+U predictions: The effect of oxygen vacancy on the structural, electronic and photocatalytic properties of Mn-doped BiOCl. *Comput. Mater. Sci.* **2013**, *71*, 135–145. [CrossRef]
21. Shivakumara, C.; Saraf, R.; Halappa, P. White luminescence in Dy^{3+} doped BiOCl phosphors and their Judd-Ofelt analysis. *Dyes Pigments* **2016**, *126*, 154–164. [CrossRef]
22. Zhang, L.; Han, Z.K.; Wang, W.Z.; Li, X.M.; Su, Y.; Jiang, D.; Lei, X.L.; Sun, S.M. Solar-light-driven pure water splitting with ultrathin BiOCl nanosheets. *Chem. Eur. J.* **2015**, *21*, 18089–18094. [CrossRef] [PubMed]
23. Didaskalou, C.; Buyuktiryaki, S.; Kecili, R.; Fonte, C.P.; Szekely, G. Valorisation of agricultural waste with adsorption/nanofiltration hybrid process: From materials to sustainable process design. *Green. Chem.* **2017**, *19*, 3116–3125. [CrossRef]
24. Macaskie, L.E.; Mikheenko, I.P.; Omajai, J.B.; Stephen, A.J.; Wood, J. Metallic bionanocatalysts: Potential applications as green catalysts and energy materials. *Microb. Biotechnol.* **2017**, *10*, 1171–1180. [CrossRef] [PubMed]
25. Schaepertoens, M.; Didaskalou, C.; Kim, J.F.; Livingston, A.G.; Szekely, G. Solvent recycle with imperfect membranes: A semi-continuous workaround for diafiltration. *J. Membr. Sci.* **2016**, *514*, 646–658. [CrossRef]
26. Didaskalou, C.; Kupai, J.; Cseri, L.; Barabas, J.; Vass, E.; Holtzl, T.; Szekely, G. Membrane-grafted asymmetric organocatalyst for an integrated synthesis-separation platform. *ACS. Catal.* **2018**, *8*, 7430–7438. [CrossRef]

© 2018 by the authors. Licensee MDPI, Basel, Switzerland. This article is an open access article distributed under the terms and conditions of the Creative Commons Attribution (CC BY) license (http://creativecommons.org/licenses/by/4.0/).

Article

Preparation of Magnetic Fe₃O₄/MIL-88A Nanocomposite and Its Adsorption Properties for Bromophenol Blue Dye in Aqueous Solution

Yi Liu [†], Yumin Huang [†], Aiping Xiao, Huajiao Qiu * and Liangliang Liu *

Institute of Bast Fiber Crops, Chinese Academy of Agricultural Sciences, Changsha 410205, China; lyi.keai@yahoo.com (Y.L.); YuminHuang123@gmail.com (Y.H.); aipingxiao@yahoo.com (A.X.)

* Correspondence: qiuhuajiao@caas.cn (H.Q.); liuliangliang@caas.cn (L.L.); Tel.: +86-731-88998558 (H.Q.); +86-731-88998525 (L.L.)

† These authors contributed equally to this work.

Received: 30 November 2018; Accepted: 27 December 2018; Published: 2 January 2019

Abstract: Metal-organic frameworks (MOFs) are considered as good materials for the adsorption of many environmental pollutants. In this study, magnetic Fe_3O_4/MIL-88A composite was prepared by modification of MIL-88A with magnetic nanoparticles using the coprecipitation method. The structures and magnetic property of magnetic Fe_3O_4/MIL-88A composite were characterized and the adsorption behavior and mechanism for Bromophenol Blue (BPB) were evaluated. The results showed that magnetic Fe_3O_4/MIL-88A composite maintained a hexagonal rod-like structure and has good magnetic responsibility for magnetic separation (the maximum saturation magnetization was 49.8 emu/g). Moreover, the maximum adsorption amount of Fe_3O_4/MIL-88A composite for BPB was 167.2 mg/g and could maintain 94% of the initial adsorption amount after five cycles. The pseudo-second order kinetics and Langmuir isotherm models mostly fitted to the adsorption for BPB suggesting that chemisorption is the rate-limiting step for this monomolecular-layer adsorption. The adsorption capacity for another eight dyes (Bromocresol Green, Brilliant Green, Brilliant Crocein, Amaranth, Fuchsin Basic, Safranine T, Malachite Green and Methyl Red) were also conducted and the magnetic Fe_3O_4/MIL-88A composite showed good adsorption for dyes with sulfonyl groups. In conclusion, magnetic Fe_3O_4/MIL-88A composite could be a promising adsorbent and shows great potential for the removal of anionic dyes containing sulfonyl groups.

Keywords: adsorption; bromophenol blue; magnetic nanoparticles; metal-organic frameworks; wastewater

1. Introduction

The textile industry is one of the most chemically intensive industries on Earth and the major polluter of potable water. During various stages of textile processing, huge quantities of dyes are generated in the form of wastewater [1]. Dyes usually have complex aromatic molecular structures which make them more stable and more difficult to biodegrade. As the diversity of textile products increases, different dyestuffs with highly varying chemical characteristics are used in industry, complicating further treatments of textile wastewater [2]. The direct discharge of colored and toxic wastewater into the environment affects its ecological status by causing various undesirable changes [3]. Sulfonated azo dyes, one of the aromatic sulfonates, can be easily found in the textile industry. Due to a high mobility within the aquatic system, they can easily pass through the water treatment process and cause pollution of surface water [4]. There is an urgent need for the development of effective processes to remove the dyes from wastewater.

Various physicochemical and biological methods for treating dye effluents have been reported, such as adsorption, precipitation, chemical degradation, advanced oxidation processes, biodegradation and chemical coagulation [5]. Although these methods have been widely applied, they have some disadvantages. Owing to the undesired reactions in treated water, chemical coagulation causes large amounts of sludge and extra pollution [6]. As for biological methods, it is inadequate for most textile wastewaters because of highly structured polymers with low biodegradability [7]. In the past decade, the removal of dye from aqueous solutions via adsorption has attracted much attention because of economic feasibility, simplicity, and high efficiency [8]. Traditional absorbents have some limitations such as low adsorption capacity and difficulty in separation of absorbents after reaction. Hence, it is necessary to design low cost and high-efficiency absorbents that can also be easily separated from the contaminated media [9,10]. Graphene oxide [11], metal oxides nanoparticles [12], agricultural waste peels [13], bionanomaterials [11], metal-organic frameworks (MOFs) and many kinds of materials with various modifications constantly attract researchers' attention [14].

MOFs are a class of crystalline materials made by linking metal clusters or ions and organic linkers through covalent bonds. Owing to their highly ordered structures, high porosity and large surface areas, MOFs have attracted intensive attention in gas storage [15], molecular sensing [16], catalysis [17], energy [18], and water remediation [14]. Recently, many kinds of MOF-based materials such as rod-like metal-organic framework nanomaterial and MOF composites have been successfully synthesized and are widely used to remove dyes from wastewater [19–22]. Magnetic materials gained immense attention as adsorbents as well due to their strong magnetic response, low cost and good biocompatibility represented by ferroferric oxide (Fe_3O_4) nanoparticles. Fe_3O_4 nanoparticles could be easily separated from reaction liquids by the use of an external magnetic field. Consequently, they were widely applied in separation, catalysis and environmental remediation. The combination of Fe_3O_4 nanoparticles and other nanomaterials could apparently simplify procedures, save time, and improve efficiency in adsorption and separation fields [23].

Bromophenol blue (BPB) and its structurally related derivatives have been extensively applied in many industries like food, cosmetic, textile, printing inks and laboratory indicators [24]. The present study reports the successful synthesis of magnetic composite Fe_3O_4/MIL-88A and its use for the adsorption of BPB in order to evaluate its feasibility as a novel adsorbent in environmental remediation. MIL-88A was a 3D structured framework built up from trimers of Fe^{3+} octahedra linked to fumarate dianions. This structure exhibited a pore-channel system along the c axis and cages (5–7 Å) [25]. In addition, MIL-88A exhibited a flexible framework and possessed active iron metal sites, which were applied as a photocatalyst [26]. The synthesized Fe_3O_4/MIL-88A composite was characterized with transmission electron microscopy (TEM), scanning electron microscopy (SEM), X-ray powder diffraction (XRD), thermogravimetric analysis (TGA) and vibration sample magnetometer (VSM). The adsorption properties for BPB were investigated in terms of the effects of contact time, adsorbent dosage and initial dye concentration on removal efficiency of BPB and the kinetic and isotherm of adsorption process. As a superior adsorbent material, Fe_3O_4/MIL-88A showed proper magnetic response for shortening reaction time and excellent adsorption ability for the removal of dyes.

2. Materials and Methods

2.1. Materials

The chemicals, sodium acetate, fumaric acid, ethylene glycol, ferric chloride, ethanol, ferrous sulfate and ammonium hydroxide were of analytical grade and obtained from Sigma-Aldrich Chemicals (St. Louis, MO, USA). The dyes, Bromophenol Blue (BPB), Bromocresol Green, Brilliant Green, Brilliant Crocein, Amaranth, Fuchsin Basic, Safranine T, Malachite Green and Methyl Red were obtained commercially from Sinopharm Chemical Reagent Co., Ltd. (Shanghai, China). Ultrapure water (18.2 MΩ cm resistivity) was obtained from an ELGA water purification system (ELGA Berkefeld, Veolia, Germany). All other chemicals were also analytical grade and purchased from Sinopharm Chemical Reagent Co., Ltd. (Shanghai, China).

2.2. Synthesis of Magnetic Fe$_3$O$_4$/MIL-88A Composite

The MIL-88A was prepared according to the previous synthesis customs with some modifications in the solution concentration and reaction time [26]. Typically, 10 mmol of FeCl$_3$·6H$_2$O and 10 mmol of fumaric acid were first dissolved in 25 mL of water, and then the homogeneous solution was transferred into a 120 mL Teflon-lined stainless steel autoclave and heated to 65 °C for 12 h. After cooling to room temperature, the product was dispersed in water under ultrasonic waves for several minutes and centrifuged. The liquid supernatant was decanted and the precipitate (the weight was 0.89 g after drying) was re-dispersed in 100 mL of water for further use.

The magnetic Fe$_3$O$_4$/MIL-88A composite was prepared by coprecipitation method [27]. 3 mmoL of FeCl$_3$·6H$_2$O and 1.5 mmol of FeSO$_4$·7H$_2$O were mixed in 200 mL of water to form an aqueous solution. The solution was transferred into a round bottom flask containing 100 mL of MIL-88A aqueous solution under mechanical stirring in water bath at 75 °C. While mechanical stirring, 3 mL of ammonium hydroxide was added dropwise into the flask and the color of the solution became black indicating precipitate formation. The mixture was vigorously stirred for 30 min at 75 °C and this continued for 90 min at room temperature. After the reaction, the Fe$_3$O$_4$/MIL-88A composite was magnetically separated using a magnet and washed with water and ethanol three times. Finally, it was dried in a vacuum oven at 45 °C for 12 h (1.20 g after drying).

2.3. Characterizations

In order to confirm the morphology and structure of the final products, the synthesized magnetic Fe$_3$O$_4$/MIL-88A composites were characterized by means of TEM, field emission scanning electron microscopy (FESEM), and XRD. Specifically, TEM images of magnetic Fe$_3$O$_4$/MIL-88A composites were recorded on a Tecnai-G20 transmission electron microscope (FEI, Hillsboro, OR, USA). FESEM images were recorded on a JSM-7500F Field Emission Scanning Electron Microscope (JEOL, Tokyo, Japan). The XRD spectra were recorded using a powder X-ray Diffractometer (Rigaku RINT 2500, Rigaku Corporation, Tokyo, Japan) with Cu/Kα radiation at 30 mA and 40 kV. TGA was performed in nitrogen atmosphere from 40 to 800 °C with a heating rate of 10 °C/min with a simultaneous thermal analyzer (Netzsch STA 449F3, Ahlden, Germany). Moreover, the magnetic properties of Fe$_3$O$_4$/MIL-88A composites were measured at room temperature on a vibration sample magnetometer VSM7407 (Lake Shore, Westerville, OH, USA).

2.4. Adsorption Experiments

The adsorption rate experiments were performed by immersing 0.2 g of Fe$_3$O$_4$/MIL-88A powder into 50 mL of 1.2 mg/mL of dye aqueous solutions in a 100 mL conical flask with cover. The flask was shaken using a mechanical shaker (SHA-CA, Changzhou, China) at 27 °C and 200 rpm for 135 min. At each period of time, about 2.0 mL of the solution was picked up and filtrated through a syringe filter to measure the concentration of BPB using an ultraviolet-visible (UV-Vis) spectrophotometer (UV-2700, Shimadzu, Kyoto, Japan) at a wavelength of 590 nm. Different process variables such as initial concentration (0.3–1.5 mg/mL) and doses (0.05–0.4 g) were also investigated. Percentage removal of dyes was determined using the following equation [28]:

$$\text{Removal efficiency (\%)} = \frac{(C_0 - C_t)}{C_0} \times 100\% \quad (1)$$

where C_0 represents the initial concentration of dye and C_t represents the concentration of dye after t minutes. The equilibrium amount of adsorption (q_e) and the amount of adsorption (q_t) at given time were calculated according to the following equation [29]:

$$q_e = \frac{(C_0 - C_e) \times V}{W} \quad (2)$$

$$q_t = \frac{(C_0 - C_t) \times V}{W} \qquad (3)$$

where C_e is the equilibrium concentration of dye (mg/mL), V is the solution volume (mL), and W is the adsorbent mass (g).

3. Results and Discussion

3.1. Characterization of Magnetic Fe_3O_4/MIL-88A Composite

3.1.1. Transmission Electron Microscopy (TEM) and Field Emission Scanning Electron Microscopy (FESEM)

The MIL-88A and magnetic Fe_3O_4/MIL-88A composite were characterized by TEM and FESEM to visually observe the morphologies changes during synthesis processes. TEM image (Figure 1a) showed the prepared MIL-88A were crystallized hexagonal microrods of over 5 μm in length and about 500 nm in diameter. FESEM observation (Figure 1b) confirmed the microrod shape and revealed that the size distribution of these MIL-88A was relatively uniform with some exceptions. After the combination of Fe_3O_4 nanoparticles, the TEM image (Figure 1c) showed many Fe_3O_4 nanoparticles were grown on the surface of MIL-88A and the structure of MIL-88A was retained. The diameter of Fe_3O_4 nanoparticles were about 5 to 10 nm. It could be seen that the magnetic Fe_3O_4/MIL-88A composite was successfully prepared and showed characteristics of both Fe_3O_4 nanoparticles and MIL-88A in nanostructure.

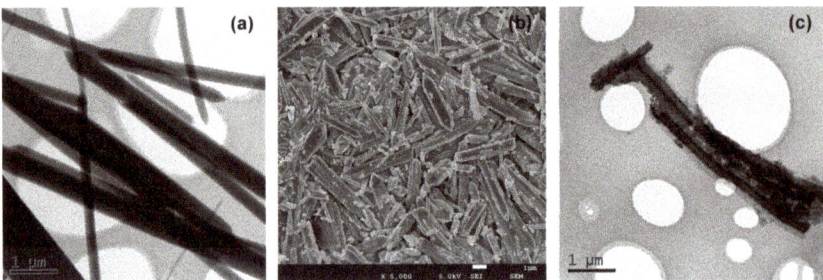

Figure 1. The transmission electron microscopy (TEM) (**a**) and field emission scanning electron microscopy (FESEM) (**b**) images of MIL-88A and the TEM (**c**) image of magnetic Fe_3O_4/MIL-88A composite.

3.1.2. X-Ray Powder Diffraction (XRD)

The structures of MIL-88A and Fe_3O_4/MIL-88A composite were analyzed by XRD and the spectra were compared with that of Fe_3O_4 nanoparticles. As shown in Figure 2a, the spectrum of MIL-88A showed peaks at 8.14°, 10.42° and 12.98°, which was accordance with the reported information [30]. Meanwhile, the spectrum of Fe_3O_4 nanoparticles also showed characteristic peaks at 30.48°, 35.72°, 43.32°, 57.56° and 62.86° corresponding the indices (220), (311), (400), (511) and (440). This pattern was in agreement with previously reported Fe_3O_4 crystal XRD data [31]. Finally, the spectrum of magnetic Fe_3O_4/MIL-88A composite exhibited some characteristic peaks of Fe_3O_4 nanoparticles at 35.72° and 62.86°. However, the peaks of MIL-88A were almost missed. As shown by TEM, the MIL-88A was coated with Fe_3O_4 nanoparticles, which might interfere the diffraction peak of MIL-88A crystals. Moreover, the quantity of Fe_3O_4 nanoparticles were much more than MIL-88A, therefore the diffraction signals of Fe_3O_4 nanoparticles were much higher than those of MIL-88A and masked the signals of MIL-88A.

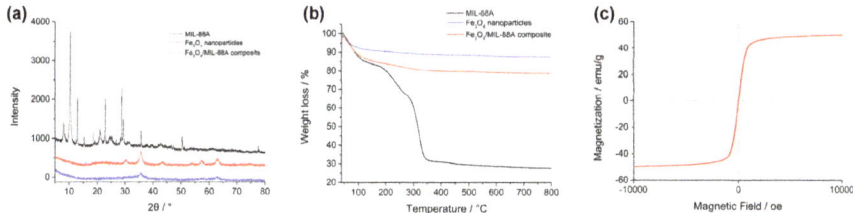

Figure 2. (a) The X-ray powder diffraction (XRD) patterns of MIL-88A (black), Fe_3O_4 nanoparticles (red) and magnetic Fe_3O_4/MIL-88A composite (blue); (b) thermogravimetric analysis (TGA) curves of MIL-88A (black), Fe_3O_4 nanoparticles (blue) and magnetic Fe_3O_4/MIL-88A composite (red); (c) The magnetization curve of magnetic Fe_3O_4/MIL-88A composite.

3.1.3. Thermogravimetric Analysis (TGA)

Figure 2b shows the weight losses of MIL-88A, Fe_3O_4 nanoparticles and magnetic Fe_3O_4/MIL-88A composite. In nitrogen below 350 °C, the weight loss of MIL-88A was attributed to the collapse of organic skeleton [32]. The weight loss of Fe_3O_4 nanoparticles below 100 °C was related to the evaporation of absorbed water, while the weight loss above 100 °C was relatively flat without obvious change. Finally, magnetic Fe_3O_4/MIL-88A composite showed the same tendency in weight loss as that of MIL-88A and the final weight loss was in between the former two materials. It illustrated that the combination of MIL-88A and Fe_3O_4 nanoparticles was effective.

3.1.4. Vibration Sample Magnetometer (VSM)

As a kind of magnetic nanomaterials, the magnetic property of magnetic Fe_3O_4/MIL-88A composite was evaluated by VSM as well. The magnetization curves of magnetic Fe_3O_4/MIL-88A composite was shown in Figure 2c. It could be found that the maximum saturation magnetization reached 49.8 emu/g. This value was less than that of bare Fe_3O_4 (about 65.0 emu/g) due to the existence of MOF without magnetic response. However, the prepared Fe_3O_4/MIL-88A composite was sufficient for magnetic separation in experiments and could be separated from solution within two minutes.

3.1.5. Adsorption Ability

The adsorption ability of magnetic Fe_3O_4/MIL-88A composite was verified and compared with that of MIL-88A and Fe_3O_4 nanoparticles. Under the same adsorption conditions, the adsorption amount of magnetic Fe_3O_4/MIL-88A composite was 141.5 mg/g with removal efficiency of 26.5%. While, the adsorption amounts of MIL-88A and Fe_3O_4 nanoparticles were 140.6 mg/g and 13.6 mg/g, respectively. It could be seen that MIL-88A and magnetic Fe_3O_4/MIL-88A composite had considerable adsorption abilities for BPB. However, the adsorption ability of Fe_3O_4 nanoparticles rather poor. The combination of two kinds of materials gave the magnetic responsibility to MIL-88A and maintained the adsorption ability. Based on these, the prepare of magnetic Fe_3O_4/MIL-88A composite could be considered successful.

3.2. Effects of Parameters on Dye Adsorption

3.2.1. Effect of Contact Time

The effect of contact time (15–135 min) on the removal efficiency of BPB was shown in Figure 3a. At the initial stage of adsorption, an increasing adsorption could be observed. However, the increase of adsorption slowed down as the adsorption proceeded, and finally the adsorption reached saturation. The maximum adsorption amount (141.2 mg/g) was achieved at 60 min with removal efficiency of 26.2%. After 135 min of adsorption, no significant increase in adsorption amount was observed

(141.2 mg/g to 147.9 mg/g). Similar patterns could be obtained in many adsorption experiments of dyes [33].

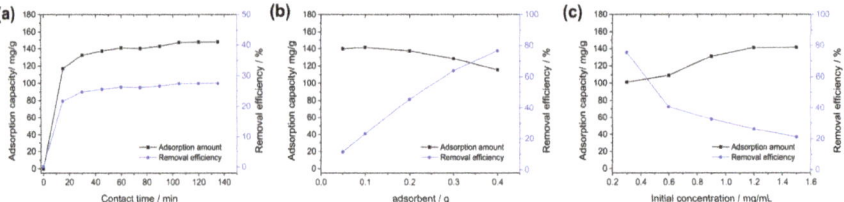

Figure 3. Effects of contact time (**a**), adsorbent dosage (**b**) and initial concentration (**c**) on the adsorption amount of Fe_3O_4/MIL-88A composite (black) and removal efficiency (blue).

3.2.2. Effect of Adsorbent Dosage

The effect of adsorbent dosage was investigated by addition of various amounts of magnetic Fe_3O_4/MIL-88A composite in 50 mL of dyes solution (1.2 mg/L) at room temperature for 60 min. As shown in Figure 3b, the adsorption amount of Fe_3O_4/MIL-88A composite decreased and removal efficiency increased with increasing dosage. The removal efficiency of BPB increased from 11.6% to 76.7%, which might due to the increase of adsorption sites on Fe_3O_4/MIL-88A surface were available for adsorption to dyes [34]. However, the adsorption capacity of Fe_3O_4/MIL-88A composite of decreased from 140.2 mg/g to 115.6 mg/g and showed maximum of 141.7 mg/g at 0.1 g. This kind of trend was commonly shown in many adsorption researches, which was caused by the agglomeration of adsorbent at high dosage, resulting in the reduction of effective sites on the adsorbent surface [35].

3.2.3. Effects of Initial Dye Concentration

The initial concentrations have great influences in this system, because it provides driving force to overcome mass transfer resistance between dye ion and solid phase [36]. The effect of initial dye concentrations (0.3–1.5 mg/mL) on the adsorption on magnetic Fe_3O_4/MIL-88A composite was shown in Figure 3c. It could be found that the removal efficiency of BPB decreased from 75.6% to 21.2% accompanied by the increase of adsorption capacity of Fe_3O_4/MIL-88A composite (101.2 mg/g to 141.7 mg/g) with initial dye concentration increased. The increase of adsorption capacity of absorbent could also be observed in the adsorption of R-250 dye on starch/poly(alginic acid-*cl*-acrylamide), direct orange 34 on natural clay and crystal violet on polyaniline nanoparticles [36–38]. However, the reduction of removal efficiency with increasing initial dye concentration might be attributed to relatively limited number of active sites for dyes compared with the increasing dye molecules.

3.3. Adsorption Kinetics

In order to study the mechanism of adsorption kinetics, two kinds of commonly used kinetic models, pseudo-first-order and pseudo-second-order, were applied in this research to study the adsorption behavior of BPB on magnetic Fe_3O_4/MIL-88A composite. Especially, the pseudo-first-order model was the simplest model and widely exploited for investigating the adsorption behavior. The pseudo-first-order model was expressed as follows [39]:

$$\ln(q_e - q_t) = \ln q_e - k_1 t \quad (4)$$

where q_t (mg/g) and q_e (mg/g) are the amounts of adsorbed dyes at a certain time and at equilibrium status respectively, t is contact time (min) and k_1 ($\min^{-1}$) is the pseudo-first order rate constant. The pseudo-second-order model could be represented as:

$$\frac{t}{q_t} = \frac{1}{k_{ad} q_e^2} + \frac{1}{q_e} t \quad (5)$$

where k_{ad} (g/mg/min) is the pseudo-second-order rate constant.

Figure 4 and Table 1 illustrated the linear plots of first and second order models for the adsorption of BPB on magnetic Fe$_3$O$_4$/MIL-88A composite. To estimate the suitability of two models, the corresponding correlation coefficients (R^2) were obtained by linear regression methods and higher R^2 value indicated more applicable model for describing the kinetics of BPB adsorption. As a result, the higher R^2 value for pseudo-second-order model (0.999) indicated this model was in good agreement with the experimental values and was more suitable for this adsorption. However, the R^2 for pseudo-first-order (0.902) is much lower than that offered by the pseudo-second-order model, which indicated the pseudo-first-order model was not suitable for the adsorption of BPB. This result reflected the rate limiting step for this adsorption might be chemisorption, involving valence force via sharing or exchanging electron between adsorbent and adsorbate [40].

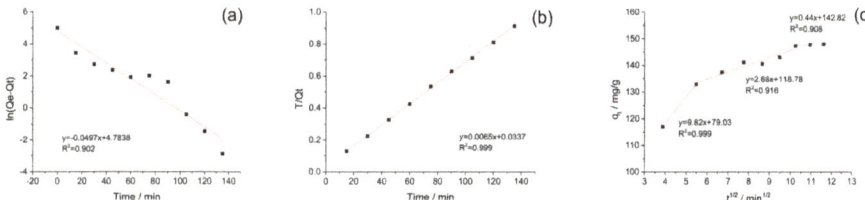

Figure 4. Adsorption kinetics plots (black dot) and the fitting curve (line) for the adsorption of Bromophenol Blue (BPB) on magnetic Fe$_3$O$_4$/MIL-88A composite at 27 °C: pseudo-first-order kinetic models (**a**) and pseudo-second-order kinetic models (**b**) and intraparticle diffusion model (**c**).

Table 1. Adsorption kinetic parameters for BPB on magnetic Fe$_3$O$_4$/MIL-88A composite.

Exp	Pseudo-First-Order Model			Pseudo-Second-Order Model			Intraparticle Diffusion		
q_{exp}	q_e	k_1	R^2	q_e	k_{ad}	R^2	k_1	k_2	k_3
148.0	119.56	0.0497	0.902	153.85	0.00125	0.999	9.82	2.68	0.44

In order to determine the adsorption process mechanism, an intraparticle diffusion model was also used to determine the rate-limiting step during the adsorption process [41]. The expression of this model is shown as the following equation:

$$q_t = k_i t^{0.5} + C \qquad (6)$$

where C (mg/g) is the intercept in intraparticle diffusion plot, and k_i (mg/g/min) is the intraparticle diffusion rate constant. Figure 4c showed linear plots in three sections, implying that three steps were involved in the adsorption with decreasing rates: (a) surface adsorption; (b) intraparticle diffusion; (c) adsorption close to equilibrium [42]. The rate constants k_i decreased and C values increased from step (a) to step (c) showed the increased contribution of the boundary layer to the adsorption rate. This kind of evolution was reported in other dye/adsorbent systems [43].

3.4. Adsorption Isotherms

The analysis on adsorption equilibrium could reveal types of adsorbate layers formed on the adsorbent surface. Three isotherm models were used in this study including the Langmuir model, Freundlich model and Temkin model. The Langmuir model assumed that uptake occurs on a homogeneous surface by monolayer adsorption without interaction between the absorbed materials, which could be expressed as following [44]:

$$\frac{C_e}{q_e} = \frac{1}{bq_m} + \frac{C_e}{q_m} \qquad (7)$$

where C_e (mg/mL) and q_e (mg/g) represent the equilibrium concentration of dye and adsorption capacity at equilibrium, q_m (mg/g) represents the maximum adsorption capacity and b represents the equilibrium adsorption constant (mL/mg).

The Freundlich model described the formation of multilayers by adsorbate molecules on the adsorbent surface because of different affinities for various active sites on adsorbent surface [45]. The equation was expressed as following:

$$\log q_e = \log k + \frac{1}{n} \log C_e \tag{8}$$

where k (mL/mg) and n are the Freundlich constants.

The Temkin model assumed that adsorbent-adsorbate interactions could not be neglected during the adsorption mechanism, and the heat of adsorption decreases linearly with the adsorbate coverage due to the interaction [46]. This model could be represented by the following equation:

$$q_e = B\log A_T + B\log C_e \tag{9}$$

where B (mg/g) and A_T (mL/mg) are the Temkin isotherm equilibrium binding constant.

The experimental data were fitted to the Langmuir, Freundlich and Temkin models as described in Figure 5 and the detail parameters were shown in Table 2. Through the comparison on the R^2 values (0.984 for Langmuir, 0.954 for Freundlich and 0.975 for Temkin, respectively), the Langmuir model appeared to be the most suitable model in describing adsorption of BPB on magnetic Fe_3O_4/MIL-88A composite. Thus, the adsorption of BPB was typical monomolecular-layer adsorption.

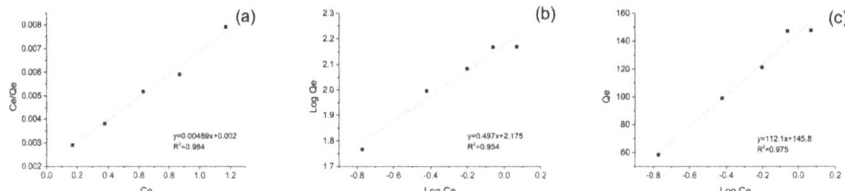

Figure 5. Adsorption isotherms plots (Black dot) and the fitting curve (Red line) for the adsorption of BPB dye on magnetic Fe_3O_4/MIL-88A composite at 27 °C: Langmuir model (a), Freundlich model (b) and Temkin model (c).

Table 2. The adsorption isotherm parameters for BPB on magnetic Fe_3O_4/MIL-88A composite.

Langmuir			Freundlich			Temkin		
q_m	b	R^2	$1/n$	k	R^2	B	A_T	R^2
204.50	2.445	0.984	0.497	149.62	0.954	112.1	19.98	0.975

3.5. Comparison Study

Eight dyes, Bromocresol Green, Brilliant Green, Brilliant Crocein, Amaranth, Fuchsin Basic, Safranine T, Malachite Green and Methyl Red, were investigated and compared at the same adsorption conditions (Figure 6). 0.1 g of Fe_3O_4/MIL-88A powders were transferred into 50 mL of 1.0 mg/mL of dye solutions, the mixture was shaken at 27 °C for 60 min. The adsorption amount was calculated through monitoring the change of absorbance for each dye solution. Among these dyes, Bromocresol Green, Brilliant Green, Brilliant Crocein, Amaranth, and BPB all contain sulfonyl groups. However, there is no sulfonyl group in the structure of the other four dyes, Fuchsin Basic, Safranine T, Malachite Green and Methyl Red. As a result, five dyes containing sulfonyl groups showed much better adsorption amounts than that of dyes without sulfonyl group. In particular, there was nearly no adsorption for Fuchsin Basic and Methyl Red. Some computed and experimental properties were listed in Table 3. Topological polar surface area (TPSA) values were obtained on the Pubchem website (https://pubchem.ncbi.nlm.nih.gov/compound/) and pKa values were obtained on the Chemicalbook website (https://www.chemicalbook.com/). The TPSA was defined as the sum of surfaces of polar atoms in a molecule. This property has been shown to correlate with the human intestinal absorption and blood–brain barrier penetration. Herein, BPB, Bromocresol Green, Brilliant Green, Brilliant Crocein and Amaranth with higher adsorption amounts showed relative higher TPSA values and lower pKa values. Based on these results, it might be assumed that magnetic Fe_3O_4/MIL-88A composite could effectively adsorbed dyes with sulfonyl groups, and the polar surface and pKa values of molecules might affect the adsorption. However, the detailed mechanism, especially in the surface charge of absorbent and dyes, still needs more in-depth and systematic studies in future [47].

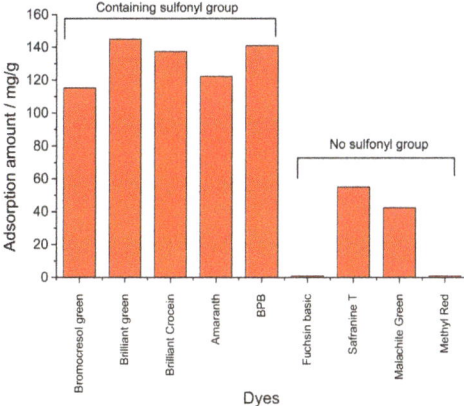

Figure 6. Comparison of the adsorption amount among various dyes on magnetic Fe_3O_4/MIL-88A composite.

Table 3. The chemical properties of investigated dyes.

Dyes	TPSA/A^2	pKa
BPB	92.2	3.85
Bromocresol Green	92.2	4.7
Brilliant Green	92.1	2.5
Brilliant Crocein	197	-[a]
Amaranth	238	-
Fuchsin Basic	75.9	-
Safranine T	68.8	6.4
Malachite Green	6.2	6.9
Methyl Red	65.3	4.95

[a] means the value could not be found in the website.

Adsorption capacities of various adsorbents for BPB as reported in literature were presented in Table 4. The comparison between this work and other reported data showed that magnetic Fe_3O_4/MIL-88A composite was a satisfied adsorbent for BPB compared to other adsorbents, as the adsorption capacity of magnetic Fe_3O_4/MIL-88A composite was higher than that of most reported materials. Therefore, it could be safely concluded that the materials prepared in this work exhibited considerable ability for adsorbing BPB from aqueous solutions.

Table 4. A comparison of adsorption of BPB by various reported adsorbents.

Adsorbent	Adsorbate	Adsorption Capacity (mg/g)	Reference
α-Chitin nanoparticles	BPB	22.72	[48]
Activated charcoal	BPB	About 9.0×10^{-3}	[49]
$SiO_2 \cdot Bth^+ \cdot PF_6^-$ ionic liquids	BPB	238.10	[24]
Modified layered silicate	BPB	184.5	[50]
Sorel's cement nanoparticles	BPB	4.88	[51]
Polymer-clay composite	BPB	About 7.5	[52]
Mesoporous MgO nanoparticles	BPB	40	[53]
Mesoporous hybrid gel	BPB	18.43	[54]
$CoFe_2O_4$ nano-hollow spheres	BPB	29.3	[55]
Graphene oxide functionalized magnetic chitosan composite	BPB	9.5	[56]
CuS-NP-AC	BPB	106.4	[57]
Fe_2O_3-ZnO-$ZnFe_2O_4$/carbon nanocomposite	BPB	90.91	[58]
Iron oxide nanoparticles	BPB	About 110	[59]
Fe_3O_4/MIL-88A	BPB	141.9–167.2	This study

3.6. Recycling of Fe_3O_4/MIL-88A Composite

The reuse of adsorbent is an important aspect for practical application in economic aspect. To evaluate the reusability of magnetic Fe_3O_4/MIL-88A composite, the adsorbed composite was desorbed with ethanol solution and used for next adsorption cycles. The adsorption capacity of each cycle was monitored and the relative adsorption capacity was calculated by comparing with the first run in percentage form (adsorption capacity defined as 100%). Five cycles' reuse of magnetic Fe_3O_4/MIL-88A composite was shown in Figure 7. It could be seen that Fe_3O_4/MIL-88A maintained high adsorption capacity (94%) without significant loss after five cycles. The result demonstrated that Fe_3O_4/MIL-88A composite could be applied in practical application owing to their high adsorption capacity and good reusability.

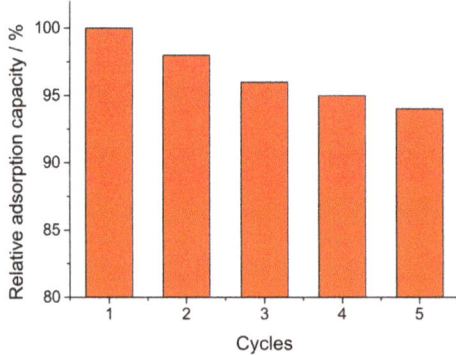

Figure 7. Reusability of magnetic Fe3O4/MIL-88A composite.

4. Conclusions

The magnetic Fe_3O_4/MIL-88A composite was prepared and characterized by TEM, FESEM, XRD, TGA and VSM. The characterizations showed the preparation was successful and sufficient for magnetic separation. The magnetic Fe_3O_4/MIL-88A composite showed good adsorption ability for BPB and other dyes containing sulfonyl groups. The adsorption amount of magnetic Fe_3O_4/MIL-88A composite was higher than many reported materials for BPB and could be maintained during five cycles. The results illustrated that the magnetic Fe_3O_4/MIL-88A composite has promising application in dye-contaminated wastewater treatment, especially for anionic dyes containing sulfonyl groups.

Author Contributions: Data curation, Y.H.; Investigation, Y.L. and L.L.; Methodology, Y.H. and H.Q.; Project administration, A.X. and H.Q.; Supervision, A.X.; Writing—original draft, L.L.; Writing—review and editing, L.L.

Funding: This research received no external funding.

Acknowledgments: This work was supported by the risk assessment of agricultural products quality and safety project (GJFP2018010) and the Natural Science Foundation of Hunan Province (2017JJ3348).

Conflicts of Interest: The authors declare no conflict of interest.

References

1. Ji, Y.; Ma, C.; Li, J.; Zhao, H.; Chen, Q.; Li, M.; Liu, H. A magnetic adsorbent for the removal of cationic dyes from wastewater. *Nanomaterials* **2018**, *8*, 710. [CrossRef] [PubMed]
2. Matmin, J.; Affendi, I.; Ibrahim, S.; Endud, S. Additive-free rice starch-assisted synthesis of spherical nanostructured hematite for degradation of dye contaminant. *Nanomaterials* **2018**, *8*, 702. [CrossRef] [PubMed]
3. Pajootan, E.; Arami, M.; Mahmoodi, N.M. Binary system dye removal by electrocoagulation from synthetic and real colored wastewaters. *J. Taiwan Inst. Chem. Eng.* **2012**, *43*, 282–290. [CrossRef]
4. Loos, R.; Niessner, R. Analysis of aromatic sulfonates in water by solid-phase extraction and capillary electrophoresis. *J. Chromatogr. A* **1998**, *822*, 291–303. [CrossRef]
5. Pan, L.; Wang, Z.; Yang, Q.; Huang, R. Efficient removal of lead, copper and cadmium ions from water by a porous calcium alginate/graphene oxide composite aerogel. *Nanomaterials* **2018**, *8*, 957. [CrossRef] [PubMed]
6. Gao, P.; Chen, X.; Shen, F.; Chen, G. Removal of chromium(VI) from wastewater by combined electrocoagulation–electroflotation without a filter. *Sep. Purif. Technol.* **2005**, *43*, 117–123. [CrossRef]
7. Can, O.T.; Bayramoglu, M.; Kobya, M. Decolorization of reactive dye solutions by electrocoagulation using aluminum electrodes. *Ind. Eng. Chem. Res.* **2003**, *42*, 3391–3396. [CrossRef]
8. Carmalin Sophia, A.; Lima, E.C. Removal of emerging contaminants from the environment by adsorption. *Ecotoxicol. Environ. Saf.* **2018**, *150*, 1–17. [CrossRef]
9. Jabbari, V.; Veleta, J.M.; Zarei-Chaleshtori, M.; Gardea-Torresdey, J.; Villagrán, D. Green synthesis of magnetic MOF@GO and MOF@CNT hybrid nanocomposites with high adsorption capacity towards organic pollutants. *Chem. Eng. J.* **2016**, *304*, 774–783. [CrossRef]
10. Li, W.J.; Gao, S.Y.; Liu, T.F.; Han, L.W.; Lin, Z.J.; Cao, R. In situ growth of metal-organic framework thin films with gas sensing and molecule storage properties. *Langmuir* **2013**, *29*, 8657–8664. [CrossRef]
11. Tan, K.B.; Vakili, M.; Horri, B.A.; Poh, P.E.; Abdullah, A.Z.; Salamatinia, B. Adsorption of dyes by nanomaterials: Recent developments and adsorption mechanisms. *Sep. Purif. Technol.* **2015**, *150*, 229–242. [CrossRef]
12. Santhosh, C.; Velmurugan, V.; Jacob, G.; Jeong, S.K.; Grace, A.N.; Bhatnagar, A. Role of nanomaterials in water treatment applications: A review. *Chem. Eng. J.* **2016**, *306*, 1116–1137. [CrossRef]
13. Bhatnagar, A.; Sillanpää, M.; Witek-Krowiak, A. Agricultural waste peels as versatile biomass for water purification—A review. *Chem. Eng. J.* **2015**, *270*, 244–271. [CrossRef]
14. Hasan, Z.; Jhung, S.H. Removal of hazardous organics from water using metal-organic frameworks (MOFs): Plausible mechanisms for selective adsorptions. *J. Hazard. Mater.* **2015**, *283*, 329–339. [CrossRef] [PubMed]
15. Li, J.-R.; Kuppler, R.J.; Zhou, H.-C. Selective gas adsorption and separation in metal–organic frameworks. *Chem. Soc. Rev.* **2009**, *38*, 1477–1504. [CrossRef] [PubMed]
16. Allendorf, M.D.; Bauer, C.A.; Bhakta, R.K.; Houk, R.J.T. Luminescent metal–organic frameworks. *Chem. Soc. Rev.* **2009**, *38*, 1330–1352. [CrossRef]
17. Lee, J.; Farha, O.K.; Roberts, J.; Scheidt, K.A.; Nguyen, S.T.; Hupp, J.T. Metal–organic framework materials as catalysts. *Chem. Soc. Rev.* **2009**, *38*, 1450–1459. [CrossRef]
18. Manos, G.; Dunne, L. Predicting the features of methane adsorption in large pore metal-organic frameworks for energy storage. *Nanomaterials* **2018**, *8*, 818. [CrossRef]
19. Wu, Y.; Pang, H.; Yao, W.; Wang, X.; Yu, S.; Yu, Z.; Wang, X. Synthesis of rod-like metal-organic framework (MOF-5) nanomaterial for efficient removal of U(VI): Batch experiments and spectroscopy study. *Sci. Bull.* **2018**, *63*, 831–839. [CrossRef]
20. Ke, F.; Qiu, L.G.; Yuan, Y.P.; Jiang, X.; Zhu, J.F. Fe_3O_4@MOF core–shell magnetic microspheres with a designable metal–organic framework shell. *J. Mater. Chem.* **2012**, *22*, 9497–9500. [CrossRef]

21. Askari, H.; Ghaedi, M.; Dashtian, K.; Mha, A. Rapid and high-capacity ultrasonic assisted adsorption of ternary toxic anionic dyes onto MOF-5-activated carbon: Artificial neural networks, partial least squares, desirability function and isotherm and kinetic study. *Ultrason. Sonochem.* **2017**, *37*, 71. [CrossRef] [PubMed]
22. Li, H.; Cao, X.; Zhang, C.; Yu, Q.; Zhao, Z.; Niu, X.; Sun, X.; Liu, Y.; Ma, L.; Li, Z. Enhanced adsorptive removal of anionic and cationic dyes from single or mixed dye solutions using MOF PCN-222. *RSC Adv.* **2017**, *7*, 16273–16281. [CrossRef]
23. Zhao, X.; Liu, S.; Zhi, T.; Niu, H.; Cai, Y.; Wei, M.; Wu, F.; Giesy, J.P. Synthesis of magnetic metal-organic framework (MOF) for efficient removal of organic dyes from water. *Sci. Rep.* **2015**, *5*, 11849. [CrossRef] [PubMed]
24. Liu, J.; Yao, S.; Wang, L.; Zhu, W.; Xu, J.; Song, H. Adsorption of bromophenol blue from aqueous samples by novel supported ionic liquids. *J. Chem. Technol. Biotechnol.* **2014**, *89*, 230–238. [CrossRef]
25. Chalati, T.; Horcajada, P.; Gref, R.; Couvreur, P.; Serre, C. Optimisation of the synthesis of MOF nanoparticles made of flexible porous iron fumarate MIL-88A. *J. Mater. Chem.* **2011**, *21*, 2220–2227. [CrossRef]
26. Xu, W.-T.; Ma, L.; Ke, F.; Peng, F.-M.; Xu, G.-S.; Shen, Y.-H.; Zhu, J.-F.; Qiu, L.-G.; Yuan, Y.-P. Metal–organic frameworks MIL-88A hexagonal microrods as a new photocatalyst for efficient decolorization of methylene blue dye. *Dalton Trans.* **2014**, *43*, 3792–3798. [CrossRef]
27. Fang, J.; Wang, H.; Xue, Y.; Wang, X.; Lin, T. Magnet-induced temporary superhydrophobic coatings from one-pot synthesized hydrophobic magnetic nanoparticles. *ACS Appl. Mater. Interfaces* **2010**, *2*, 1449–1455. [CrossRef]
28. Azha, S.F.; Sellaoui, L.; Shamsudin, M.S.; Ismail, S.; Bonilla-Petriciolet, A.; Ben Lamine, A.; Erto, A. Synthesis and characterization of a novel amphoteric adsorbent coating for anionic and cationic dyes adsorption: Experimental investigation and statistical physics modelling. *Chem. Eng. J.* **2018**, *351*, 221–229. [CrossRef]
29. Gao, Y.; Deng, S.-Q.; Jin, X.; Cai, S.-L.; Zheng, S.-R.; Zhang, W.-G. The construction of amorphous metal-organic cage-based solid for rapid dye adsorption and time-dependent dye separation from water. *Chem. Eng. J.* **2019**, *357*, 129–139. [CrossRef]
30. Wu, H.; Ma, M.-D.; Gai, W.-Z.; Yang, H.; Zhou, J.-G.; Cheng, Z.; Xu, P.; Deng, Z.-Y. Arsenic removal from water by metal-organic framework MIL-88A microrods. *Environ. Sci. Pollut. Res.* **2018**, *25*, 27196–27202. [CrossRef]
31. Liu, L.; Ma, Y.; Chen, X.; Xiong, X.; Shi, S. Screening and identification of BSA bound ligands from *Puerariae lobata* flower by BSA functionalized Fe_3O_4 magnetic nanoparticles coupled with HPLC–MS/MS. *J. Chromatogr. B* **2012**, *887–888*, 55–60. [CrossRef] [PubMed]
32. Wang, Y.; Guo, X.; Wang, Z.; Lü, M.; Wu, B.; Wang, Y.; Yan, C.; Yuan, A.; Yang, H. Controlled pyrolysis of MIL-88A to Fe_2O_3@C nanocomposites with varied morphologies and phases for advanced lithium storage. *J. Mater. Chem. A* **2017**, *5*, 25562–25573. [CrossRef]
33. Hamza, W.; Dammak, N.; Hadjltaief, H.B.; Eloussaief, M.; Benzina, M. Sono-assisted adsorption of cristal violet dye onto tunisian smectite clay: Characterization, kinetics and adsorption isotherms. *Ecotoxicol. Environ. Saf.* **2018**, *163*, 365–371. [CrossRef] [PubMed]
34. Wang, X.; Jiang, C.; Hou, B.; Wang, Y.; Hao, C.; Wu, J. Carbon composite lignin-based adsorbents for the adsorption of dyes. *Chemosphere* **2018**, *206*, 587–596. [CrossRef] [PubMed]
35. Tang, C.-Y.; Yu, P.; Tang, L.-S.; Wang, Q.-Y.; Bao, R.-Y.; Liu, Z.-Y.; Yang, M.-B.; Yang, W. Tannic acid functionalized graphene hydrogel for organic dye adsorption. *Ecotoxicol. Environ. Saf.* **2018**, *165*, 299–306. [CrossRef] [PubMed]
36. Chaari, I.; Moussi, B.; Jamoussi, F. Interactions of the dye, c.I. Direct orange 34 with natural clay. *J. Alloys Compd.* **2015**, *647*, 720–727. [CrossRef]
37. Sharma, G.; Naushad, M.; Kumar, A.; Rana, S.; Sharma, S.; Bhatnagar, A.; Florian, J.S.; Ghfar, A.A.; Khan, M.R. Efficient removal of coomassie brilliant blue r-250 dye using starch/poly(alginic acid-cl-acrylamide) nanohydrogel. *Process Saf. Environ. Prot.* **2017**, *109*, 301–310. [CrossRef]
38. Saad, M.; Tahir, H.; Khan, J.; Hameed, U.; Saud, A. Synthesis of polyaniline nanoparticles and their application for the removal of crystal violet dye by ultrasonicated adsorption process based on response surface methodology. *Ultrason. Sonochem.* **2017**, *34*, 600–608. [CrossRef]
39. Al-Hussain, S.; Atta, A.; Al-Lohedan, H.; Ezzat, A.; Tawfeek, A. Application of new sodium vinyl sulfonate-*co*-2-acrylamido-2-me[thylpropane sulfonic acid sodium salt-magnetite cryogel nanocomposites for fast methylene blue removal from industrial waste water. *Nanomaterials* **2018**, *8*, 878. [CrossRef]
40. Tekin, N.; Şafaklı, A.; Bingöl, D. Process modeling and thermodynamics and kinetics evaluation of basic yellow 28 adsorption onto sepiolite. *Desalin. Water Treat.* **2015**, *54*, 2023–2035. [CrossRef]

41. Mouni, L.; Belkhiri, L.; Bollinger, J.-C.; Bouzaza, A.; Assadi, A.; Tirri, A.; Dahmoune, F.; Madani, K.; Remini, H. Removal of methylene blue from aqueous solutions by adsorption on kaolin: Kinetic and equilibrium studies. *Appl. Clay Sci.* **2018**, *153*, 38–45. [CrossRef]
42. Ghosal, P.S.; Gupta, A.K. Determination of thermodynamic parameters from langmuir isotherm constant-revisited. *J. Mol. Liq.* **2017**, *225*, 137–146. [CrossRef]
43. Tanhaei, B.; Ayati, A.; Lahtinen, M.; Sillanpää, M. Preparation and characterization of a novel chitosan/Al$_2$O$_3$/magnetite nanoparticles composite adsorbent for kinetic, thermodynamic and isotherm studies of methyl orange adsorption. *Chem. Eng. J.* **2015**, *259*, 1–10. [CrossRef]
44. Wang, X.; Zhang, Z.; Zhao, Y.; Xia, K.; Guo, Y.; Qu, Z.; Bai, R. A mild and facile synthesis of amino functionalized CoFe$_2$O$_4$@SiO$_2$ for Hg(II) removal. *Nanomaterials* **2018**, *8*, 673. [CrossRef] [PubMed]
45. Carazo, E.; Borrego-Sánchez, A.; Sánchez-Espejo, R.; García-Villén, F.; Cerezo, P.; Aguzzi, C.; Viseras, C. Kinetic and thermodynamic assessment on isoniazid/montmorillonite adsorption. *Appl. Clay Sci.* **2018**, *165*, 82–90. [CrossRef]
46. Wong, S.; Tumari, H.H.; Ngadi, N.; Mohamed, N.B.; Hassan, O.; Mat, R.; Saidina Amin, N.A. Adsorption of anionic dyes on spent tea leaves modified with polyethyleneimine (PEI-STL). *J. Clean. Prod.* **2019**, *206*, 394–406. [CrossRef]
47. Heimann, S.; Ndé-Tchoupé, A.I.; Hu, R.; Licha, T.; Noubactep, C. Investigating the suitability of Fe^0 packed-beds for water defluoridation. *Chemosphere* **2018**, *209*, 578–587. [CrossRef] [PubMed]
48. Dhananasekaran, S.; Palanivel, R.; Pappu, S. Adsorption of methylene blue, bromophenol blue, and coomassie brilliant blue by α-chitin nanoparticles. *J. Adv. Res.* **2016**, *7*, 113–124. [CrossRef]
49. Iqbal, M.J.; Ashiq, M.N. Adsorption of dyes from aqueous solutions on activated charcoal. *J. Hazard. Mater.* **2007**, *139*, 57–66. [CrossRef]
50. Shapkin, N.P.; Maiorov, V.I.; Leont'ev, L.B.; Shkuratov, A.L.; Shapkina, V.Y.; Khal'chenko, I.G.J.C.J. A study of the adsorption properties of modified layered silicate. *Colloid J.* **2014**, *76*, 746–752. [CrossRef]
51. El-Gamal, S.M.A.; Amin, M.S.; Ahmed, M.A. Removal of methyl orange and bromophenol blue dyes from aqueous solution using sorel's cement nanoparticles. *J. Environ. Chem. Eng.* **2015**, *3*, 1702–1712. [CrossRef]
52. El-Zahhar, A.A.; Awwad, N.S.; El-Katori, E.E. Removal of bromophenol blue dye from industrial waste water by synthesizing polymer-clay composite. *J. Mol. Liq.* **2014**, *199*, 454–461. [CrossRef]
53. Ahmed, M.A.; Abou-Gamra, Z.M. Mesoporous mgo nanoparticles as a potential sorbent for removal of fast orange and bromophenol blue dyes. *Nanotechnol. Environ. Eng.* **2016**, *1*, 10. [CrossRef]
54. You, L.; Wu, Z.; Kim, T.; Lee, K. Kinetics and thermodynamics of bromophenol blue adsorption by a mesoporous hybrid gel derived from tetraethoxysilane and bis(trimethoxysilyl)hexane. *J. Colloid Interface Sci.* **2006**, *300*, 526–535. [CrossRef] [PubMed]
55. Rakshit, R.; Khatun, E.; Pal, M.; Talukdar, S.; Mandal, D.; Saha, P.; Mandal, K. Influence of functional group of dye on the adsorption behaviour of CoFe$_2$O$_4$ nano-hollow spheres. *New J. Chem.* **2017**, *41*, 9095–9102. [CrossRef]
56. Sohni, S.; Gul, K.; Ahmad, F.; Ahmad, I.; Khan, A.; Khan, N.; Bahadar Khan, S. Highly efficient removal of acid red-17 and bromophenol blue dyes from industrial wastewater using graphene oxide functionalized magnetic chitosan composite. *Polym. Compos.* **2018**, *39*, 3317–3328. [CrossRef]
57. Mazaheri, H.; Ghaedi, M.; Asfaram, A.; Hajati, S. Performance of cus nanoparticle loaded on activated carbon in the adsorption of methylene blue and bromophenol blue dyes in binary aqueous solutions: Using ultrasound power and optimization by central composite design. *J. Mol. Liq.* **2016**, *219*, 667–676. [CrossRef]
58. Mohammadzadeh, A.; Ramezani, M.; Ghaedi, A.M. Synthesis and characterization of Fe$_2$O$_3$–ZnO–ZnFe$_2$O$_4$/carbon nanocomposite and its application to removal of bromophenol blue dye using ultrasonic assisted method: Optimization by response surface methodology and genetic algorithm. *J. Taiwan Inst. Chem. Eng.* **2016**, *59*, 275–284. [CrossRef]
59. Saha, B.; Das, S.; Saikia, J.; Das, G. Preferential and enhanced adsorption of different dyes on iron oxide nanoparticles: A comparative study. *J. Phys. Chem. C* **2011**, *115*, 8024–8033. [CrossRef]

© 2019 by the authors. Licensee MDPI, Basel, Switzerland. This article is an open access article distributed under the terms and conditions of the Creative Commons Attribution (CC BY) license (http://creativecommons.org/licenses/by/4.0/).

Article

Magnetic Photocatalyst BiVO₄/Mn-Zn ferrite/Reduced Graphene Oxide: Synthesis Strategy and Its Highly Photocatalytic Activity

Taiping Xie [1,2], Hui Li [3], Chenglun Liu [1,3,*], Jun Yang [4], Tiancun Xiao [5] and Longjun Xu [1,*]

1. State Key Laboratory of Coal Mine Disaster Dynamics and Control, Chongqing University, Chongqing 400044, China; deartaiping@163.com
2. Chongqing Key Laboratory of Extraordinary Bond Engineering and Advanced Materials Technology (EBEAM), Yangtze Normal University, Chongqing 408100, China
3. College of Chemistry and Chemical Engineering, Chongqing University, Chongqing 401331, China; lihui@163.com
4. Chongqing Key Laboratory of Environmental Materials & Remediation Technologies Chongqing University of Arts and Sciences, Yongchuan 402160, China; yangjun@163.com
5. Inorganic Chemistry Laboratory, University of Oxford, Oxford OX1 3QR, UK; xiaotiancun@chem.ox.ac.uk
* Correspondence: xlclj@cqu.edu.cn (C.L.); xulj@cqu.edu.cn (L.X.); Tel.: +86-138-8370-2103 (C.L.); +86-153-1060-6337 (L.X.)

Received: 23 April 2018; Accepted: 28 May 2018; Published: 29 May 2018

Abstract: Magnetic photocatalyst BiVO₄/Mn-Zn ferrite ($Mn_{1-x}Zn_xFe_2O_4$)/reduced graphene oxide (RGO) was synthesized by a simple calcination and reduction method. The magnetic photocatalyst held high visible light-absorption ability with low band gap energy and wide absorption wavelength range. Electrochemical impedance spectroscopies illustrated good electrical conductivity which indicated low charge-transfer resistance due to incorporation of $Mn_{1-x}Zn_xFe_2O_4$ and RGO. The test of photocatalytic activity showed that the degradation ratio of rhodamine B (RhB) reached 96.0% under visible light irradiation after only 1.5 h reaction. The photocatalytic mechanism for the prepared photocatalyst was explained in detail. Here, the incorporation of RGO enhanced the specific surface area compared with BiVO4/$Mn_{1-x}Zn_xFe_2O_4$. The larger specific surface area provided more active surface sites, more free space to improve the mobility of photo-induced electrons, and further facilitated the effective migration of charge carriers, leading to the remarkable improvement of photocatalytic performance. Meanwhile, RGO was the effective acceptor as well as transporter of photo-generated electron hole pairs. $\bullet O_2^-$ was the most active species in the photocatalytic reaction. BiVO₄/$Mn_{1-x}Zn_xFe_2O_4$/RGO had quite a wide application in organic contaminants removal or environmental pollution control.

Keywords: BiVO₄; RGO; Mn–Zn ferrite; magnetic photocatalyst; magnetic performance; photocatalytic mechanism

1. Introduction

In the recent decade, composite semiconductor materials are considered extraordinarily attractive in the field of solar energy and pollution control engineering. Many kinds of photocatalytic composite materials with superior optical properties and high photo-induced activity have been synthesized and studied [1,2]. However, the utilization efficiency of visible light for some photocatalysts is very low, owing to their large intrinsic band gap energy, which impels scientists to explore new photocatalytic compounds with high visible light-driven photocatalytic activity. Bismuth-based composites with

n-type junctions exhibited excellent photocatalytic activity and high stability [3]. Among them, monoclinic crystal $BiVO_4$, due to its relatively lower band gap energy has been of much interest in the photocatalysis field. Nevertheless, single component $BiVO_4$ has poor absorption ability for visible light, leading to low quantum efficiency.

Meanwhile, the difficulty in separation and recovery for bismuth-contained photocatalysts greatly restricts their industrial application. Therefore, magnetic composite photocatalysts are vitally important in photocatalysis materials science, due to their simple recovery via an external magnet after reaction. Magnetic compounds, such as Fe_3O_4 and $ZnFe_2O_4$, have been extensively studied, due to their interesting properties, including photoactivity and stability. There are synthesis strategies and property studies for magnetic composite catalysts [4–8]. However, the recovery rate and the photocatalytic activity of these composites do not meet the need of industrial applications yet. Comparing magnetization and stability, $Mn_{1-x}Zn_xFe_2O_4$ is superior to Fe_3O_4 and $ZnFe_2O_4$.

Reduced graphene oxide (RGO) possessing several good properties (e.g., electrical conductivity, optical transparency and carrier mobility) has been paid considerable attention [9–11]. Single layer graphene sheet is composed of sp^2-hybridized carbon atoms in the two-dimension honeycomb lattice, which donates high mobility for electron carriers. The distinctive structure of RGO determines that its band gap energy is zero [12,13]. There are reports on the preparation method for RGO-composed catalysts and their activity [14–16]. It is reasonable to mingle RGO with $BiVO_4$, which is aimed at enhancement of the migration rate of photo-produced electrons and holes of $BiVO_4$. Previous investigation showed that $BiVO_4$–RGO composite possessed photocatalytic performance and redox ability [17]. Unexpectedly, a larger visible light photocatalytic activity could not be observed in $BiVO_4$–RGO system under visible light irradiation [18,19].

Here, fabrication of $BiVO_4/Mn_{1-x}Zn_xFe_2O_4/RGO$ was a continuation of our research about the syntheses and application of $BiVO_4/Mn_{1-x}Zn_xFe_2O_4$ [20]. The RhB degradation reaction using $BiVO_4/Mn_{1-x}Zn_xFe_2O_4$ as photocatalyst was slow (take 3 h). The incorporation of RGO could boost the photocatalytic reaction kinetics. Here, the photocatalytic activity and mechanism are deeply investigated with RhB degradation and the radical capturing experiments using $BiVO_4/Mn_{1-x}Zn_xFe_2O_4/RGO$ as photocatalyst.

2. Experimental Procedures

2.1. Preparation of $BiVO_4/Mn_{1-x}Zn_xFe_2O_4/RGO$

$BiVO_4/Mn_{1-x}Zn_xFe_2O_4$ was prepared according to our previous report [20]. Graphene oxide (GO) was fabricated with improving Hummars method [21].

GO (36.0 mg) and 1.2 g $BiVO_4/Mn_{1-x}Zn_xFe_2O_4$ were dispersed in deionized water with ultrasonication and stirring for 2 h. GO was reduced into RGO with NH_3 H_2O–N_2H_4 H_2O solution (1.0 mL–3.0 mL), then filtered and washed four times with deionized water and ethanol before placing at 80 °C for 2 h. $BiVO_4/Mn_{1-x}Zn_xFe_2O_4/RGO$ was obtained after drying at 60 °C for 24 h.

2.2. Materials Characterization

The phase and structure of samples were determined by X-ray Diffractometer (Shimadzu, XRD-6000, Kyoto, Japan), Fourier transform infrared spectroscopy (FTIR, Perkin-Elmersystem 2000, Akron, OH, USA), and INVIA Raman microprobe (Renishaw Instruments, Wotton-under-Edge, UK). The light absorption, magnetization, and surface performances of samples were examined by ultraviolet–visible diffuse reflectance spectrophotometer (UV–vis DRS, TU1901, Beijing, China), vibrating sample magnetometer (VSM 7410, LakeShore, Carson, CA, USA), Brunauer–Emmett–Teller (BET, ASAP-2020, Micromeritics, Norcross, GA, USA). The electrochemical workstation (PGSTAT30) was employed to measure electrochemical impedance spectroscopy (EIS) of the as-prepared samples. The test parameters of EIS were the following, $K_3[Fe(CN)_6]/K_4[Fe(CN)_6]$ (1:1)—KCl electrolyte solution was employed. The work electrode content contained the as-produced catalyst, acetylene black, and polytetrafluoroethylene (mass ratio, 85.0:10:5), the counter electrode was platinum foils, and the reference electrode was the saturated calomel electrode (SCE), setting AC voltage amplitude of 5 mV and a frequency range of 1×10^5–1×10^{-2} Hz.

2.3. Photocatalytic Activity, Stability, and Corresponding Mechanism

The photocatalytic activity of $BiVO_4/Mn_{1-x}Zn_xFe_2O_4/RGO$ was investigated by the rhodamine B (RhB) degradation under visible light irradiation [22]. Ninety milligrams of composite photocatalyst (named fresh photocatalyst) was put into 5.0 mg/L RhB solution (100.0 mL). The solution was placed for 0.5 h with stirring in the dark to reach the adsorption–desorption equilibrium. A 500 W Xe lamp was used as the visible light source, equipped with ultraviolet (UV) light cut-off filter ($\lambda \geq 400$ nm). At given irradiation time intervals, a series of the reaction solution was sampled and the absorption spectrum was measured.

The stability for the photocatalyst were assessed by cycling tests. After each cycle, the photocatalyst was separated and recovered by means of an external magnet. The recovered catalyst was respectively washed with ethanol and deionized water, then dried at the end of each cycle.

The photocatalytic mechanism of $BiVO_4/Mn_{1-x}Zn_xFe_2O_4/RGO$ was explored by holes-radical trapping experiments with *p*-benzoquinone (BZQ,) (•O_2^- radical scavenger), Na_2-EDTA (hole scavenger), and *tert*-butanol (*t*-BuOH) (•OH radical scavenger) in photocatalytic reaction.

3. Results and Discussion

3.1. Optimal Synthesis Condition

The $Mn_{1-x}Zn_xFe_2O_4$, prepared in advance, had a strong magnetization. In order to completely form $BiVO_4$ precursor and reduce the impurity, $Mn_{1-x}Zn_xFe_2O_4$ was put into the precursor instead of $Bi(NO_3)_3$ solution, in other words, $BiVO_4$ precursor was already formed before magnetic substance was added. $Mn_{1-x}Zn_xFe_2O_4/BiVO_4$ was assembled via calcination at only 450 °C. This temperature was lower than that of $Mn_{1-x}Zn_xFe_2O_4$ formation (1200 °C), as well as $BiVO_4$ formation (500 °C). Therefore, the calcination approach was indeed low-cost and economical.

GO was dispersed in $BiVO_4/Mn_{1-x}Zn_xFe_2O_4$ with deionized water under room temperature, RGO was produced by $NH_3 H_2O + N_2H_4 H_2O$ reduction of GO without heating. This in situ synthesis method was simple and with low-energy consumption.

3.2. Structure and Phase Identification

The XRD spectra of the obtained samples were shown in Figure 1. The characteristic spectra (Figure 1b–d) of monoclinic crystal $BiVO_4$ was well indexed with the standard card (JCPDS card No: 14-0688) [17], corresponding to the diffraction phases of (110), (011), (121), (040), (200), (002), (211), (150), (132), and (042). The diffraction pattern in Figure 1a of $Mn_{1-x}Zn_xFe_2O_4$ was fully matched with the standard card (JCPDS card No: 74-2400), agreeing with the result of the literature report [20]. The diffraction peaks of $Mn_{1-x}Zn_xFe_2O_4$ patterns were hardly observed in Figure 1c,d. Not only was the amount (15.0%) of the magnetic matrix low, but also, the diffraction patterns location of $Mn_{1-x}Zn_xFe_2O_4$ overlapped with the domain diffraction patterns of $BiVO_4$. The diffraction peak of GO (Figure 1e) was observed at 10.8° (crystal plane (001)) [23]. However, the peak (Figure 1d) disappeared after GO was mostly reduced to RGO under the reduction of $NH_3 H_2O$ and $N_2H_4 H_2O$ [24]. Moreover, the amount (3%, w/w) of RGO was not enough to be detected in X-ray diffraction. In short, it was deduced that the prepared samples totally exhibited good crystallinity.

The peak–intensity ratio (I_D/I_G) of D band (~1364.0 cm^{-1}, originating from disorder-activated Raman mode) and G band (~1598.0 cm^{-1}, corresponding to sp^2 hybridized carbon) in RGO was usually used to assess the reduction extent. Figure 2 showed the Raman spectra of the above-obtained samples. It was seen that G-band of RGO was shifted from 1598.0 cm^{-1} to 1589.0 cm^{-1}, while the D-band shorted from 1364.0 cm^{-1} to 1352 cm^{-1} after the thermal reduction finished. The I_D/I_G ratio of GO was 1.10, and that of $BiVO_4/Mn_{1-x}Zn_xFe_2O_4/RGO$ decreased to 0.84. The relative low I_D/I_G ratio of RGO implied high reduction efficiency in $BiVO_4/Mn_{1-x}Zn_xFe_2O_4/RGO$ [17]. Typical Raman bands of $BiVO_4$ were located at 120.0, 210.0, 324.0, 366.0, and 826.0 cm^{-1} in Figure 2. The two bands at 324.0 cm^{-1} and 366.0 cm^{-1} changed into one wide band in $BiVO_4/Mn_{1-x}Zn_xFe_2O_4$, as well as in $BiVO_4/Mn_{1-x}Zn_xFe_2O_4/RGO$. The result was also consistent with the results of the previous report [25].

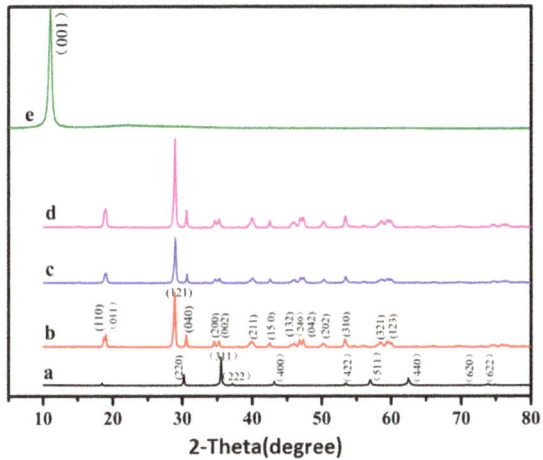

Figure 1. XRD patterns of (a) $Mn_{1-x}Zn_xFe_2O_4$; (b) $BiVO_4$; (c) $BiVO_4/Mn_{1-x}Zn_xFe_2O_4$; (d) $BiVO_4/Mn_{1-x}Zn_xFe_2O_4/RGO$ (reduced graphene oxide); (e) GO (graphene oxide).

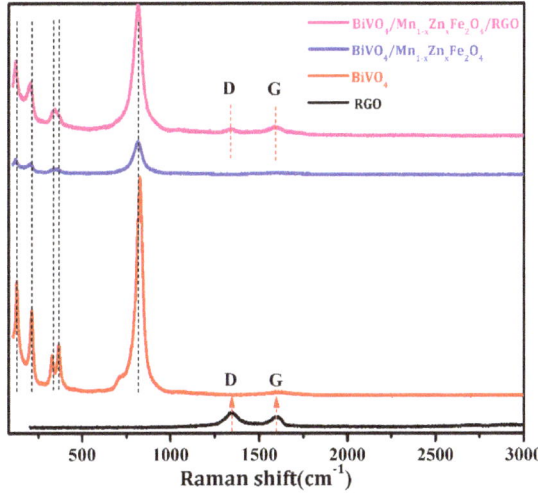

Figure 2. Raman spectra of RGO, $BiVO_4$, $Mn_{1-x}Zn_xFe_2O_4/BiVO_4$, and $Mn_{1-x}Zn_xFe_2O_4/BiVO_4/RGO$.

To investigate the valence state and the surface property of $BiVO_4/Mn_{1-x}Zn_xFe_2O_4/RGO$, XPS spectrum characterization was employed. As displayed in Figure 3a, the spectrum intensity of C 1s in $BiVO_4/Mn_{1-x}Zn_xFe_2O_4/RGO$ was larger than that in $BiVO_4/Mn_{1-x}Zn_xFe_2O_4$, namely, the introduction of RGO brought the intensity increase of C 1s. The spectrum intensity of oxygen-containing functional groups in Figure 3b was larger than that in Figure 3c, meaning the decrease of GO and the increase of RGO in BiVO$_4$/Mn$_{1-x}$Zn$_x$Fe$_2$O$_4$/RGO sample. This feature confirmed the efficient reduction of GO and the valence states for various elements in BiVO$_4$/Mn$_{1-x}$Zn$_x$Fe$_2$O$_4$/RGO.

Figure 4 was the transmission electron microscopy (TEM) images of the as-synthesized BiVO$_4$/Mn$_{1-x}$Zn$_x$Fe$_2$O$_4$/RGO. In detail, there were the black core of Mn$_{1-x}$Zn$_x$Fe$_2$O$_4$ and the gray shell of BiVO$_4$, and RGO sheets had good interfacial contact with BiVO$_4$/Mn$_{1-x}$Zn$_x$Fe$_2$O$_4$ spherical particle. In other words, there was an overlap between BiVO$_4$/Mn$_{1-x}$Zn$_x$Fe$_2$O$_4$ and RGO. At the same time, energy dispersive spectroscopy (EDS) of the composite revealed the presence of Fe, Bi, V, O, and C elements in Mn$_{1-x}$Zn$_x$Fe$_2$O$_4$, BiVO$_4$, and RGO, which was in good agreement with XPS investigation.

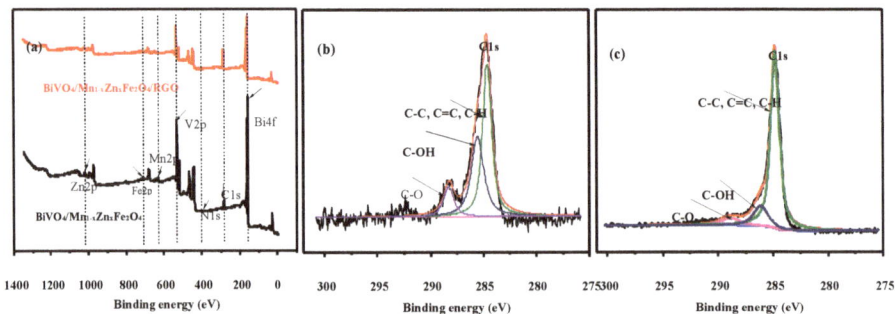

Figure 3. XPS survey spectra of (**a**) full range scan of BiVO$_4$/Mn$_{1-x}$Zn$_x$Fe$_2$O$_4$ and BiVO$_4$/Mn$_{1-x}$Zn$_x$Fe$_2$O$_4$/RGO; (**b**) C1s peaks in BiVO$_4$/Mn$_{1-x}$Zn$_x$Fe$_2$O$_4$ (**c**) C1s peaks in BiVO$_4$/Mn$_{1-x}$Zn$_x$Fe$_2$O$_4$/RGO.

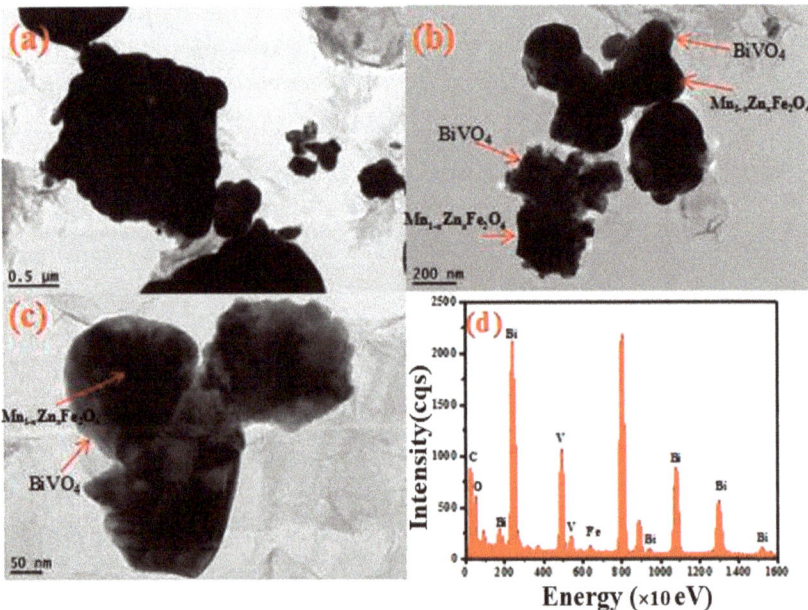

Figure 4. TEM images of Mn$_{1-x}$Zn$_x$Fe$_2$O$_4$/BiVO$_4$/RGO with different resolution (**a**–**c**) and (**d**) EDS of Mn$_{1-x}$Zn$_x$Fe$_2$O$_4$/BiVO$_4$/RGO.

Specific surface area of the as-obtained compounds was determined with the adsorption instrument, and the result was shown in Figure 5. The adsorption–desorption isotherms in Figure 5 were the typical isotherm III, agreeing with the reference report [26]. The discrete curve of $BiVO_4/Mn_{1-x}Zn_xFe_2O_4/RGO$ was in p/p_0 range of 0.45–0.55, and the pore diameter distribution was mainly 2–10 nm, and the most probable distribution was located in 4 nm. It was deduced that the introduction of RGO caused the mesopore increase and the macropore decrease. Thus, there was the uniform surface structure in the ternary composite. Calculating with the data in Figure 6, the specific surface area of $BiVO_4/Mn_{1-x}Zn_xFe_2O_4/RGO$ was 8.84 m^2/g, and that of $BiVO_4/Mn_{1-x}Zn_xFe_2O_4$ was only 2.22 m^2/g. The incorporation of RGO enhanced the specific surface area compared with $BiVO4/Mn_{1-x}Zn_xFe_2O_4$. The larger specific surface area provided more active surface sites, more free space to improve the mobility of photo-induced electrons, and further facilitated the effective migration of charge carriers, leading to the remarkable improvement of photocatalytic performance [27]. The surface structure characterization could demonstrate, in advance, the photocatalytic activity of $BiVO_4/Mn_{1-x}Zn_xFe_2O_4/RGO$ to some extent.

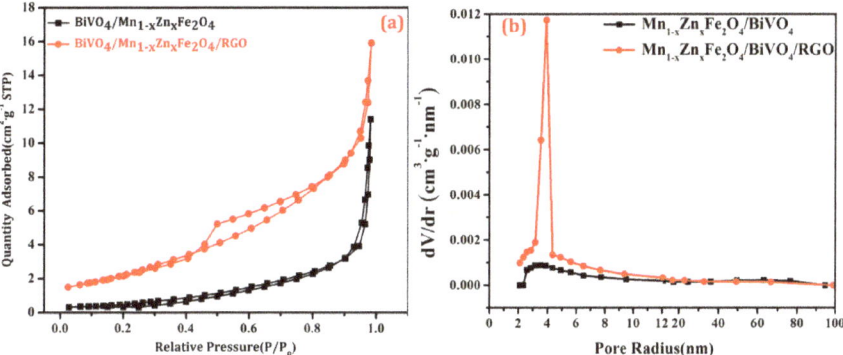

Figure 5. The adsorption–desorption isotherms of compounds (**a**), and the pore size distribution curves of compounds (**b**).

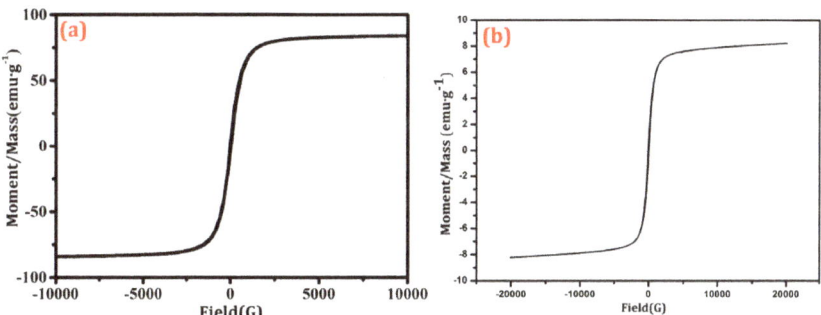

Figure 6. Hysteresis loops of products (**a**) $Mn_{1-x}Zn_xFe_2O_4$; (**b**) $BiVO_4/Mn_{1-x}Zn_xFe_2O_4/RGO$.

3.3. Magnetic Performance and Optical Properties

The magnetic hysteresis loops of the samples were displayed in Figure 6. The saturation magnetization (Ms) of $Mn_{1-x}Zn_xFe_2O_4$ and $BiVO_4/Mn_{1-x}Zn_xFe_2O_4/RGO$ were 84.03 and 8.21 emu g^{-1}, respectively. Ms of the compounds was lower than that of the pure $Mn_{1-x}Zn_xFe_2O_4$, owing to the amount decrease of the magnetic substance quantity in per unit composite. It was obvious that the prepared composite $BiVO_4/Mn_{1-x}Zn_xFe_2O_4/RGO$ had a soft-magnetic feature like pure

$Mn_{1-x}Zn_xFe_2O_4$, which further confirmed than the synthesized composite must be comprised of $Mn_{1-x}Zn_xFe_2O_4$ component [20,26].

It was worth noting that Ms was no attenuation after $BiVO_4/Mn_{1-x}Zn_xFe_2O_4/RGO$ was employed after five rounds of recycling, indicating the stable magnetism of the as-prepared composite photocatalyst. More importantly, the compound exhibited outstanding paramagnetism because both coercivity (Hc) and remnant magnetization (Mr) were near to zero. Obviously, the excellent magnetic property ensured the high recovery ratio of $BiVO_4/Mn_{1-x}Zn_xFe_2O_4/RGO$ using an external magnet after reaction.

The light absorption ability of the as-prepared samples was investigated with UV–vis DRS, and the diffuse reflectance spectra were recorded in Figure 7. It was seen from Figure 8a that the maximum absorption wavelength (λ_{max}) of pure $BiVO_4$ was about 500 nm. The further insights revealed the absorbance (at λ_{max} = 500 nm) of the compounds was higher than that of $BiVO_4$. The band gap energy (E_g) was estimated from $(Ah\upsilon)^{1/2}$ ~hv plots [5] (Figure 7b). E_g of $BiVO_4$, $BiVO_4/Mn_{1-x}Zn_xFe_2O_4$, and $BiVO_4/Mn_{1-x}Zn_xFe_2O_4/RGO$ were approximately 2.36 eV, 2.36 eV, and 2.27 eV, respectively. The introduction of $Mn_{1-x}Zn_xFe_2O_4$ did not extend the absorbance light range of $BiVO_4$ [20]. However, the introduction of RGO could be conducive to lessen E_g, leading to the enhancement of visible light absorbance for $BiVO_4/Mn_{1-x}Zn_xFe_2O_4$. It is true that the great light absorption was closely related to good photocatalytic activity of catalysts [26].

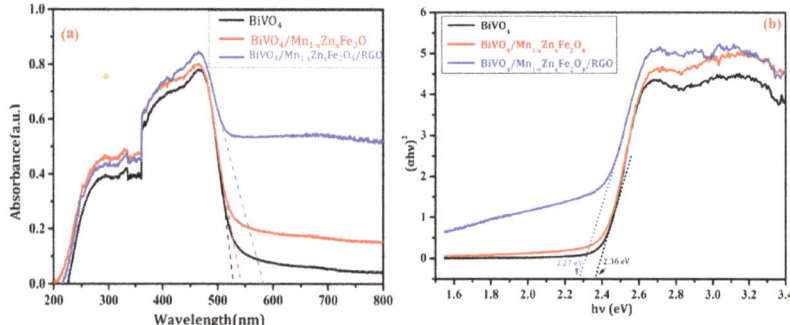

Figure 7. UV–vis diffuse reflectance spectra of the as-prepared products (**a**) and corresponding the plot of $(Ah\upsilon)^{1/2}$ versus hυ (**b**).

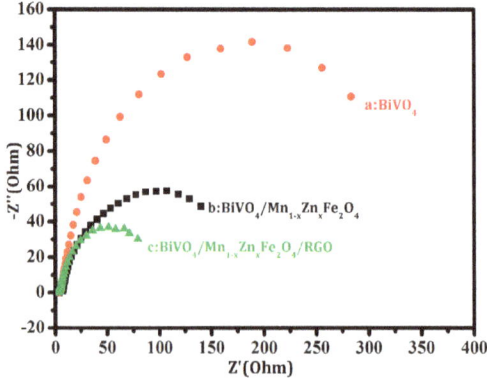

Figure 8. EIS of the work electrode containing $BiVO_4$ (**a**); $BiVO_4/Mn_{1-x}Zn_xFe_2O_4$ (**b**) and $BiVO_4/Mn_{1-x}Zn_xFe_2O_4/RGO$ (**c**).

3.4. Electrochemical Performance

Electrochemical impedance spectroscopy (EIS) was an effective approach to evaluate electron transfer ability in the interface between solid phase electrodes and electrolyte solution [28]. The typical impedance spectra of the samples were displayed with Nyquist plots. The semicircle diameter in Figure 8 became small when RGO inserted in the work electrode contained the compound. This change implied the resistance decrease and the conductivity increase in the test interface. The charge-transfer resistance (R_{ct}) of the samples was gained by fitting the data from Figure 8. R_{ct} of $BiVO_4$, $BiVO_4/Mn_{1-x}Zn_xFe_2O_4$, and $BiVO_4/Mn_{1-x}Zn_xFe_2O_4/RGO$ were 351.0 Ω cm^2, 206.0 Ω cm^2, and 103.0 Ω cm^2, respectively. It was clear that R_{ct} of the ternary composite was the lowest.

The good electron accepting and transporting properties of RGO could contribute to the prevention of charge recombination. It was reasonable that the introduction of RGO was beneficial to the efficient charge separation and transportation in the compound interface. The electrochemical behavior brought high conductivity of the comprising electrode. As a result, the enhancement of conductivity promoted the improvement of photocatalytic activity for $BiVO_4/Mn_{1-x}Zn_xFe_2O_4/RGO$.

3.5. Photocatalytic Activity, Stability, and Corresponding Mechanism

The photocatalytic activity was probed with photodegradation of RhB dye, and the result was shown in Figure 9. It was found from Figure 9 that the degradation ratio of RhB with $BiVO_4/Mn_{1-x}Zn_xFe_2O_4/RGO$ under visible light irradiation reached to 96.0% after only 1.5 h reaction. It is worth noting that the self-degradation of RhB was very weak in the comparative test. It took about 3 h to get the same degradation ratio (96.0%) with pure $BiVO_4$ as well as $BiVO_4/Mn_{1-x}Zn_xFe_2O_4$ under identical conditions. Significantly, $BiVO_4/Mn_{1-x}Zn_xFe_2O_4/RGO$ exhibited more excellent photocatalytic activity than that of $BiVO_4/Mn_{1-x}Zn_xFe_2O_4$. Moreover, the activity of $BiVO_4/Mn_{1-x}Zn_xFe_2O_4/RGO$ was greatly superior to that of $SrFe_{12}O_{19}/BiVO_4$, as well as $BiVO_4/RGO$, in the literature [19,22]. The high photocatalytic property of the as-produced compound $BiVO_4/Mn_{1-x}Zn_xFe_2O_4/RGO$ was explained as follows: the graphene owned two-dimensional $\pi-\pi$ conjugate structure was not only a good electron acceptor, but also a good electronic vector. RGO excited electrons in $BiVO_4/Mn_{1-x}Zn_xFe_2O_4$ and prompted the transferring of the conduction band in itself [29]. It was more interesting that the photocatalytic activity of $BiVO_4/Mn_{1-x}Zn_xFe_2O_4/RGO$ was obviously better than that of $Mn_{1-x}Zn_xFe_2O_4/\beta-Bi_2O_3$ in previous our group's report [20].

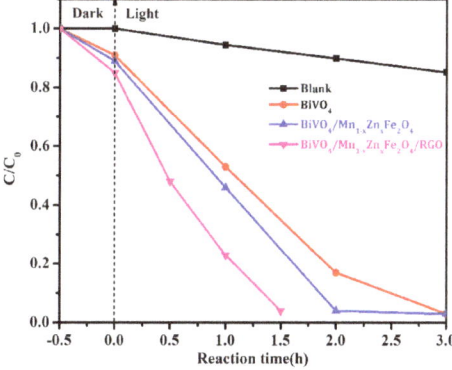

Figure 9. Degradation ratios of rhodamine B (RhB) with photocatalysts.

The stability was a key property in the industrial application of catalytic materials. Each recycling experiment was operated in triplicate, and average values and standard deviations were also shown in Figure 10. The degradation ratio of RhB in the fifth recycling was still 85.0% after 1.5 h of reaction under the same test parameters. The photocatalytic activity was only reduced a little within five recycles. The result revealed good stability of BiVO$_4$/Mn$_{1-x}$Zn$_x$Fe$_2$O$_4$/RGO.

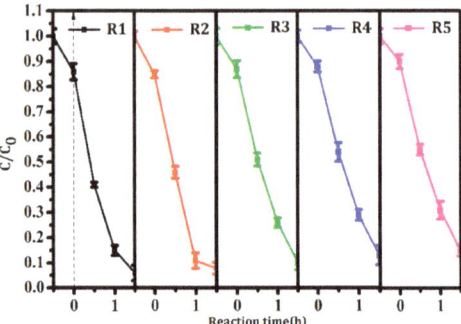

Figure 10. Cycling test of BiVO$_4$/Mn$_{1-x}$Zn$_x$Fe$_2$O$_4$/RGO in RhB photodegradation (five recycles).

Figure 11 was FTIR spectra of the compounds. The peaks at 473.7 cm^{-1} and 412.4 cm^{-1} in Figure 11a,b were assigned to Zn–O and Fe–O vibrations in Mn$_{1-x}$Zn$_x$Fe$_2$O$_4$. Characteristic patterns of V–O symmetric and asymmetric stretching vibrations spectra in BiVO$_4$ were present at 734.3 cm^{-1} and 823.4 cm^{-1}. The abovementioned peak location and intensity of Mn$_{1-x}$Zn$_x$Fe$_2$O$_4$ and BiVO$_4$ were not varied, demonstrating a high stability of BiVO$_4$/Mn$_{1-x}$Zn$_x$Fe$_2$O$_4$ during the photocatalytic reaction. The peak at 1625.9 cm^{-1} in Figure 11 belonged to C=C stretch of aromatic group in RGO. The peak at 1629.5 cm^{-1} in Figure 11b was weaker than that in Figure 11a, due to a little loss of RGO quantity after the fifth cycle. The absorption peaks located at around 1400.0 cm^{-1} and 1065.0 cm^{-1} illustrated the functional group of RGO [30]. By comparing pattern (a) and (b) in Figure 11, the typical peaks were detected in the spectra of the fresh, as well as the recovered compound. Thereby, it was concluded that the structure of BiVO$_4$/Mn$_{1-x}$Zn$_x$Fe$_2$O$_4$/RGO was stable in the process of RhB photocatalytic degradation.

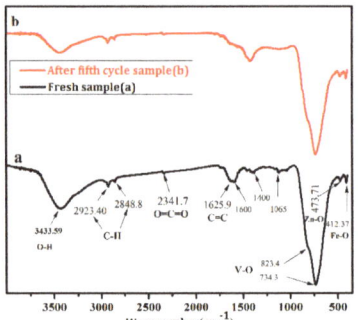

Figure 11. FTIR spectra of Mn$_{1-x}$Zn$_x$Fe$_2$O$_4$/BiVO$_4$/RGO (a) the fresh sample; (b) the recovered sample.

The photocatalytic mechanism of BiVO$_4$/Mn$_{1-x}$Zn$_x$Fe$_2$O$_4$/RGO was probed with radical scavengers [31]. t-BuOH (•OH scavenger), EDTA-Na$_2$ (h$^+$ scavenger), and BZQ (•O$_2^-$ scavenger) were employed to ascertain the dominant radical species in the photocatalytic degradation of RhB with the as-synthesized compound. Degradation ratios of RhB under these scavengers were given in Figure 12. The photodegradation ratio of RhB was 72.0% and 65.0% only when 5.0 mM t-BuOH and

1.0 mM EDTA-Na$_2$ were added into the reaction system. Namely, h$^+$ or •OH scavenger brought about the decrease of degradation ratios. Thus, the photocatalytic activity of the compound greatly decreased. The inactivation of the photocatalytic test was evidently proven when 1 mM BZQ was added in the same reaction system. The above results demonstrated that the most active species was •O$_2^-$, though •OH and h$^+$ also contributed to the photocatalytic activity of BiVO$_4$/Mn$_{1-x}$Zn$_x$Fe$_2$O$_4$/RGO.

Figure 13 described the electron–hole pairs forming process under the light irradiation. The potentials of conduction band (CB) and valence band (VB) of BiVO$_4$ were 0.46 eV and 2.86 eV, respectively (referring to hydrogen electrode, NHE). The electrons in VB were excited to CB under visible light irradiation, forming driven-electrons (e$^-$) and holes (h$^+$). The VB potential of BiVO$_4$ was close to E$^\theta$ of •OH/H$_2$O, and the CB potential was larger than E$^\theta$ of O$_2$/•O$_2^-$. This meant that electrons were able to directly reduce O$_2$ molecules into superoxide O$_2^-$. Besides, RhB molecules were directly oxidized by the holes on VB of BiVO$_4$. In addition, there were much more active adsorption centers and photocatalytic reaction sites in RGO with a large surface area. These active centers and sites were beneficial to the improvement of the photocatalytic activity. As a good electron acceptor and electronic vector, RGO facilitates the transmission of photo-produced electrons, which was conducive to the separation of photo-produced electrons and holes, and further promoted the formation of •O$_2^-$. It was ensured that •O$_2^-$ played the main role for the RhB photodegradation, though •OH and h$^+$ had a collaborative oxidation role in the photocatalytic reaction of BiVO$_4$/Mn$_{1-x}$Zn$_x$Fe$_2$O$_4$/RGO.

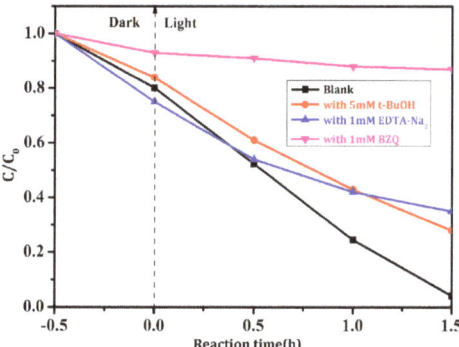

Figure 12. Photodegradation ratios of RhB with BiVO$_4$/Mn$_{1-x}$Zn$_x$Fe$_2$O$_4$/RGO under scavengers.

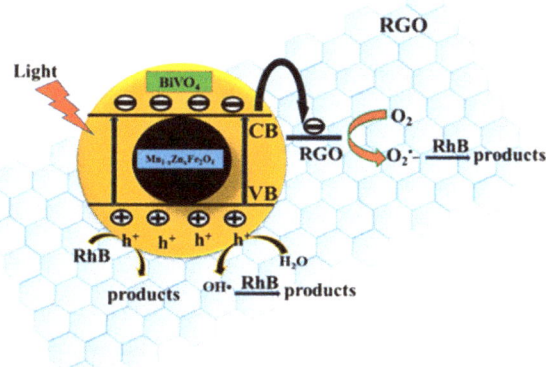

Figure 13. Photocatalytic scheme of BiVO$_4$/Mn$_{1-x}$Zn$_x$Fe$_2$O$_4$/RGO under visible light irradiation.

In fact, since our group found that the magnetic composite $ZnFe_2O_4/SrFe_{12}O_{19}$ had a highly photocatalytic activity in 2013 [27], we firmly thought that the stable magnetic field from $SrFe_{12}O_{19}$ itself could promote the separation of photo-generated electrons and holes, and furthermore, that the photocatalyst could produce more photo-generated electrons and holes under identical light irradiation. Thus, the photoelectric transformation efficiency would be boosted. Our group the studied $BiVO_4/SrFe_{12}O_{19}$, $Bi_2O_3/SrFe_{12}O_{19}$, and $BiOCl/SrFe_{12}O_{19}$ [22,32,33] magnetic heterojunction to confirm previous speculation. However, these studies about its photoelectron transfer mechanism were not enough. In future work, our group will continue to attempt to confirm our speculation via experimental and theoretical calculations. Of course, this work was carried out in order to compare with $SrFe_{12}O_{19}$ functions.

4. Conclusions

Magnetic photocatalyst $BiVO_4/Mn_{1-x}Zn_xFe_2O_4/RGO$ was synthesized with the simple and economical roasting-reduction approach. The photocatalyst exhibited excellent photocatalytic activity and stability. The degradation ratio of RhB reached 96.0% under visible light irradiation after only 1.5 h reaction with the photocatalyst. The degradation ratio of RhB was still maintained at 85.0% after five cycles of photocatalytic reaction. Here, the incorporation of RGO enhanced the specific surface area compared with $BiVO4/Mn_{1-x}Zn_xFe_2O_4$. The larger specific surface area provided more active surface sites, more free space to improve the mobility of photo-induced electrons, and further facilitated the effective migration of charge carriers, leading to the remarkable improvement of photocatalytic performance. RGO was the effective acceptor as well as transporter of photo-generated electron–hole pairs. $•O_2^-$ was the most active species in this photocatalytic reaction. We hope this photocatalyst has a wide application in organic contaminants removal or environmental pollution control in practical.

Author Contributions: Conceptualization, T.X. and T.X.; Methodology, C.L.; Formal Analysis, H.L. and T.X.; Investigation, H.L. and J.Y.; Resources, C.L.; Data Curation, H.L. and T.X.; Writing-Original Draft Preparation, H.L. and T.X.; Writing-Review & Editing, T.X., C.L., and L.X.; Supervision, C.L. and L.X.

Funding: The work was financially supported by Chongqing Basic Science and Advanced Technology Research Program (CSTC2015jcyjBX0015) and Chongqing Bureau of Land Resources Housing Management, PR China (No. CQGT-KJ-2014012).

Acknowledgments: We authors want to thank Jiankang Wang for his partial data processing.

Conflicts of Interest: The authors declare that they have no conflict of interest.

References

1. Willkomm, J.; Katherine, L.O.; Reynal, A. Dye-sensitised semiconductors modified with molecular catalysts for light-driven H_2 production. *Chem. Soc. Rev.* **2016**, *45*, 9–23. [CrossRef] [PubMed]
2. Kudo, A.; Ueda, K.; Kato, H.; Mikami, I. Photocatalytic O_2 Evolution under Visible Light Irradiation on $BiVO_4$ in Aqueous $AgNO_3$ Solution. *Catal. Lett.* **1998**, *53*, 229–230. [CrossRef]
3. Zhu, G.Q.; Hojamberdievc, M.; Que, W.X.; Liu, P. Hydrothermal synthesis and visible-light photocatalytic activity of porous peanut-like $BiVO_4$ and $BiVO_4/Fe_3O_4$ submicron structures. *Ceram. Int.* **2013**, *39*, 9163–9172. [CrossRef]
4. Zhang, L.S.; Lian, J.S.; Wu, L.Y.; Duan, Z.R.; Jiang, J.; Zhao, L.J. Synthesis of a Thin-Layer MnO_2 Nanosheet-Coated Fe_3O_4 Nanocomposite as a Magnetically Separable Photocatalyst. *Langmuir* **2014**, *30*, 7006–7013. [CrossRef] [PubMed]
5. Zhang, W.Q.; Wang, M.; Zhao, W.J.; Wang, B.Q. Magnetic composite photocatalyst $ZnFe_2O_4/BiVO_4$: Synthesis, characterization, and visible-light photocatalytic activity. *Dalton Trans.* **2013**, *42*, 15464–15474. [CrossRef] [PubMed]
6. Laohhasurayotin, K.; Pookboonmee, S.; Viboonratanasri, D.; Kangwansupamonkon, W. Preparation of magnetic photocatalyst nanoparticles—TiO_2/SiO_2/Mn–Zn ferrite—And its photocatalytic activity influenced by silica interlayer. *Mater. Res. Bull.* **2012**, *47*, 1500–1507. [CrossRef]

7. Dong, X.L.; Shao, Y.; Zhang, X.X.; Ma, H.C.; Zhang, X.F.; Shi, F.; Ma, C.; Xue, M. Synthesis and properties of magnetically separable $Fe_3O_4/TiO_2/Bi_2O_3$ photocatalysts. *Res. Chem. Intermed.* **2014**, *40*, 2953–2961. [CrossRef]
8. Ma, P.C.; Jiang, W.; Wang, F.H.; Li, F.S.; Shen, P.; Chen, M.D.; Wang, Y.J.; Liu, J.; Li, P.Y. Synthesis and photocatalytic property of $Fe_3O_4@TiO_2$ core/shell nanoparticles supported by reduced graphene oxide sheets. *J. Alloys Compd.* **2013**, *578*, 501–506. [CrossRef]
9. Dong, S.Y.; Cui, Y.R.; Wang, Y.F.; Li, Y.K.; Hu, L.M.; Sun, J.Y.; Sun, J.H. Designing three-dimensional acicular sheaf shaped $BiVO_4$/reduced graphene oxide composites for efficient sunlight-driven photocatalytic degradation of dye wastewater. *Chem. Eng. J.* **2014**, *249*, 102–110. [CrossRef]
10. Gupta, B.; Melvin, A.A.; Matthews, T.; Dhara, S.; Dash, S.; Tyagi, A.K. Facile gamma radiolytic methodology for TiO2-rGO synthesis: Effect on photo-catalytic H_2 evolution. *Int. J. Hydrog. Energy* **2015**, *40*, 5815–5823. [CrossRef]
11. Aiga, N. Electron−Phonon Coupling Dynamics at Oxygen Evolution Sites of Visible-Light-Driven Photocatalyst: Bismuth Vanadate. *J. Phys. Chem. C* **2013**, *117*, 9881–9886. [CrossRef]
12. Boruah, P.K.; Borthakur, P.; Darabdhara, G.; Kamaja, C.K.; Karbhal, I.; Shelke, M.V.; Phukan, P.; Saikiad, D.; Das, M.R. Sunlight assisted degradation of dye molecules and reduction of toxic Cr(VI) in aqueous medium using magnetically recoverable Fe_3O_4/reduced graphene oxide nanocomposite. *RSC Adv.* **2016**, *6*, 11049–11063. [CrossRef]
13. Li, J.Q.; Guo, Z.Y.; Liu, H.; Du, J.; Zhu, Z.F. Two-step hydrothermal process for synthesis of F-doped $BiVO_4$ spheres with enhanced photocatalytic activity. *J. Alloys Compd.* **2013**, *581*, 40–45. [CrossRef]
14. Xu, L.; Huang, W.Q.; Wang, L.L.; Tian, Z.A.; Hu, W.Y.; Ma, Y.M.; Wang, X.; Pan, A.L.; Huang, G.F. Insights into Enhanced Visible-Light Photocatalytic Hydrogen Evolution of $g-C_3N_4$ and Highly Reduced Graphene Oxide Composite: The Role of Oxygen. *Chem. Mater.* **2015**, *27*, 1612–1621. [CrossRef]
15. Wu, X.F.; Wen, L.L.; Lv, K.L.; Deng, K.J.; Tang, D.G.; Ye, H.P.; Du, D.Y.; Liu, S.L.; Li, M. Fabrication of ZnO/graphene flake-like photocatalyst with enhanced photoreactivity. *Appl. Surf. Sci.* **2015**, *358*, 130–136. [CrossRef]
16. Yang, X.F.; Qin, J.L.; Li, Y.; Zhang, R.X.; Tang, H. Graphene-spindle shaped TiO_2 mesocrystal composites: Facile synthesis and enhanced visible light photocatalytic performance. *J. Hazard. Mater.* **2013**, *261*, 342–350. [CrossRef] [PubMed]
17. Wang, A.L.; Shen, S.; Zhao, Y.B.; Wu, W. Preparation and characterizations of $BiVO_4$/reduced graphene oxide nanocomposites with higher visible light reduction activities. *J. Colloid Interface Sci.* **2015**, *445*, 330–336. [CrossRef] [PubMed]
18. Yan, Y.; Suna, S.F.; Song, Y.; Yan, X.; Guan, W.S.; Liu, X.L.; Shi, W.D. Microwave-assisted in situ synthesis of reduced graphene oxide-$BiVO_4$ composite photocatalysts and their enhanced photocatalytic performance for the degradation of ciprofloxacin. *J. Hazard. Mater.* **2013**, *250–251*, 106–114. [CrossRef] [PubMed]
19. Lia, Y.K.; Dong, S.Y.; Wang, Y.F.; Sun, J.Y.; Li, Y.F.; Pi, Y.Y.; Hu, L.M.; Sun, J.H. Reduced graphene oxide on a dumbbell-shaped $BiVO_4$ photocatalyst for an augmented natural sunlight photocatalytic activity. *J. Mol. Catal. A Chem.* **2014**, *387*, 138–146. [CrossRef]
20. Xie, T.P.; Xu, C.L.L.L.J.; Li, H. New Insights into $Mn_xZn_{1−x}Fe_2O_4$ via Fabricating Magnetic Photocatalyst Material $BiVO_4/Mn_xZn_{1−x}Fe_2O_4$. *Materials* **2018**, *11*, 335. [CrossRef] [PubMed]
21. Liu, C.L.; He, C.L.; Xie, T.P.; Yang, J. Reduction of Graphite Oxide Using Ammonia Solution and Detection Cr(VI) with Graphene-Modified Electrode. *Fuller. Nanotub. Carbon Nanostruct.* **2015**, *23*, 125–130. [CrossRef]
22. Liu, C.L.; Li, H.; Ye, H.P.; Xu, L.J. Preparation and Visible-Light-Driven Photocatalytic Performance of Magnetic $SrFe_{12}O_{19}/BiVO_4$. *J. Mater. Eng. Perform.* **2015**, *24*, 771–777.
23. Cui, H.Y.; Yang, X.F.; Gao, Q.X. Facile synthesis of grapheme oxide-enwrapped Ag_3PO_4 composites with highly efficient visible light photocatalytic performance. *Mater. Lett.* **2013**, *93*, 28–31. [CrossRef]
24. Fu, Y.S.; Sun, X.Q.; Wang, X. $BiVO_4$–graphene catalyst and its high photocatalytic performance under visible light irradiation. *Mater. Chem. Phys.* **2011**, *131*, 325–330. [CrossRef]
25. Yu, Q.Q.; Tang, Z.R.; Xu, Y.J. Synthesis of $BiVO_4$ nanosheets-graphene composites toward improved visible light photoactivity. *J. Energy Chem.* **2015**, *23*, 564–574. [CrossRef]
26. Xie, T.P.; Xu, L.J.; Liu, C.L.; Wang, Y. Magnetic composite $ZnFe_2O_4/SrFe_{12}O_{19}$: Preparation, characterization, and photocatalytic activity under visible light. *Appl. Surf. Sci.* **2013**, *273*, 684–691. [CrossRef]

27. Wang, B.; Ru, Q.; Su, C.Q.; Cheng, S.K.; Liu, P.; Guo, Q.; Hou, X.H.; Su, S.C.; Ling, C.C. $Ni_{12}P_5$ Nanoparticles Hinged by Carbon Nanotubes as 3D Mesoporous Anodes for Lithium-Ion Batteries. *ChemElectroChem* **2018**. [CrossRef]
28. Sun, L.L.; Wu, X.L.; Meng, M.; Zhu, X.B.; Chu, P.K. Enhanced Photodegradation of Methyl Orange Synergistically by Microcrystal Facet Cutting and Flexible Electrically-Conducting Channels. *J. Phys. Chem. C* **2014**, *118*, 28063–28068. [CrossRef]
29. Zhang, H.; Lv, X.J.; Li, Y.M.; Wang, Y.; Li, J.H. P25-Graphene Composite as a High Performance Photocatalyst. *J. Am. Chem. Soc.* **2010**, *4*, 380–386. [CrossRef] [PubMed]
30. Yang, X.; Chen, W.; Huang, J.; Zhou, Y.; Zhu, Y.; Li, C. Sonochemical synthesis of nanocrystallite Bi_2O_3 as a visible-light-driven photocatalyst. *Sci. Rep.* **2015**, *5*, 10632–10641. [CrossRef] [PubMed]
31. Chen, X.J.; Dai, Y.Z.; Wang, X.Y.; Guo, J.; Liu, T.H.; Li, F.F. Synthesis and characterization of Ag_3PO_4 immobilized with graphene oxide (GO) for enhanced photocatalytic activity and stability over 2,4-dichlorophenol under visible light irradiation. *J. Hazard. Mater.* **2015**, *292*, 9–18. [CrossRef] [PubMed]
32. Xie, T.P.; Liu, C.L.; Xu, L.J.; Yang, J.; Zhou, W. Novel Heterojunction Bi_2O_3/$SrFe_{12}O_{19}$ Magnetic Photocatalyst with Highly Enhanced Photocatalytic Activity. *J. Phys. Chem. C* **2013**, *117*, 24601–24610. [CrossRef]
33. Xie, T.P.; Xu, L.J.; Liu, C.L.; Yang, J.; Wang, M. Magnetic composite BiOCl–$SrFe_{12}O_{19}$: A novel p-n type heterojunction with enhanced photocatalytic activity. *Dalton Trans.* **2014**, *43*, 2211–2220. [CrossRef] [PubMed]

© 2018 by the authors. Licensee MDPI, Basel, Switzerland. This article is an open access article distributed under the terms and conditions of the Creative Commons Attribution (CC BY) license (http://creativecommons.org/licenses/by/4.0/).

MDPI
St. Alban-Anlage 66
4052 Basel
Switzerland
Tel. +41 61 683 77 34
Fax +41 61 302 89 18
www.mdpi.com

Nanomaterials Editorial Office
E-mail: nanomaterials@mdpi.com
www.mdpi.com/journal/nanomaterials

www.ingramcontent.com/pod-product-compliance
Lightning Source LLC
LaVergne TN
LVHW071952080526
838202LV00064B/6726